Theory and Application of Hydraulic Modeling

This edited volume from Japan's Research Subcommittee on Methodology for Dealing with Geomaterials in Hydraulic Model Experiments presents readers with a state-of-the-art overview of experimental and computational methods used to address similarity scaling incompatibilities present in fluid–sediment flows.

Readers will gain an understanding of complex phenomena in the boundary fields of hydraulics and geotechnical engineering. Chapter contributors focus on the phenomena that are affected by the interactions between fluid wave and ground in a complex field, which for many years have been challenging to process and model. In addition to describing the implementation of model tests and the concept of the law of similarity, this book contrasts these phenomena with the laws of similarity, describes models and numeral analysis methods, and explains important considerations using experimental case studies. Each chapter is written by leading researchers in Japan who are members of the Research Subcommittee on Methodology for Dealing with Geomaterials in Hydraulic Model Experiments. The chapters are closely linked but are written so that each can be read individually. Readers will be able to apply this knowledge to their work and to create models that more accurately simulate the interactions between wave and ground, allowing them to better understand these phenomena and devise more appropriate strategies for defense and so on when necessary.

This collection provides information that can be used by young researchers and postgraduate students in the boundary fields of hydraulics and geotechnical engineering who aim at becoming civil engineers, and it will be of particular value to practicing engineers of all experience levels who must regularly analyze complex interactions between fluids and ground.

Taro Arikawa is a Professor specializing in coastal engineering and tsunami disaster prevention at Chuo University Tokyo, Japan. He obtained his BS, MS, and Ph.D. from the University of Tokyo in 1995, 1997, and 2000, respectively. He acts as Secretary and Chair for subcommittees of several societies, including the Japan Society of Civil Engineers, Architectural Institute of Japan, Japan Association for Earthquake Engineering, and the Engineering Academy of Japan.

Theory and Application of Hydraulic Modeling

Interaction between Fluid and Ground under Water Waves

Edited by

Taro Arikawa

Research Subcommittee on Methodology for Dealing with Geomaterials in Hydraulic Model Experiments,
Coastal Engineering Committee,
Japan Society of Civil Engineers

CRC Press
Taylor & Francis Group
Boca Raton London New York

CRC Press is an imprint of the Taylor & Francis Group, an **informa** business

JSCE

CEC 土木学会海岸工学委員会
Coastal Engineering Committee

Note: This book is an English translation of *Suiri mokei jikken no riron to ouyo – hadou to jiban no sougosayou*, which was first published in Japanese in 2021 by the Japanese Society of Civil Engineers.

Designed cover image: Large Hydro Geo Flume, Port and Airport Research Institute

First edition published [2024]
by CRC Press
2385 NW Executive Center Drive, Suite 320, Boca Raton FL 33431

and by CRC Press
4 Park Square, Milton Park, Abingdon, Oxon, OX14 4RN

CRC Press is an imprint of Taylor & Francis Group, LLC

ISBN: 978-1-032-54413-7 (hbk)
ISBN: 978-1-032-54414-4 (pbk)
ISBN: 978-1-003-42475-8 (ebk)

DOI: 10.1201/9781003424758

Typeset in Palatino
by KnowledgeWorks Global Ltd.

Contents

Publication........ ix
Members List of Research Subcommittee on Methodology for Dealing with Geomaterials in Hydraulic Model Experiments........ xi
List of Contributors........ xiii

1 Introduction........ **1**
Yasuo Kotake

2 Methodology of Model Tests........ **5**
Tatsuya Matsuda, Iwao Hasegawa, Shin-ichi Kubota, Yutaka Sawada, Hidenori Takahashi, Tetsuo Tobita, and Tomoaki Nakamura
2.1 Purpose of Model Tests........ 5
2.2 Hydraulic Model Test........ 6
2.2.1 Model Test Plan........ 6
2.2.2 Classification and Target of Model Tests........ 9
2.2.3 Target and Method of Measurement........ 12
2.2.4 Model Test Procedure........ 15
2.2.5 Typical Examples of Hydraulic Model Tests........ 24
2.3 Geotechnical Model Test........ 29
2.3.1 Planning of Model Test........ 29
2.3.2 Classification and Target of Model Tests........ 32
2.3.3 Target and Method of Measurement........ 42
2.3.4 Model Test Procedure........ 47
2.3.5 Typical Model Test Example........ 54
2.4 Review on the Handling of Geomaterials and Organization of Issues in Japan (Published in Japanese)........ 60
2.4.1 Examples of Past Experiments Dealing with a Movable Bed in Hydraulic Model Tests........ 60
2.4.2 Extraction and Organization of Issues Related to the Handling of Geomaterials in Hydraulic Model Experiments........ 63
References........ 65

3 Concept of the Law of Similarity in Model Tests........ **72**
Hidenori Takahashi and Masaaki Ikeno
3.1 Role of Law of Similarity........ 72
3.1.1 Law of Similarity........ 72
3.1.2 History of Law of Similarity........ 74
3.2 Law of Similarity for Fluids........ 77
3.2.1 Derivation of Law of Similarity........ 77
3.2.2 Model Tests Involving Breaking Waves........ 79

3.2.3 Model Tests in Earth's Gravity Field 80
3.2.4 Model Tests in Centrifugal Gravity Field 81
3.2.5 Model Tests on Fluid–Structure Interaction 81
3.3 Law of Similarity for Ground 82
3.3.1 Derivation of the Law of Similarity 82
3.3.2 Model Tests in Earth's Gravitational Field 85
3.3.3 Model Tests in Centrifugal Gravity Field 87
3.3.4 Model Tests on Soil–Structure Interaction 88
3.3.5 Law of Similarity for Unsaturated Ground 89
3.4 Law of Similarity in Combined Fluid and Ground Model Tests 93
3.4.1 Sand Drift 93
3.4.2 Seabed Topography Changes 98
3.4.3 Scouring Phenomenon below Permeable Structure 100
3.4.4 Combined Wave and Ground Model Test 102
References 105

4 Case Study of Model Experiments 110
Susumu Araki, Kojiro Suzuki, Junji Miyamoto, Takenori Shimozono, Fuminori Kato, Takayuki Suzuki, and Kenichi Maeda
4.1 Viewpoint Concerning the Case Study 110
4.2 Case (1) Scouring Phenomenon under Wave-dissipating Blocks 111
4.2.1 Subsidence Phenomena of Wave-dissipating Blocks 111
4.2.2 Method of Reproducing Scouring Phenomenon with Small-to-Medium-Scale Model Experiments 113
4.2.3 Mechanism of Subsidence of Blocks Based on a Large-scale Experiment (1/16 Scale of the Site) 115
4.2.4 Reproduction of Subsidence of Blocks by Super-large-scale Experiment (1/4 Scale of the Site) 120
4.2.5 Small-to-medium-scale Model Experiment (1/14 of Super-large-scale Experiment and 1/55.8 of the Site): Institution A 123
4.2.6 Small-to-medium-scale Model Experiment (1/14 of Super-large-scale Experiment and 1/55.8 of the Site): Institution B 126
4.2.7 Small-to-medium-scale Model Experiment (1/25 of Super-large-scale Experiment and 1/100 of the Site): Institution C 130
4.2.8 Centrifuge Model Tests (1/38 of Super-large-scale Experiment and 1/152 of the Site): Institution D 131
4.2.9 Dean's Law (Dean Number) and Results of Experiments of Each Scale 137
4.2.10 Conclusion 140
4.3 Case (2) Erosion of Beach Nourishment Filling 141
4.3.1 Introduction 141
4.3.2 Existing Knowledge Regarding Recession Mechanism of Sand Dune and Beach Scarp 142
4.3.3 Test Method 146

4.3.4 Test Result ... 149
4.3.5 Comparison between the Test Result and the Actual Phenomenon ... 156
4.3.6 Conclusion ... 158
References ... 159

5 Numerical Calculation Methodology for Wave–Seabed Interaction ... 162
Tomoaki Nakamura, Taro Arikawa, Kazunori Fujisawa, and Eiji Harada
5.1 Role of Numerical Calculation ... 162
5.2 Development of Numerical Calculation for Wave–Seabed Interaction ... 163
5.2.1 Grid Method ... 164
5.2.2 Particle Method ... 172
5.2.3 Coupling of Grid Method and Particle Method ... 176
5.3 Treatment of Interface between Fluid and Porous Domains ... 177
5.3.1 Role of Interfacial Condition ... 181
5.3.2 Examples of Numerical Analyses ... 183
5.4 Bottom Sediment Transport by Particle Method ... 187
5.4.1 Multiphase Flow Analysis ... 188
5.4.2 Example of Numerical Analysis ... 195
5.4.3 Conclusion ... 204
5.5 Wave-induced Seabed Response by Grid Method ... 204
5.5.1 Wave Model ... 204
5.5.2 Seabed Model ... 211
5.5.3 Coupling of Wave Model and Seabed Model ... 212
5.5.4 Examples of Numerical Analysis ... 214
References ... 220

6 Future Challenges ... 228
Taro Arikawa

Index ... 231

Publication

In this publication, the research activities conducted over six years by the Research Subcommittee on Methodology for Dealing with Geomaterials in Hydraulic Model Experiments that was established by the Japan Society of Civil Engineers (JSCE) Coastal Engineering Committee in June 2016 are summarized. In summarizing the results of the activities, we would like to describe the background.

Model tests in the field of hydraulics are difficult to carry out on an actual scale. Hence, it is customary to use Froude's law of similarity when specifying the model scale. Nevertheless, if Froude's law of similarity is strictly applied as the geometric scale when using a geomaterial in a model test, it may not reproduce the local phenomenon adequately. For example, if clay must be used in a model as opposed to the local material, which is sand, and the experiment is carried out as it is, then the experimental results obtained will not be a proper reproduction of the local phenomenon because of the difference in material properties. Thus, in the field of hydraulics, a technique that uses a homogeneous ground material of the smallest possible size is generally used. Nevertheless, in this case, the difference between the geometrical law of similarity and the mechanical law of similarity of the materials must be considered when converting the results of the experiment to the scale of the field.

Conversely, in the field of geotechnical engineering, a technique for conducting model tests using a centrifugal loading device that has been used conventionally was developed for experiments that involve the behavior of water. Examples of this application have also been reported at coastal engineering lectures organized by the JSCE Coastal Engineering Committee. In hydraulic model tests using a centrifugal loading device, a law of similarity to match the Froude number is used. However, in the field of hydraulics, the restoring force due to gravity is emphasized and Froude's law of similarity, which assumes a gravitational force field of $1g$, is used. In a centrifuge model test, emphasis is placed on matching the pressure and stress fields between the actual site and model, so a law of similarity in which the gravitational acceleration is changed using centrifugal force is used. Hence, even when dealing with the same phenomenon in a field combining fluid and ground, a shared understanding of the consistency of the experiments and the handling of results, including the law of similarity, had not been formed among researchers in the fields of hydraulics and geotechnical engineering.

Amid such an environment, the Great East Japan Earthquake that occurred in March 2011 revealed several issues that were difficult to elucidate in the respective fields of study alone; thus, an interdisciplinary approach in implementation and clarification became a pressing issue. Under such circumstances, besides a shared understanding of the problem, the problem of the law of similarity was closely examined in the practical use of the interpreted results of each model test.

Against this background, the subcommittee aims to (1) review the handling of ground materials in previous studies in the fields of hydraulics and geotechnical engineering, (2) sort out the consistency of the laws of similarity used in both fields and the points to note when converting experimental results to local site conditions, and (3) present a fixed direction on methodology for dealing with geomaterials in hydraulic model experiments. Established under the JSCE Coastal Engineering Committee, the term of the subcommittee has been extended twice to vigorously tackle these issues. I am writing the preface to this book as I was appointed subcommittee chairman when the committee was first established.

Nevertheless, during the implementation period of the subsequent case study, Associate Professor Susumu Araki (Osaka University) was appointed the subcommittee chairman. Also, during the full-scale activity period related to this publication, Prof. Taro Arikawa (Chuo University) was appointed the subcommittee chairman. They took on the role with much enthusiasm. This book summarizes the achievements of the last six years. We hope that it will contribute to the future efforts of researchers and engineers in this field.

Finally, besides the Subcommittee Chairman Arikawa, former Subcommittee Chairman Araki, Vice Chairman Kotake (Toyo Construction) who acted as a coordinator, and all the members who participated enthusiastically in the activities, we would like to express our gratitude to the JSCE Coastal Engineering Committee for supporting the activities of the subcommittee.

Norimi Mizutani
Professor at Graduate School of Nagoya University
First Chairman of the Research Subcommittee on Methodology for Dealing with Geomaterials in Hydraulic Model Experiments
Nagoya, Japan, September 2021

Members List of Research Subcommittee on Methodology for Dealing with Geomaterials in Hydraulic Model Experiments

(As of April 1, 2021)

Norimi Mizutani	Graduate School of Engineering, Nagoya University (First Subcommittee Chairman)
Susumu Araki	Graduate School of Engineering, Osaka University(Second Subcommittee Chairman)
Taro Arikawa	Department of Civil and Environmental Engineering, Faculty of Science and Engineering, Chuo University (Third Subcommittee Chairman)
Yasuo Kotake	Technical Research Institute of Naruo, Technical Research Institute, Toyo Construction Co., Ltd. (Vice Chairman)
Katsuya Ikeno	Institute of Technology, Penta-Ocean Construction Co., Ltd.
Masaaki Ikeno	Meteorology and Fluid Science Division Sustainable System Research Laboratory, Central Research Institute of Electric Power Industry
Kazunori Ito	Innovation Strategy Department and Infrastructure Technology Research Department, Technology Center, Taisei Corporation
Yuri Ogata	Infrastructure Development Department, Civil Engineering Center, Tokyo Electric Power Services Co., Ltd.
Fuminori Kato	Coast Division, River Department, National Institute for Land and Infrastructure Management
Shin-ichi Kubota	Maritime and Coastal Hydraulics Group, Technical Research Institute, Fudo Tetra Corporation
Takaaki Kobayashi	Analytical Technology Department, Civil Engineering Design Division, Kajima Corporation
Yutaka Sawada	Graduate School of Agricultural Science, Kobe University
Shinji Sassa	Port and Airport Research Institute, National Institute of Maritime, Port and Aviation Technology
Kojiro Suzuki	Maritime Structures Group, Coastal and Ocean Engineering Department, Port and Airport Research Institute
Takayuki Suzuki	Graduate School of Urban Innovation, Yokohama National University
Takenori Shimozono	School of Engineering, The University of Tokyo
Hidenori Takahashi	Soil Stabilization Group, Geotechnical Engineering Department, Port and Airport Research Institute
Tetsuo Tobita	Faculty of Environmental and Urban Engineering, Kansai University
Tomoaki Nakamura	Graduate School of Engineering, Nagoya University
Eiji Harada	Graduate School of Engineering, Kyoto University
Kazunori Fujisawa	Division of Environmental Science and Technology, Graduate School of Agriculture, Kyoto University
Kenichi Maeda	Department of Civil Engineering and Systems Management, Nagoya Institute of Technology
Tatsuya Matsuda	Department of Architecture and Civil Engineering, Toyohashi University of Technology
Junji Miyamoto	Technical Research Institute of Naruo, Technical Research Institute, Toyo Construction Co., Ltd.

Contributors

Susumu Araki
Graduate School of Engineering, Osaka University, Suita, Japan (Second Subcommittee Chairman)

Taro Arikawa
Department of Civil and Environmental Engineering, Faculty of Science and Engineering, Chuo University, Bunkyo-ku, Japan (Third Subcommittee Chairman)

Kazunori Fujisawa
Division of Environmental Science and Technology, Graduate School of Agriculture, Kyoto University, Kyoto, Japan

Eiji Harada
Graduate School of Engineering, Kyoto University, Kyoto, Japan

Iwao Hasegawa
Ecoh Corporation, Taito-ku, Japan

Katsuya Ikeno
Institute of Technology, Penta-Ocean Construction Co., Ltd., NasuShiobara, Japan

Masaaki Ikeno
Meteorology and Fluid Science Division Sustainable System Research Laboratory, Central Research Institute of Electric Power Industry, Abiko, Japan

Kazunori Ito
Innovation Strategy Department and Infrastructure Technology Research Department, Technology Center, Taisei Corporation, Yokohama, Japan

Fuminori Kato
Coast Division, River Department, National Institute for Land and Infrastructure Management, Tsukuba, Japan

Takaaki Kobayashi
Analytical Technology Department, Civil Engineering Design Division, Kajima Corporation Minato-ku, Japan

Yasuo Kotake
Technical Research Institute of Naruo, Technical Research Institute, Toyo Construction Co., Ltd., Nishinomiya, Japan (Vice Chairman)

Shin-ichi Kubota
Maritime and Coastal Hydraulics Group, Technical Research Institute, Fudo Tetra Corporation, Tsuchiura, Japan

Kenichi Maeda
Department of Civil Engineering and Systems Management, Nagoya Institute of Technology, Nagoya, Japan

Tatsuya Matsuda
Department of Architecture and Civil Engineering, Toyohashi University of Technology, Toyohashi, Japan

Junji Miyamoto
Technical Research Institute of Naruo, Technical Research Institute, Toyo Construction Co., Ltd, Nishinomiya, Japan

Norimi Mizutani
Graduate School of Engineering, Nagoya University, Nagoya, Japan (First Subcommittee Chairman)

Tomoaki Nakamura
Graduate School of Engineering, Nagoya University, Nagoya, Japan

Yuri Ogata
Infrastructure Development Department, Civil Engineering Center, Tokyo Electric Power Services Co., Ltd., Koto-ku, Japan

Shinji Sassa
Port and Airport Research Institute, National Institute of Maritime, Port and Aviation Technology, Yokosuka, Japan

Yutaka Sawada
Graduate School of Agricultural Science, Kobe University, Kobe, Japan

Takenori Shimozono
School of Engineering, The University of Tokyo, Bunkyo-ku, Japan

Kojiro Suzuki
Maritime Structures Group, Coastal and Ocean Engineering Department, Port and Airport Research Institute, Yokosuka, Japan

Takayuki Suzuki
Graduate School of Urban Innovation, Yokohama National University Yokohama, Japan

Hidenori Takahashi
Soil Stabilization Group, Geotechnical Engineering Department, Port and Airport Research Institute, Yokosuka, Japan

Tetsuo Tobita
Faculty of Environmental and Urban Engineering, Kansai University, Suita, Japan

1

Introduction

Yasuo Kotake

This book focuses on "phenomena that are affected by the interaction between fluid (wave) and ground in a complex field," whose difficulty in handling has been highlighted for many years in the field of coastal engineering. Besides sorting out the ways of implementing model tests and the concept of the law of similarity which ought to be noted, it contrasts the phenomena with the laws of similarity and explains the points to note on the basis of the experimental cases. In the future, we will introduce the current status of numerical analysis techniques that can complement or replace model tests. By providing information that can be used by young researchers who aim at becoming civil engineers and engineers who have had difficulty in handling complex fields involving fluid (wave) and ground in practice, we hope that this publication will serve as a stepping stone to address current issues and aid future development besides fostering a shared understanding of complex phenomena in the boundary fields of hydraulics and geotechnical engineering.

Figure 1.1.1 shows schematically a typical phenomenon affected by interaction in a complex field consisting of fluid (wave) and ground that is covered in this book. Here, the complex field of or interaction between fluid (wave) and ground refers to the phenomenon in which the wave motion of the fluid that acts as an external force causes the ground to erode and soil to be deposited, which results in changes to the characteristics of the fluid field that further causes the behavior of the ground to change. Traditionally, in such interdisciplinary fields, research has been conducted independently by fields that specialize in geotechnical engineering or hydraulics. For example, in the geotechnical engineering approach, research has been conducted by treating ground as an aggregate of soil particles and focusing on the deformation characteristics of the ground as a "mass of soil (soil element)" with consideration to the effects of pore water such as saturation, unsaturation, and infiltration. Conversely, in the hydraulics approach, research has been conducted focusing on the phenomenon in which soil particles move because of the shear force acting on the surface of the ground on the assumption that the fluid motion caused by waves acts as an external force. Results that can be used in practice have been reported in each field, but the current situation is that research is being conducted on the basis of certain theoretical hypotheses and empirical models in each field and not all mechanisms in complex fields have been elucidated. Additionally, to shed light on the mechanism in the future, it has become increasingly necessary to merge independent fields and achieve a shared understanding. Thus, as introduced in the "Publication," the JSCE Coastal Engineering Committee established a Research Subcommittee on Methodology for Dealing with Geomaterials in Hydraulic Model

DOI: 10.1201/9781003424758-1

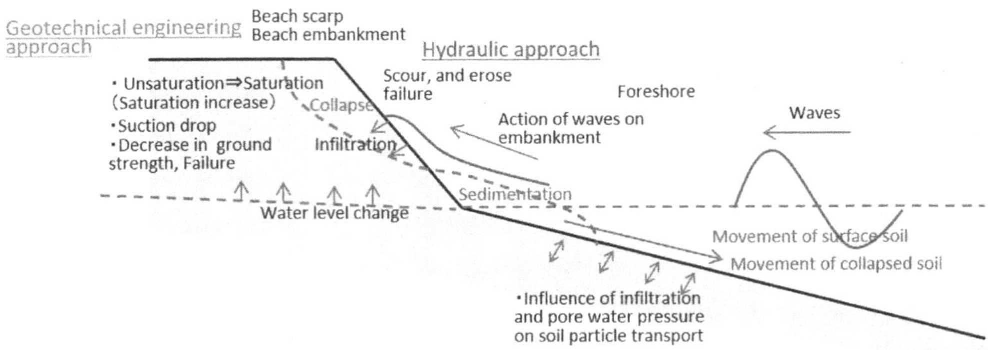

FIGURE 1.1.1
Typical phenomenon affected by the interaction of fluid (wave) and ground in a complex field.

Experiments, which is composed of researchers who are working on this theme in the fields of geotechnical engineering and hydraulics and promoted discussions toward the clarification of phenomena affected by the interaction between a fluid (wave) and the ground in a complex field.

In each field, besides theoretical analysis, model tests for elucidating phenomena have been actively conducted in the past. Recently, with the advancement of computers, computational techniques that can replace model tests have been proposed and it is hoped that mechanisms will be clarified by numerical calculations in the future. Nonetheless, model tests will continue to play an important role in the development of computational models.

In a model test, engineering-wise, how the scale is considered and how to apply the experimental result to the actual phenomenon are very important. Chapter 2 provides the history of the model tests conducted in each field and details of specific experimental methods, whereas Chapter 3 describes in detail the laws of similarity that are important in considering the model scale. Kindly read these chapters before proceeding further. Here, we will briefly introduce the laws of similarity and related dimensionless numbers used in Chapters 2 and 3, as well as in Chapter 4, which introduces specific examples.

Table 1.1.1 summarizes the main laws of similarity used in a fluid field and their related dimensionless numbers. Particularly, Froude's law is often used as a law of similarity for reproducing general coastal hydraulic phenomena in model tests, whereas Reynolds's

TABLE 1.1.1
Main Laws of Similarity and Dimensionless Numbers Used in Fluid Fields

Main Laws of Similarity		Corresponding Dimensionless Numbers	
Notation in This Book	**For Use in Fluid Fields**	**Term**	**Definition**
Froude's law	Gravitational dominant phenomena	Froude number	Ratio of inertia to gravity
Reynolds's law	Viscous dominant phenomena	Reynolds number	Ratio of inertia to viscous force

law is used for phenomena in which viscosity is dominant. These laws of similarity are used for conversion between the model and actual phenomenon by matching the dimensionless numbers listed in the table as corresponding dimensionless numbers on the model scale and actual scale. These are sometimes referred to as "Froude law of similarity," "Froude's law of similarity," "Reynolds law of similarity," or "Reynolds's law of similarity," but in this book, they are referred to as "Froude's law" and "Reynolds's law."

Next, Table 1.1.2 introduces the Shields number and the Dean number as the main dimensionless numbers used for complex phenomena involving fluids and ground. Shields number has traditionally been used as a dimensionless number of the friction at the bottom to express sand movement on the seabed in relation to local drift sand and scouring phenomena, and it is often referred to as a dimensionless tractive force in the area of rivers. Conversely, the Dean number, as described in the definition, represents the ratio of the horizontal wave power acting on the sand particles to the gravity in the vertical direction and is a dimensionless number used when terrain changes are subject to study. As it has a shorter history compared with the Shields number, the Dean number appears to be often written in alphabetical notation in past references. In this book, it is referred to as the "Dean number" to make it more readable.

The contents of each chapter are introduced below. Historically, in many cases, the same parameters are expressed by different symbols in the fields of hydraulics (or coastal engineering) and geotechnical engineering. For this reason, although explanations of the symbols are provided as close to the text as possible in this publication, note that the same symbol may refer to different parameters depending on the chapter or section.

First, as mentioned earlier, Chapter 2 summarizes model test methods adopted in the fields of hydraulics and geotechnical engineering. The experimental technique is described in a concise and very detailed manner that also includes past examples, so it may be used as an introduction to those who are going to start conducting experiments in the future. Conventional techniques in the fields of hydraulics (or coastal engineering) and geotechnical engineering are introduced, and a section on the review on the handling of ground materials and organization of issues is also included. It is particularly recommended for readers doing research in the field of boundary.

Next, Chapter 3 introduces the law of similarity used in each field. Current issues are organized in an easy to understand manner by unraveling the history of the law of similarity, introducing the laws of similarity that have been applied so far for phenomena of interest in hydraulics and geotechnical engineering, and applying such laws to

TABLE 1.1.2

Major Dimensionless Numbers in Combined Fluid and Ground Fields

Major Dimensionless Numbers	Definition	Situations for Use in Combined Fields
Shields number	Ratio of shear force (bottom friction force) acting on the seafloor surface to the static shear resistance of the bottom sediment	Movement of sand particles due to shear force of fluid acting on the seafloor surface
Dean number	Ratio of the force acting on a sand particle (horizontal wave force) to gravity (vertical force)	Topographic change with sedimentation of sand particles in sand drift and scour

complex problems of fluid (wave) and ground. By reading Chapters 2 and 3 in order, you will probably be able to understand the concept of model tests and the points to keep in mind when translating the results of model tests into those on the scale of the site. Nevertheless, please note that some of the latest findings introduced here must be verified while accumulating various data in the future.

Case studies are provided in Chapter 4 on "phenomena that are affected by the interaction between a fluid (wave) and the ground in a complex field," which is the most important among the phenomena covered in this book. The first case is a phenomenon in which the section below a permeable structure composed of rubble and wave-dissipating concreter blocks is scoured and the embankment sinks and is an example where the structure is involved. In this case, the "Research Subcommittee on Methodology for Dealing with Geomaterials in Hydraulic Model Experiments" has set up a full-scale experiment close to the actual scale as a basic model for a disaster that actually occurred in a coastal area. The results of the study conducted at the research facility of the institution to which the members belong will be utilized for studying differences in scale and experimental materials and to compare hydraulic model tests conducted in a centrifugal force field using a geotechnical centrifuge device that is adopted in geotechnical engineering and hydraulic model tests conducted in a typical gravitational force field that is adopted in the field of hydraulics. The second example is a phenomenon in which beach nourishment sand shaped like an embankment is eroded and forms beach scarps, as shown in Figure 1.1.1, and an experiment was conducted to develop a technology that will contribute to effective and efficient beach nourishment. These two cases are based on experiments conducted on actual disaster cases that occurred in coastal areas, and the design concepts, including the points to note when planning for the structure, are also carefully explained. You will probably be able to get a better understanding by mentally visualizing the actual phenomenon in your mind as you read it.

Finally, information on cutting-edge numerical calculations is introduced in Chapter 5. In the past, efforts were made to elucidate phenomena by making full use of model tests that consider the law of similarity, but as problems become more and more complex, there is a growing number of model tests that are similar to those presented in previous chapters and events for which the law of similarity cannot be applied easily. Conversely, numerical analysis technology has become more sophisticated because of improvements in the performance of computers, and techniques having a performance that can replace model tests have also been proposed. Nevertheless, the numerical analysis model covered in this book that deals with the boundary region between the fluid and ground and interconnects between the two is still under development at the moment. Thus, in this chapter, information on the latest numerical analysis that is expected to be utilized in this field in the future is also described.

As introduced so far, each chapter is written by researchers of their respective fields. Although the chapters are generally related to one another, they are written such that each of them can be read individually. This book is structured such that readers can get a sufficient understanding even if they start reading from the chapter that they are interested in. We hope that it will be used by not only young researchers aiming to pursue research in civil engineering but also engineers who already have a lot of practical experience.

2

Methodology of Model Tests

Tatsuya Matsuda, Iwao Hasegawa, Shin-ichi Kubota, Yutaka Sawada, Hidenori Takahashi, Tetsuo Tobita, and Tomoaki Nakamura

2.1 Purpose of Model Tests

In civil engineering, it is necessary to work out appropriate and optimal solutions for complicated conditions and unknown phenomena, such as the construction of structures, the creation of earth structures, the clarification of natural disaster mechanisms, and measures against such disasters. For example, to clarify a phenomenon that may occur, useful information can be obtained if the phenomenon can be examined using the actual object or a large-scale model test. However, when conducting an experiment on the prototype or a large-scale model test, various problems may arise in the cost, equipment, reproducibility, and safety of the experiment.[2.1.1]

Thus, stringent theories, model tests, and numerical analyses have been used to solve problems in civil engineering regarding phenomena that may occur. This method is adopted in engineering in general, but one of the characteristics in civil engineering is that scale model tests play a very important role in the clarification of phenomena and the demand is still high at present.

The appeal of model tests is that the result (output) is always a physical quantity relative to the set conditions (input), and the phenomenon can be visually observed as an image. What is important here is how the model test is conducted. Even if the input conditions are correct, depending on how the experiment is carried out, a result completely different from the correct answer may be obtained. Thus, in conducting a model test, how the way of reproducing a phenomenon that may occur with the prototype is devised is important.

In the field of coastal engineering, model tests have been conducted for many years to clarify the mechanism of various phenomena occurring at the coast, such as waves, currents, and beach deformation. Additionally, model tests are positioned as a general technique for checking the functions of structures against waves, determining the optimal structures, and so on in the design of various structures, and many experiments have been conducted so far. Initially, various observations and measurements were conducted by generating regular waves in wave tank, but subsequently, as the importance of irregularity in the waves acting on coastal structures became more and more recognized, experiments using irregular waves advanced rapidly. In line with this, experimental methods unique to irregular waves were developed, and at present, multidirectional spectral waves are reproduced in wave tank. Even now, after numerical

DOI: 10.1201/9781003424758-2

calculation techniques have advanced significantly, the importance of model experiments remains high, and experiments with a good level of accuracy are required to determine various coefficients necessary for numerical calculations and to verify the validity of calculation models.

In the field of geotechnical engineering,[2.1.2, 2.1.3] model tests have been carried out to elucidate the mechanical properties of the soils and to clarify soil failure and interaction between the soils and structures. According to Mikasa,[2.1.2] model tests on earth pressure theory, bearing capacity theory, etc. have been conducted in the past, treating the geomaterial as an elastic body, to elucidate the mechanical properties of the geomaterial. Later, mechanical tests and in situ tests were mainly conducted after Terzaghi[2.1.4] pointed out the importance of the physical and mechanical properties of soil. However, as the behavior of the geomaterial is complicated, it is necessary to clarify the mechanism of slip surface and crack generation, progressive failure,[2.1.2] behavior as a granular body, and so on. To date, model tests are conducted as part of the technique used for clarification. In model tests in geotechnical engineering, scaling law have been derived[2.1.5] based on constituent factors that govern the mechanical properties of geomaterial and earth's gravity model tests have been conducted.

Conversely, when considering problems about the stability of a soil structure, bearing capacity related to the interaction between the structure and soil, etc., the condition of stress on the soil in the model must be equivalent to that of the prototype as stress on the soil because of its own weight is important. Thus, to reproduce a stress field that is equivalent to the prototype in the model, centrifuge model tests have been conducted by obtaining a law of similarity related to the experiments under the condition in which centrifugal force is applied to the model.[2.1.6]

In this chapter, we will explain in concrete detail the information necessary for conducting experiments based on the methods and measurement developed so far in the various fields of coastal engineering and geotechnical engineering. Particularly, by explaining the information with examples of experiments in each field, we tried to provide information useful for those who are going to carry out experiments in the future. At the end of this chapter, examples of past experiments on complex fluid–ground problems, which is an important theme of this book, are given and issues related to the handling of geomaterials are identified and sorted out.

2.2 Hydraulic Model Test

2.2.1 Model Test Plan

2.2.1.1 Setting the Model Scale

In the field of coastal engineering, various phenomena caused by external forces such as waves and tides are subject to testing. Examples of such phenomena include the wave force acting on a breakwater, stability of wave-dissipating concrete blocks, wave height distribution in a harbor, and changes in shorelines and seafloor topographies. The experiment is called a hydraulic model test and is carried out using a scale model. In an experiment using a scale model, a mechanically similar relationship between the actual object and the model must be maintained, and what determines the relationship is known as the law of similarity. For example, when the scale of the length is determined, the law of similarity can be said to indicate how other physical quantities should be scaled down.[2.2.1]

The following three forces are the main forces that cause hydraulic phenomena. The first force is the inertia force. This is the force generated by water motion and is expressed as the product of mass and acceleration. The second force is gravitational force, which is expressed as the product of mass and gravitational acceleration. The third force is the viscous force, which is the force generated by the viscosity of the fluid. The three forces cannot have the same scale in the model unless the experiment is carried out with the actual object.[2.2.1] Thus, in an experiment using a scale model, how the similarity conditions are established can be said to be a problem and an opportunity for using engineering ingenuity. Details of the law of similarity are given in Chapter 3, but in hydraulic model tests, the motion of a body due to the action of waves and currents is often targeted; thus, it can be said that the inertial force needs to be made similar first. Thus, generally, experiments are conducted by matching either the ratio of the inertia force and gravitational force (Froude's law) or the ratio of the inertial force and viscous force (Reynolds's law). In the field of coastal engineering, especially in hydraulic model tests on structures, gravitational force is often superior to viscous force; thus, experiments are often performed according to Froude's law.[2.2.1, 2.2.2]

As for the specific scale in the experiment, the maximum value of the scale is first determined by the size of the wave tank used and the wave-making capability. That is, restrictions are imposed on the maximum scale at which the sea conditions and topographical conditions of the experimental object can be reproduced in the wave tank. Next, in acquiring data using a measuring instrument, the minimum value of the scale should be considered to ensure the reliability of the data. Although it is ideal to set the scale as large as possible based on the above, since experiments in a larger scale cost more, in reality, the appropriate model scale is often decided after considering the cost-effectiveness as well.[2.2.1]

2.2.1.2 Experimental Facility

Depending on the shape of the wave tank used as an experimental facility, the wave tank may be a wave flume (Figure 2.2.1) or a wave basin (Figure 2.2.2). Various organizations in and outside Japan have wave tanks of the size shown in Figures 2.2.3 and 2.2.4.[2.2.3] In a wave flume, the cross-sectional characteristics of the hydraulic phenomenon can be reproduced. For example, a wave flume is often used in examining the sliding stability of a caisson of a breakwater, the stability of a wave-dissipating concrete block, the transmitted wave height, the wave reflection, and the wave overtopping rate of a revetment. In a wave basin, the three-dimensional characteristics of hydraulic phenomena can be reproduced. For example, a wave basin is often used in various studies concerning the head of a breakwater or the corner of a revetment.

A wave generator for creating waves is attached to the wave tank. Various wave generators that adopt different wave-making methods are available, but piston-type machines (that make waves by moving a vertical paddle back and forth) or flap-type machines (that make waves by moving the top end of the paddle back and forth with the lower end hinged) are often used. In general, besides regular waves, it is possible to create

FIGURE 2.2.1
Wave flume.

FIGURE 2.2.2
Wave basin.

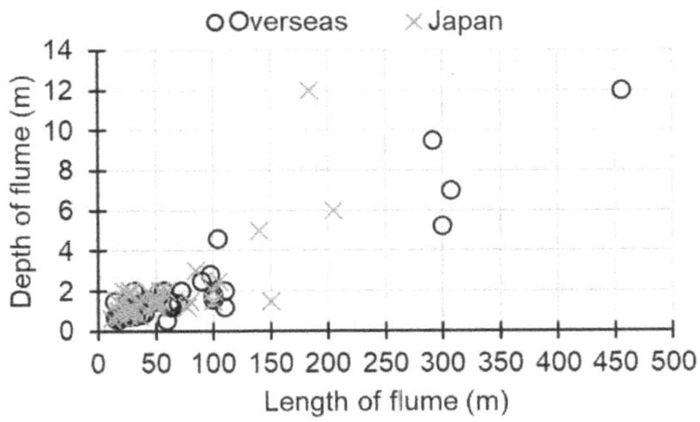

FIGURE 2.2.3
Size of wave flume.

irregular waves that reproduce the frequency spectrum of the waves. Additionally, a multidirectional irregular wave generator that can reproduce the directional spectrum of waves may sometimes be installed in a basin. Many wave generators in recent years can absorb reflected waves.

2.2.2 Classification and Target of Model Tests

Hydraulic model tests are roughly divided into wave flume experiments and wave basin experiments: an experiment using a wave flume and an experiment using a wave basin (shown in Figures 2.2.1 and 2.2.2, respectively).

Additionally, a model bed (fixed bed and movable bed) that imitate the ocean floor is set up in the wave tank according to the purpose, and the experiments are termed respectively fixed bed experiments and movable bed experiments.

2.2.2.1 Wave Flume Experiment

In a wave flume, one of the sidewalls is made of glass, allowing the observation of the wave action from the side. (Some facilities use acrylic instead of glass or have a flume with two of the sidewalls made of glass.) Generally, a wave flume is deeper than a wave

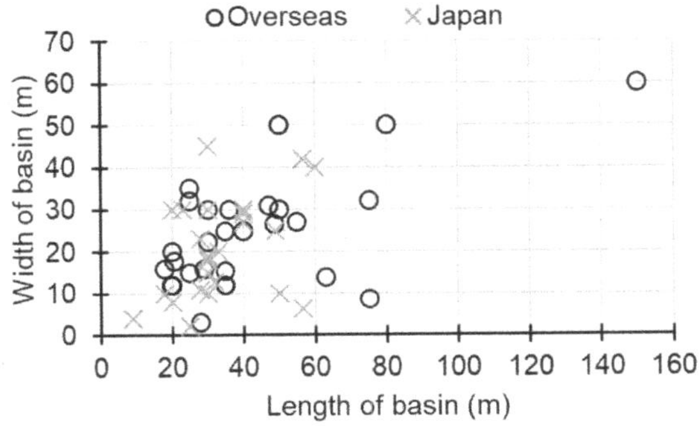

FIGURE 2.2.4
Size of wave basin.

basin and has an enhanced wave-making capability, allowing experiments to be conducted with a larger model than a wave basin. These features make it suitable for the development of new structures and the conduct of experiments for clarifying the phenomena. Besides being able to acquire data such as water level, flow velocity, and wave pressure, model tests offer the advantage of being able to visually observe the phenomenon. As shown in Figure 2.2.3, the larger wave flume in Japan has a length of 150–200 m, but the number of institutions that can use such a large flume is limited. Even in a wave flume of a typical scale, a model of the largest possible scale in terms of cost-effectiveness is adopted to reduce the influence caused by using a scale model.

Wave flume experiments are often adopted when examining standard parts of continuous structures such as beaches, coastal revetments, and breakwaters. In this case, a model having a uniform cross-sectional shape is set up in the width direction of the wave flume. Care must be taken when installing a model with a different cross-sectional shape in the width direction because water surface vibration may be created in such a direction. In some cases, experiments are conducted upon installing a column-shaped structure in the flume tank. In this case, a structure model is generally positioned at the center of the width of the wave flume. Since waves and currents pass between the structure model and the flume wall, the influence of the flume walls is included in the experiment results. In a model test for evaluating the force of tsunami waves acting on a rectangular structure installed in a wave flume, it has been emphasized that if the width of the wave flume is smaller than four times the model width, the effect of the flume wall is large, and if it is larger than eight times the model width, the effect of the flume wall is small.[2.2.4]

Several experiments have been conducted in a wave flume to confirm the stability of wave-dissipating concrete blocks and covering blocks. In a wave flume, the condition inside the flume can be seen through the glass; thus, the swaying movement of the block caused by the wave action can be observed. In an experiment using wave-dissipating concrete blocks, the porosity is made equal to the prescribed value by calculating the number of blocks that fit within a cross section of the wave-dissipating work and creating a cross section with the prescribed quantity. If the porosity is not in conformity, it will affect the hydraulic performance and stability of the wave-dissipating work. A wave-dissipating concrete block that is in contact with the flume wall is more likely to get detached than a block that is sufficiently engaged because of lack of engagement. Thus, blocks along the flume walls are often excluded from stability evaluation.

In experiments conducted under the condition that waves rising over the crown of the experimental model are transmitted to the back, the water level in front of the model may decrease and the water level behind the model may increase because of wave overtopping. If the water level change caused by this becomes a problem, the overtopped water mass must be returned to the offshore side. Examples of the methods are described below:

i. Use a circulation device installed in the wave flume.
ii. Use a pump to transmit water behind the model to the offshore side.
iii. Transmit water through the model bed by making water passage openings on the land side and the offshore side of the model bed.
iv. Set up a partitioning wall in the width direction of the wave flume, install an experimental model on the glass wall side, and use the opposite side as a circulation channel.

In Method (iv), by installing a wave-dissipating material on the circulation channel side, it is possible to prevent generated waves from affecting the back of the model through the circulation channel.

By pushing the paddle from a pulled state in one go, a solitary wave can be generated. Experiments treating this solitary wave like a tsunami are sometimes conducted. In a wave flume equipped with a wave generator and a water-circulating device that uses a pump, it is possible to study the simultaneous action of tidal currents and waves. Experiments on tsunami flow can be conducted if the pump capacity of the circulation device is large enough. In a wave flume equipped with a blower besides a wave generator, the effect of wind on the overtopping wave can be considered.

2.2.2.2 Wave Basin Experiment

Since a basin is wide in the width direction, the three-dimensional effect of seafloor topography, shape of structures, etc. can be examined. That is, it is possible to examine the wave convergence due to the seafloor topography, the influence of the corners and edges on the wave overtopping rate of the revetment, and the stability of the covering material at the head of the breakwater. Additionally, oblique incident waves that cannot be handled in a wave flume experiment can be studied. Since it is possible to change the wave making an angle with a wave generator that can create multidirectional irregular waves, it is possible to perform experiments with different incident angles without changing the installation angle of the experimental model. When conducting experiments on multidirectional irregular waves or oblique waves, ensure that the area to be tested falls within the effective wave-making area. In a basin in which multidirectional irregular wave generators are installed on two of the basin walls, the effective wave generation area can be widened.

When dealing with beach deformation in a movable bed experiment to be described later or when examining beach deformation on the basis of the measurement results of a fixed bed experiment, matters such as model layout are studied with consideration given to the effects the basin walls that do not exist in the actual location have on the coastal current.

In a wave basin equipped with a flow generator besides the wave generator, a study involving simultaneous action of tidal currents and waves can be conducted. Depending on the structure and usage of the flow generator, changes in the tide level during wave action can also be controlled.

Experiments using a wave basin often have to be smaller in scale than experiments using a wave flume. In wave height measurement such as harbor tranquility experiment, the model can be scaled down to a level at which the wave height becomes as small as several centimeters. Nevertheless, in experiments targeting the stability of structures, it is desirable to adopt a scale that will result in a wave height of at least 10 cm.

2.2.2.3 Fixed Bed Experiment and Movable Bed Experiment

A model bed simulating the seabed is installed in a wave tank. A fixed bed experiment is an experiment conducted using a model bed made of materials such as mortar, mainly to understand the fluid field, including changes in the water level and flow. Conversely, a movable bed experiment is an experiment conducted using a model bed made of sand, targeting changes in the seafloor due to waves and currents. A fixed bed experiment can be said to be aimed mainly at understanding fluid fields such as changes in the water level and flow. Experiments include measurement of the stability of structures that arise as a result of changes in these fluid fields and measurement of external forces like wave force. The movable bed experiment targets the sediment scouring phenomenon and topographical changes caused by waves and currents.

2.2.3 Target and Method of Measurement

In measuring various phenomena, a device suitable for the measurement target should be selected and the measurement must be performed appropriately. The experimental data measured using the device is stored in a storage medium as digital data by A/D conversion. Digital data are obtained by discretizing analog data that are originally continuous; hence, care must be exercised in handling the data to properly capture the target phenomenon.

2.2.3.1 Wave Height

Wave height is one of the items that is always measured in hydraulic model tests. A capacitance wave gauge or a servo-type wave gauge is often used for measurement. Figure 2.2.5 shows an example of the probe and main unit (amplifier) in a capacitance wave gauge. The probe is installed at the measurement position, and the wave height is measured by using the fact that the electrical capacitance between the capacitance line of the probe and the water varies according to the water level. A servo-type wave gauge drives a servo motor so that the resistance between the needle electrode and ground electrode is always constant and measures the position of the needle electrode with a potentiometer as a change in the water level.[2.2.1]

2.2.3.2 Flow Velocity

Propeller and impeller current meters, electromagnetic current meters, ultrasonic current meters, laser doppler velocimeter, etc. are used to measure the flow velocity.[2.2.1]

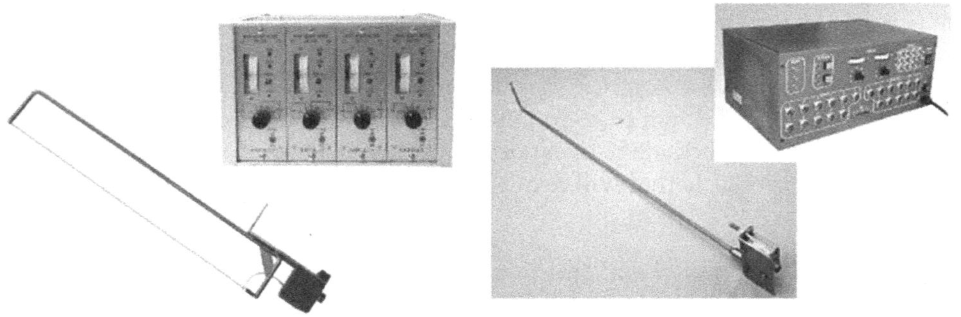

FIGURE 2.2.5
Capacitance wave gauge. With permission from (left) KENEK Co. Ltd. and (right) Mitsui E&S Systems Research Inc.

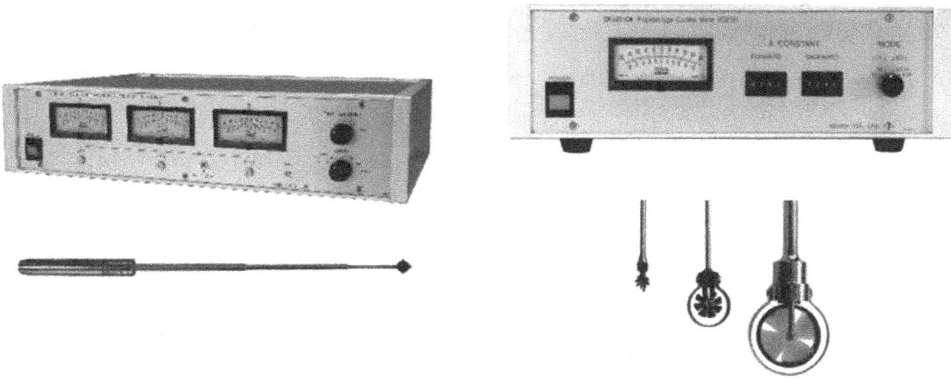

(Electromagnetic Current Meter) (Propeller and Impeller - Types Current Meter)

FIGURE 2.2.6
Current meters. With permission from KENEK Co. Ltd.

Figure 2.2.6 shows examples of a current meter. A propeller current meter can detect flow velocity components in one direction, whereas an electromagnetic current meter can detect flow velocity components in two or three directions, and an ultrasonic current meter can detect flow velocity components in three directions. The probe will be damaged by heat if the electromagnetic current meter is exposed to air while in the measurement mode. For this reason, it cannot be used unless it is constantly under water. The propeller current meter can be used under conditions in which it repeatedly goes in and out of the water. Nevertheless, noteworthily, even if the propeller is out in the air, the flow velocity will be detected when the propeller is rotated by the wind.

2.2.3.3 Wave Force, Wave Pressure

The force acting on the structure may be measured by measuring the force (wave force) acting on the entire embankment or by measuring the wave pressure at any location.[2.2.1] A multicomponent load cell (Figure 2.2.7) is used for measuring wave force. Items that can be measured include horizontal force, vertical force, and moment. Depending on the number of measurable components, there are three-component load cell, six-component load cell, etc. A wave pressure gauge (Figure 2.2.8) is used to measure the wave pressure. It is attached to the surface of a structure to measure the acting wave pressure as pressure fluctuation. Both the multicomponent load cell and the wave pressure gauge are measuring instruments that provide changes in physical quantities (wave force and wave pressure) by measuring strain.

2.2.3.4 Suspended Sediment Concentration Meter (Turbidity Meter)

The device is used to measure the concentration (turbidity) of fine particles, not just suspended sediments, present in water (Figure 2.2.9). The main ways to measure the degree of mixing of fine particles are to use the changes in the amount of transmitted light and to use the change in the backscatter of irradiated light.

FIGURE 2.2.7
Three-component load cell. With permission from Kyowa Electronic Instruments Co. Ltd.

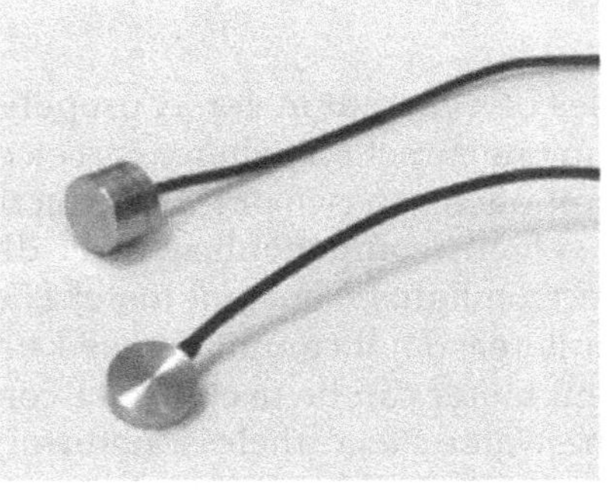

FIGURE 2.2.8
Micro pressure gauge (wave pressure gauge). With permission from SSK Co. Ltd.

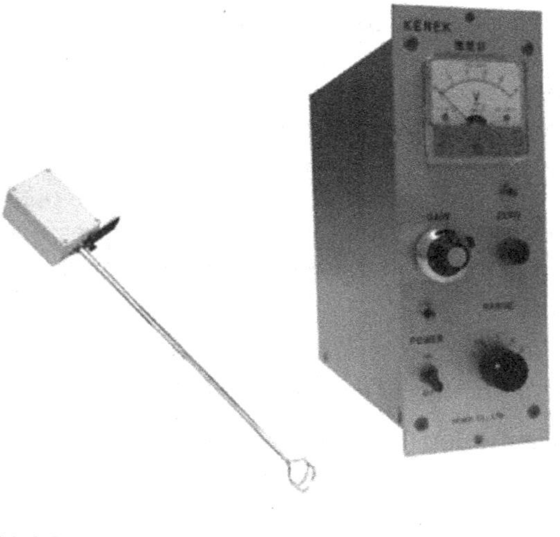

FIGURE 2.2.9
Suspended sediment concentration meter. With permission from KENEK Co. Ltd.

FIGURE 2.2.10
An example of fixed bed.

2.2.4 Model Test Procedure

2.2.4.1 Model-making

a. Making of model bed
A model bed that reproduces the sea-bottom slope and local topography is installed in the wave tank. In a wave flume, a linear sea-bottom slope is reproduced, and a structural model is often installed at the position to be examined. As mentioned in Section 2.2.2.3, the model bed has a fixed bed and a movable bed. Examples of each are shown in Figures 2.2.10 and 2.2.11, respectively. A fixed bed is a reproduction of the seabed with a fixed floor surface. Some consist

FIGURE 2.2.11
An example of movable bed.

of a floor slab installed on top of a steel or stainless steel frame, whereas others consist of a mortar floor installed on top of crushed stone. A movable bed reproduces the seabed using actual sand. The specifications of the sand used in past experiments are introduced in Section 2.4.

b. Making of structural model

Depending on the need, the structural model shall be made of acrylic, wood, etc. Models whose behaviors are subject to testing such as breakwater caisson models and wave-dissipating concrete blocks shall be manufactured so that their specific gravity is the same as the actual object. When the behavior of the caisson is the focus of the experiment, consideration should be paid to ensure that the mass of the model does not change because of the ingress and egress of water into the caisson model. A caisson model shall be made of mortar, wood, or acrylic. It is desirable to place a weight inside it to adjust the specific gravity and to make the roughness of the bottom surface the same coefficient of friction as in the field. Additionally, wave-dissipating concrete blocks are often manufactured using mortar with an adjusted specific gravity. Figures 2.2.12 and 2.2.13 show the examples of structural models.

2.2.4.2 Installation and Calibration of Measuring Instruments

Sensors have measurement limits and resolution levels, so when measuring physical quantities in various experiments, whether sensors used in the experiments can take measurements properly must be confirmed.

a. Wave gauge

A wave gauge shall be installed via a trolley or pedestal on top of the wave tank. When using a wave gauge, an elevator (a device for moving the wave gauge up and down) is often used together. The wave gauge shall be installed

FIGURE 2.2.12
A model of caisson breakwater (center: made of acrylic, both sides: made of wood).

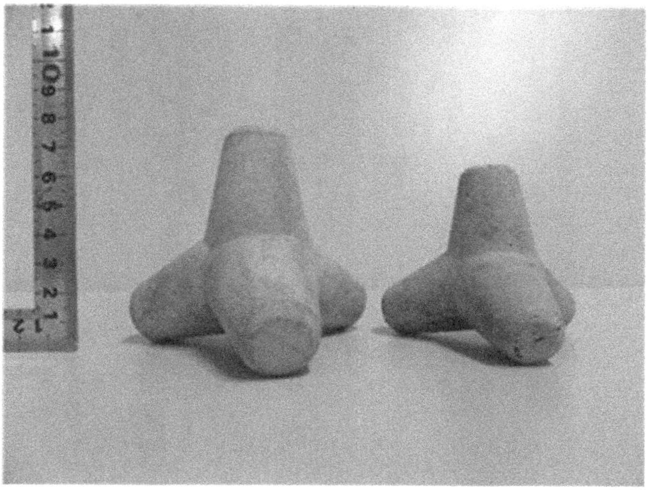

FIGURE 2.2.13
A model of wave-dissipating concrete block (made of mortar).

so that it is vertical to the water surface and that the sensor unit is located in the center of the channel as much as possible. Figure 2.2.14 shows an installation example of a capacitance wave gauge.

Calibration is the work performed to correlate the signal output from a sensor and the physical quantity of the target of measurement. For details on the calibration method, refer to Section 2.2.4.2 (e). Calibration of the wave gauge shall be performed by moving the wave gauge up and down in a filled wave tank so that the wave height subject to observation can be measured. The calibration range (maximum and minimum water levels) shall be determined so that the wave height to be observed can be measured. The response relationship between the displacement and voltage shall be obtained by moving the wave gauge up and down within the specified range with the water surface steady. This response relationship is used when converting voltage detected by the sensor into displacement. A capacitance wave gauge has several detector lengths. It is desirable to select a detector with a length appropriate for the height of the wave.

b. Current meter

Like the wave gauge, a current meter shall be installed via a trolley or pedestal on top of the wave tank (Figure 2.2.15). It shall be installed keeping in mind the mounting direction of the sensor and the direction of the flow to be measured. Bidirectional electromagnetic current meters have detectors that measure two horizontal directions and detectors that measure one horizontal direction and one vertical direction. The detector appropriate for the purpose shall be selected.

Since the electromagnetic current meter creates a magnetic field around the detection unit, if two or more detectors are installed close to each other, they will interfere with each other. Since the effect of interference differs depending on the combination of each product, a combination that does not cause

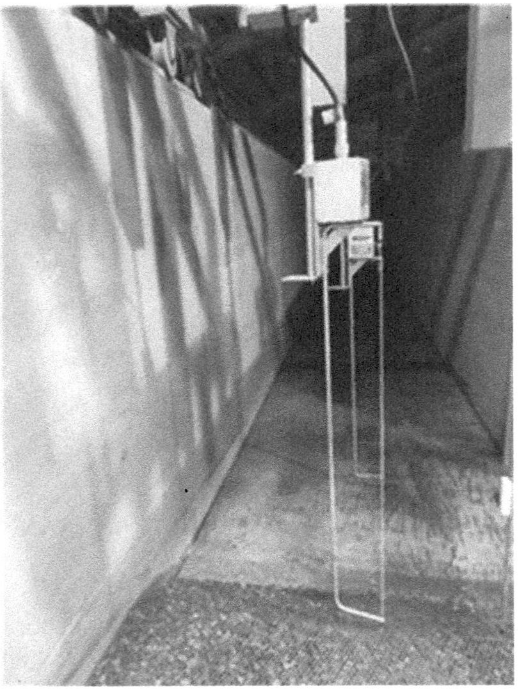

FIGURE 2.2.14
An example of capacitance wave gauges setup.

interference easily shall be selected and the installation interval that does not result in interference shall be checked.

c. Multicomponent load cell, wave pressure gauge

A multicomponent load cell may be installed directly under or behind the structure model to be measured. When installing the sensor directly underneath, a space for installing the multicomponent load cell shall be provided in the model bed and it shall be attached firmly to the bottom (Figure 2.2.16). Also, to measure the wave force acting on the model appropriately, it should be confirmed that the model is attached securely to the multicomponent load cell. The relationship between the rated load of the multicomponent load cell and the weight of the model should also be paid close attention. For example, if a 100 *N* model is installed on a multicomponent load cell with a rated load of 100 *N*, the rated load will be used up by just the weight of the model. The multicomponent load cell may be damaged just by installing the model if the weight of the model is larger than the rated load. The wave pressure gauge shall be attached to the surface of the structural model to be measured. When installing the instrument, it is often embedded so that the wave pressure measuring surface does not protrude from the surface of the structure (Figure 2.2.17). In such a case, in a structural model made of acrylic or other materials, a hole for embedment at the mounting position of the wave pressure gauge should be provided beforehand. A highly sensitive wave pressure gauge can easily malfunction if a pressure

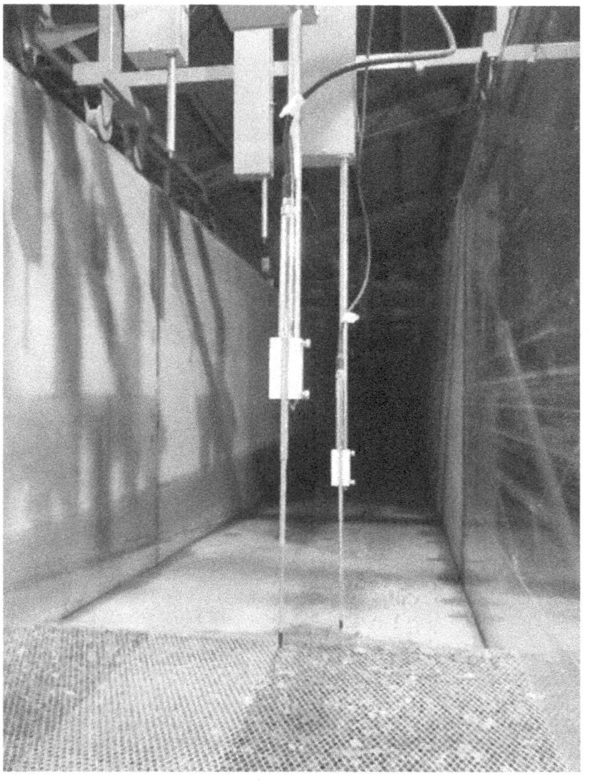

FIGURE 2.2.15
An example of electromagnetic current meters setup.

FIGURE 2.2.16
An example of three-component load cell setup.

FIGURE 2.2.17
An example of pressure gauges setup.

exceeding the rated load is applied to the pressure receiving surface; thus, care should be taken when installing the wave pressure gauge or model.

In an experiment to measure wave force and wave pressure, the components of the frequency corresponding to the natural vibration may be amplified and measured as they are affected by the natural vibration of the measurement system comprising the securing jig of the multicomponent load cell, the multicomponent load cell, and the experimental model attached to the multicomponent load cell. An impact test of the measurement system shall be performed in advance to investigate the natural frequency, and the effect of the natural vibration of the measurement system included in the measurement data shall be studied.

d. Suspended sediment concentration meter (turbidity meter)

Regardless of whether it is a laboratory experiment or field survey, when measuring suspended sediment that is directly above the bottom surface, the measuring instrument should be carefully installed so that it does not affect the suspension of the sediment. Additionally, background turbidity (regular turbidity) may be present, especially in the local measurement area; hence, care must be taken when taking measurements. The amount of transmitted light and the reflection intensity differ depending on the difference in the target fine particles (for example, the difference in the grain size of the sediment). Thus, calibration must be carried out before performing the measurement; the relationship between the actual concentration and the changes in the amount of transmitted light or reflection intensity, which are output values from the measuring instrument, should be determined; and the results must be converted into concentration after the measurement. For details on the calibration method, refer to Section 2.2.4.2 (e).

e. Calibration

Values measured by the measuring instruments (sensors) used in the experiment are not physical values (acceleration, pressure, displacement, etc.). For example, in the case of voltage-type measuring instruments, the voltage (V) is the output, whereas in the case of strain-type measuring instruments, the measured value of the change in the strain ($\mu\varepsilon$) is the output. Hence, the relationship between the measured value and the physical value should be investigated beforehand. This is known as calibration.

One-way to calibrate an instrument is to add a certain known physical quantity (referred to as a measured quantity) to the sensor and record the output value from the sensor. Several measured quantities and their output values shall be collected, and an approximate expression of a linear function ($y = a_1 x + a_2$) shall be obtained on the basis of the series of data (Figure 2.2.18). Here, a_1 and a_2 are calibration values (calibration factors) of the sensor. Normally, when the measured quantity is 0, the sensor output is assumed to be 0, so $a_2 = 0$. When you purchase a sensor, the manufacturer will calibrate it in advance and send you the calibration certificate; therefore, it is good to use this as the original calibration result to check whether the calibration that you performed yourself is valid. Additionally, there may be cases where appropriate values cannot be obtained because of distortion of the sensor resulting from the conditions and frequency of use. To that end, calibration shall be performed before the experiment to confirm that the sensor is operating normally.

2.2.4.3 Calibration of Test Waves

Before installing the structure model and other items, a calibration of test waves shall be conducted with just the model bed. It is a process for clarifying the relationship between the amplitude of the paddle (signal level to be entered into the wave generator) and the wave height in advance by generating waves under experimental conditions like water depth and period. The test is generally known as a "passing wave test" or "experimental wave test." Specifically, the wave heights at two locations at a deeper water section on the offshore side (offshore wave heights) and the wave heights at two locations sandwiching the structure installation position that is subject to testing (onshore wave heights) shall be measured. The incident wave height at each position shall be obtained by the resolution method of incident and reflected waves. Since the wave height changes drastically

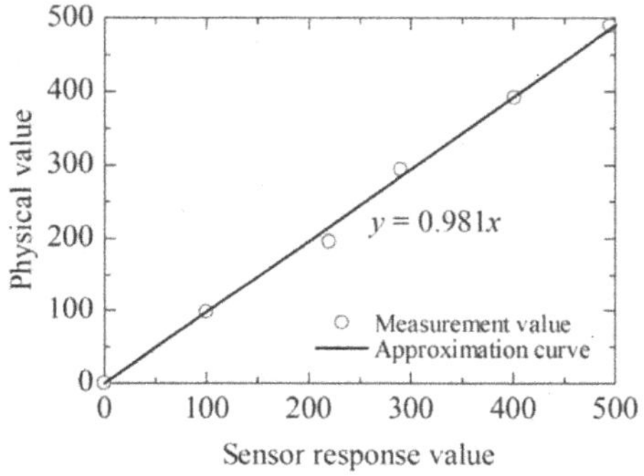

FIGURE 2.2.18
An example of calibration.

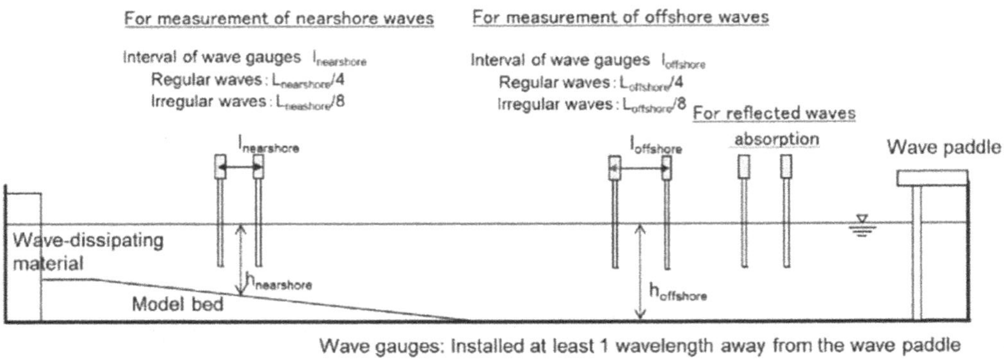

FIGURE 2.2.19
An example of wave gauges arrangement for calibration of test waves.

near the reflecting boundary such as the paddle, the wave gauge shall be installed at least one wavelength away from the paddle. See Goda et al.[2.2.5] for details on the resolution method of incident and reflected waves. Figure 2.2.19 shows an example of the installation of a wave gauge during a passing wave test. The wave-making signal is adjusted so that the significant wave period in the deeper water section matches the targeted significant wave period. Furthermore, the relationship between the wave height at a deeper water section (offshore wave height) and the wave height at the structure installation position (passing wave height) shall be obtained beforehand by changing the input wave height. Assuming that this relationship does not change even if a structural model is installed, the wave height at the structure installation position (passing wave height) shall be estimated from the wave height at a deeper water section (offshore wave height). It is desirable to install a wave-dissipating material at the end of the wave flume to suppress reflected waves from the wall at the end of the flume.

In an experiment involving re-reflected waves from a wave paddle, such as an irregular wave experiment, data such as those on waves are generally recorded after multiple reflections are formed in the wave flume. For the waiting time from the start of wave generation to the start of data recording, you may refer to Tanimoto et al.[2.2.6]

The two wave gauges at the right end of Figure 2.2.19 are wave gauges for controlling the absorption of reflected waves from the wave paddle.[2.2.7] Besides this, a method of controlling the absorption of reflected waves using a single wave gauge fixed to the wave paddle[2.2.8] is available. In an experimental facility that does not have a reflected wave absorption control function, the wave-making height may vary if the reflection coefficient in the passing wave test is different from the reflection coefficient after the experimental model is installed. When the wave reflected by the experimental model is re-reflected by the wave paddle, it reaches the model together with the wave made. If the reflection coefficient when the model is installed is larger than the experimental wave test, the re-reflected wave will be larger; thus, the height of the wave reaching the model will be larger than the wave in the passing wave test. Since the re-reflected wave travels in the same direction as the wave made by the wave generator, it cannot be removed by resolution analysis of the incident and reflected waves. Thus, in an experimental facility that does not have a reflected wave absorption control function, the height of the offshore wave in front of the paddle shall be confirmed after the model is installed. If it is different from the passing wave test, the wave paddle amplitude shall be adjusted. This method can also be applied to experiments in which regular waves are created over a long period of time.

In the case of a regular wave experiment, there is also a way to perform the measurement to eliminate the influence of reflected waves. After the waves generated by the wave paddle reach the model structure, are reflected, and return to the paddle, measurements are taken while such wave is reflected again by the wave paddle until they reach the model. Because the wave re-reflected by the wave paddle is superimposed onto the wave that ought to be created by the wave paddle, the measurement is stopped after the re-reflected wave reaches the model. Alternatively, a technique in which wave generation stops before the reflected wave from the model reaches the wave paddle or a technique in which the number of waves generated is limited to approximately ten is also available.

2.2.4.4 Conducting the Experiment

After the calibration of test waves, a structural model is installed in the wave tank, and various measuring instruments are installed before the experiment is conducted. During the experiment, besides collecting the data, the experiment situation is recorded with photos, videos, etc.

2.2.4.5 Analysis and Organization of Experimental Data

Experimental data measured using the above-mentioned equipment is stored in the storage medium as digital data. Care must be taken in acquiring the data to properly capture the target phenomenon. It is also important to confirm the reproducibility by comparing the results of repeat measurements taken under the same conditions.

a. Sampling interval

   Experimental data can be obtained as digital data by A/D conversion. Since the data are obtained by discretizing analog data that is originally continuous, care must be taken in its handling.

   For example, in measuring the wave height, to measure the wave accurately, it is necessary to take 10–20 data samples in one wave (one wave period). Additionally, when measuring phenomena with drastic changes over time such as wave force and wave pressure, the peak values should be captured by shortening the sampling interval to units of 1/1,000 s.

b. Measurement time

   The measurement time should be determined properly according to the target. For example, when determining significant wave specifications in an irregular wave experiment, at least 100–200 waves should be captured during the measurement time to obtain statistically stable data. Additionally, when examining the stability of a wave-dissipating concrete block, for example, approximately 1,000 waves or more may be required. When targeting a phenomenon that includes a long period component, an adequate measurement time must be ensured according to the long period.

c. Method of experimental data analysis

The zero up-crossing method and zero down-crossing method are ways to analyze water surface elevation data measured using the wave gauge. For general waves, the typical values of the period and height of irregular waves are almost the same regardless of the method used; hence, the zero up-crossing method is generally used. Nevertheless, the zero down-crossing method may be used at points where the water depth is small or in the surf zone.[2.2.9, 2.2.10] In the zero up-crossing method, the mean water level is calculated from all the measured data. The point at which the rising water level crosses the mean water level shall be the start of the wave, and after the water surface elevation drops and starts to rise again, the period until the point at which the rising water surface elevation crosses the mean water level again is defined as one wave. The one wave interval is the period, whereas the height difference of the water surface elevation within one wave is the wave height. The maximum wave (highest wave), the highest 1/10 of waves, significant wave (the highest 1/3 of waves, etc.), average wave, etc. are defined from the data obtained.[2.2.9] If the amount of rise in the mean water level is required, premeasurement is performed before the wave action, and the value is determined by the difference between the average value of the measurement data during wave generation and the average value of the premeasurement.

The same method used to calculate the mean water level rise from the wave gauge data is used to find the average flow velocity from the data measured by the current meter. When determining the fluctuation component of the flow velocity, the zero up-crossing method is applied similarly to the wave gauge data.

The zero up-crossing method may not be applicable to wave force and wave pressure data. For example, the concept of zero-crossing cannot be applied because there are no data on the trough side in the data of the wave pressure gauge installed at a position higher than the still water surface of the breakwater caisson. A wave gauge is installed directly in front of a structure such as a caisson to measure wave force and wave pressure, and the start and end times of each wave are determined by zero up-crossing analysis of the wave gauge data. The maximum value of the wave force and wave pressure within that time period is detected and specified as the local maximum value of each wave. For this, statistical processing is performed to sort out the maximum value, the highest 1/10 value, significant value, mean value, and so on. In wave force and wave pressure measurement, especially in cases targeting shocking phenomena, the natural vibration of the measurement system often affects the measurement, so this influence should be eliminated during analysis.[2.2.11]

2.2.5 Typical Examples of Hydraulic Model Tests

2.2.5.1 Experiment on Wave Force Acting on a Structure (Caisson)

We will introduce an example of an experiment in which the wave force acting on the breakwater caisson was measured using a wave flume. This experiment was conducted to investigate the wave force reduction effect of the wave-dissipating work on

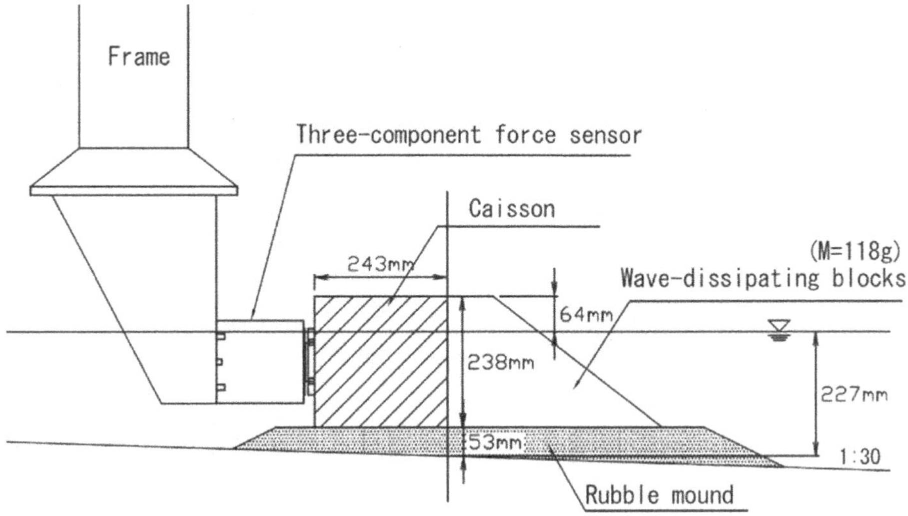

FIGURE 2.2.20
Cross-sectional view of the breakwater model.

the caisson. The wave flume used is 55 m long, 1.2 m wide, and 1.5 m high and has a piston-type wave generator. A breakwater model was installed on a fixed mortar bed with a sea-bottom slope of 1/30. Figure 2.2.20 is a cross-sectional view of the breakwater model. The installation water depth was 22.7 cm, and the caisson dimensions were 23.8 cm in height and 24.3 cm in width. The caisson model for wave force measurement is made of acrylic, and a three-component load cell was attached to the back. The three-component load cell was fixed by hanging it from a frame built into a gate shape with H-shaped steel. The caisson for wave force measurement is 40 cm in the width direction of the flume. Dummy wooden caissons are installed on both sides and a clearance of approximately 5 mm in width is provided to ensure they do not contact with the caisson for wave force measurement. The dummy caissons were fixed with sufficient weights so that they would not move due to wave action. The surface of the rubble mound was covered with a wire mesh, and a clearance of approximately 5 mm was provided so that the bottom surface of the caisson for wave force measurement does not contact with the mound. Additionally, to ensure that the load leaning on the wave-dissipating work would not act on the caisson for measuring the wave force, a wire mesh wall was installed between the wave-dissipating work and caisson so that the wave-dissipating concrete blocks would not contact with the caisson. Figure 2.2.21 shows a photograph of the installation status of the breakwater model.

Irregular waves with a modified Bretshneider–Mitsuyasu spectrum were used as the test waves. The significant wave period $T_{1/3}$ was set to 1.67 and 2.14 s and the significant wave height $H_{1/3}$ was set to six values within a range of 1.9–10.7 cm. Two types of wave signal were used, with a length of 819.2 s, which is equivalent to 491 and 383 waves when converted with the respective significant wave periods. The wave force for one cycle of the wave signal was measured 5 min after the start of wave generation. The measurement was performed at a sampling interval of 1 ms (0.001 s). Boss ends of the wave flume are connected via a pipe that is designed to suppress a

FIGURE 2.2.21
Installation status of the breakwater model.

rise in the water level behind the dike due to overtopping waves. Nevertheless, a wave gauge was also installed behind the breakwater model to compensate for changes in the water pressure due to a rise in the water level at the back when the return current is not able to keep up.

Figure 2.2.22 shows the time-series of the horizontal wave force under significant wave period $T_{1/3} = 1.67$ s and significant wave height $H_{1/3} = 8.8$ cm. This is a comparison of the differences in wave force waveforms that result from the presence and absence of wave-dissipating work and extracts the period close to the time at which the horizontal wave force is the highest under the condition without wave-dissipating work. It can be verified that the sharp peak value of the wave force is reduced by installing wave-dissipating concrete blocks. Figure 2.2.23 shows the reduction rate of the horizontal wave force by wave-dissipating work. Here, the reduction rate λ_H was obtained by dividing the experimental wave force with a wave-dissipating work by the calculated

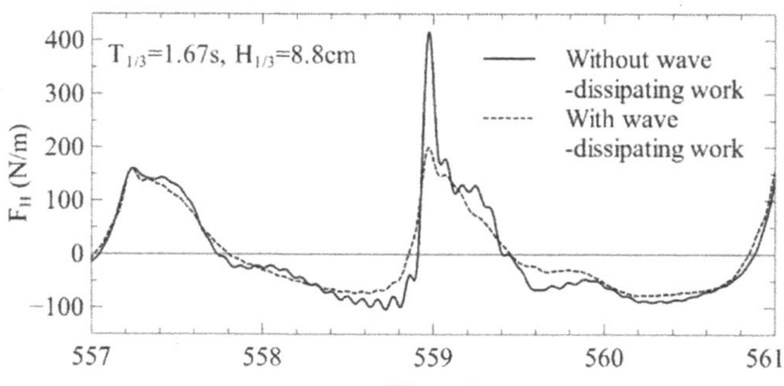

FIGURE 2.2.22
Time-series of the horizontal wave force.

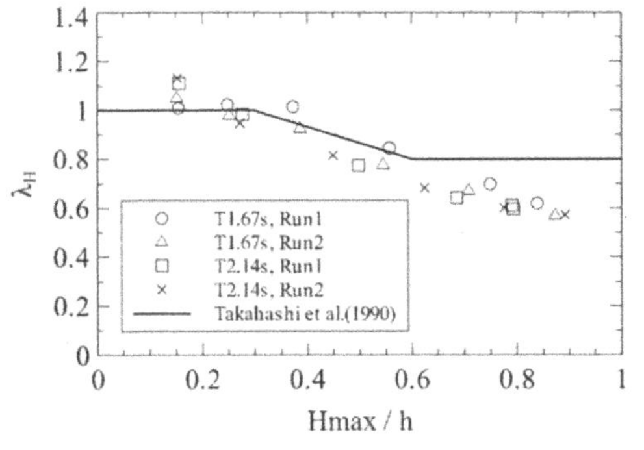

FIGURE 2.2.23
Reduction rate of the horizontal wave force by wave-dissipating work λ_H.

wave force according to the Goda formula. The reduction rate decreases as the wave height–water depth ratio H_{max}/h increases. Additionally, the value of λ proposed by Takahashi et al.[2.2.12] is also shown in the figure, which shows that it is roughly consistent with the experimental results.

2.2.5.2 Stability Experiment of Wave-Dissipating Concrete Blocks

We will introduce an example of an experiment on the hydraulic stability of wave-dissipating concrete blocks installed in front of a breakwater. This experiment was conducted to investigate the effect of the crown height of the breakwater on the stability of the wave-dissipating concrete blocks against wave attack. The wave flume used is 55 m long, 1.2 m wide, and 1.5 m high. A breakwater model was installed on a fixed mortar bed with a sea-bottom slope of 1/30. Figure 2.2.24 is a cross-sectional view of the breakwater model. The installation water depth h of the breakwater was 26.9 cm, and the water depth d at the crown of the foundation mound was 21.5 cm. There are two types of breakwater crown height h_c above the static water level, which are 7.0 and 11.7 cm. The wave height at the stability limit H_D calculated from the mass, etc. of the wave-dissipating concrete blocks is 11.8 cm, and the relative crown height h_c/H_D is 0.59 and 0.99, respectively. The breakwater is covered fully with the wave-dissipating

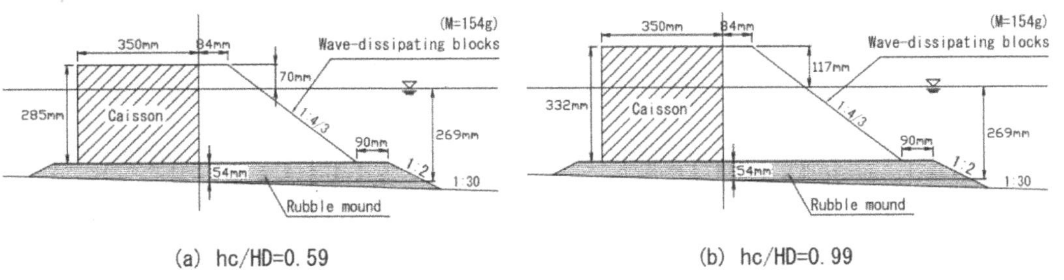

FIGURE 2.2.24
Cross-sectional view of the breakwater model.

concrete blocks, with the crown width equivalent to two blocks. The wave-dissipating concrete block model was made of mortar, and its mass and density were measured in advance. An average mass of 154 g, a density of 2.30 g/cm,[3] and a mass within ±3% of the average value were used in the experiment.

Irregular waves with a modified Bretshneider–Mitsuyasu spectrum were used as the test wave. The significant wave period $T_{1/3}$ was set to 1.5, 1.8, and 2.1 s, and the significant wave height $H_{1/3}$ was set to seven values within a range of 8.5–17.5 cm, two types of wave signals were used.

In the stability experiment, the wave height started from a small wave height rank and thereafter increased sequentially. For each wave height, 1,000 waves were produced. When increasing the wave height, the blocks were not stacked again, the number of the blocks damaged was defined by the cumulative number which is the number of blocks previously damaged plus the number of blocks newly damaged under the wave height. Since the blocks in contact with the sidewall of the flume are under different conditions, they were excluded from the observation. The damage rate was calculated by dividing the number of damaged blocks by the number of blocks subject to observation. Regarding the damage criteria, damaged blocks are those with a movement of 1/2 of the size of the block or more and a rotation of 90° or more.

Figure 2.2.25 shows the relationship between the damage rate and the K_D value. It is shown that the larger the relative crown height, the lower the damage rate of the blocks. For a relative crown height of 0.59, damage of the blocks was often observed near the crown of the wave-dissipating work. It can be concluded that the blocks near the crown are structurally susceptible to damage because there is no block to hold them down from above. Nevertheless, as the relative crown height increases, the flow velocity near the top decreases and the fluid force acting on the blocks also decreases, so the damage may be reduced.

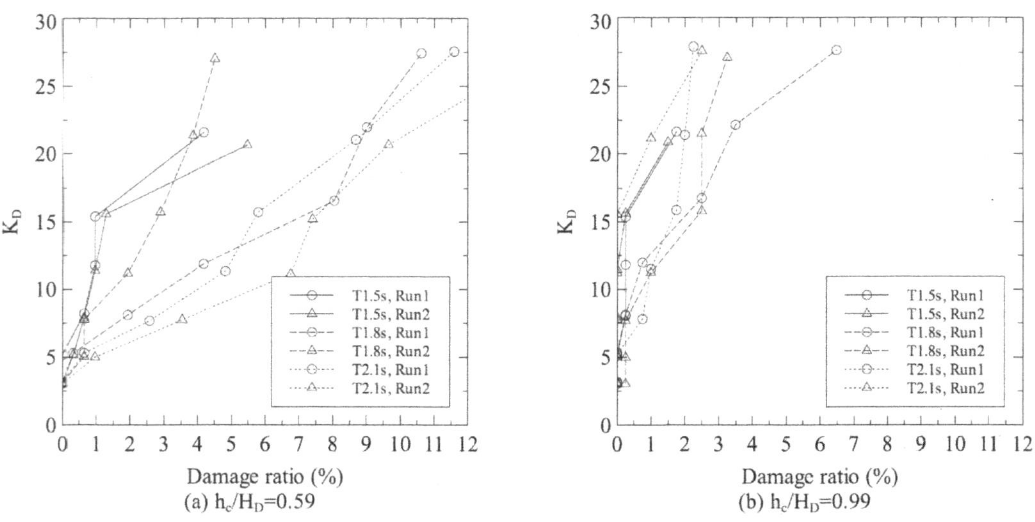

FIGURE 2.2.25
Relationship between the damage ratio and the K_D value. (a) The case of relative crown height (h_c/H_D) of 0.59 and (b) The case of h_c/H_D of 0.99.

2.3 Geotechnical Model Test

2.3.1 Planning of Model Test

Model tests need to be carried out under various constraints; thus, preliminary experiment planning is important. Particularly, the actual complicated phenomenon must be idealized or simplified, and then, under what conditions the model test should be carried out to obtain various information from the model must be decided.

In the book *Introduction to Model Tests in Geotechnical Engineering*, published by the Japanese Geotechnical Society[2.3.1], model tests are divided into categories of "Component factor of model," "Physical value of model," and "Efficiency of model test," and a schematic illustration of general terms is also shown (Figure 2.3.1). According to this, it is desirable to formulate an experimental plan by sorting out the constraints such as the purpose and detailed factors of each individual task, scale of the test equipment, cost, period, number of workers, and so on. Those responsible for the experiment struggle the most in deriving an optimal plan to achieve the purpose of the experiment within the constraints imposed, but this is important in conducting the experiment and must be carefully considered.

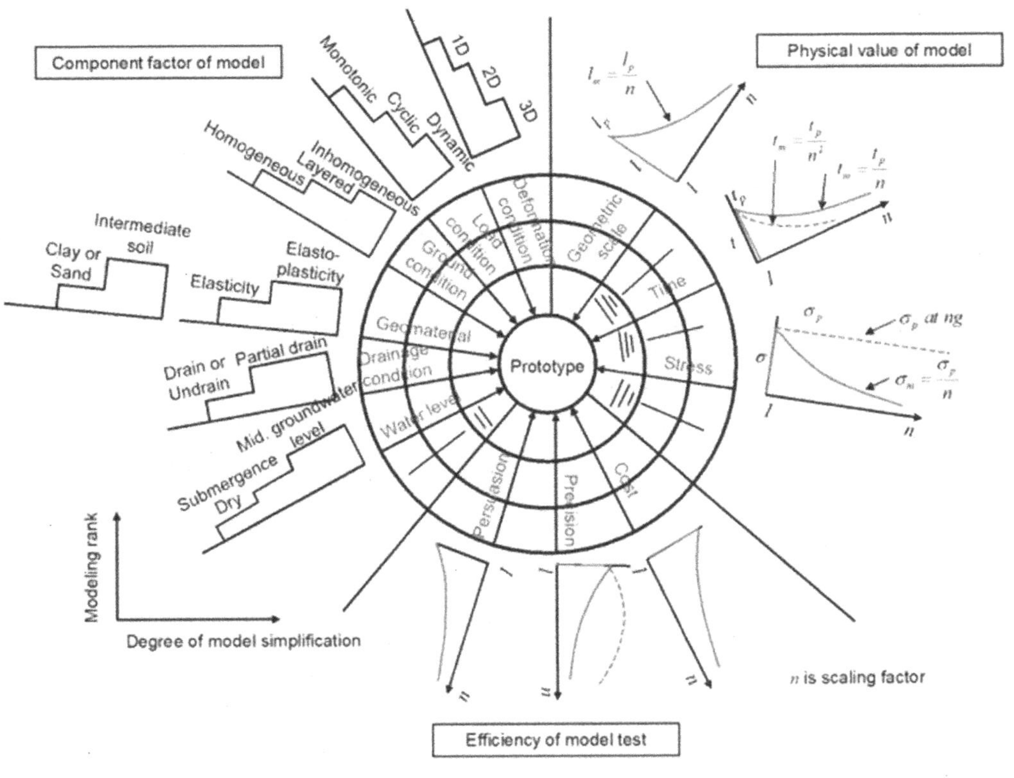

FIGURE 2.3.1
Structure of model test (adapted from note 2.3.1).

An explanation of the model test plan is given here by citing the "purpose of the experiment," "similitude ratio," and "model experiment method and design."[2.3.1]

2.3.1.1 Purpose of the Experiment

To carry out a model test, the purpose of the experiment is set after defining the phenomenon to be clarified. In the model test, as described later, the scaling law is studied for the target phenomenon. The scaling law is important in conducting a model test. For example, if an experiment is performed only under geometric similarity conditions, appropriate information cannot be obtained from the model test. For that purpose, we will consider the purpose on the basis of whether the target phenomenon can be elucidated by the model test while predicting the phenomenon that may occur in the prototype to some extent.

2.3.1.2 Scaling Ratio

Once the purpose of the experiment is determined, the conditions under which the relationship between the prototype and the model are similar regarding the target phenomenon should be determined. Here, the scaling is the three conditions of "geometric," "kinematic," and "mechanical".[2.3.2] The scaling law will be considered on the basis of these scaling conditions, but as shown in Chapter 3, the following methods can be cited as ways to determine the scaling ratios: (1) from dimensionless quantities based on physical quantities related to the phenomenon, (2) from dimensionless quantities based on the ratio of forces related to the phenomenon, and (3) from equations governing the phenomenon. See Chapter 3 for details.

The specific scale ratio of the model (for example, the geometric scaling ratio based on the model dimensions is assumed to be $1/N$) is examined on the condition that the experimental equipment can be used on the basis of the law of similarity. Additionally, the selection of the geomaterial used in the experiment is very important; thus, we will also study it.

For example, if the conditions of the target prototype and structure are fixed, the target region is clear, making it easy to determine the specific scale ratio under the constraint conditions for conducting the experiment. Nevertheless, if there are no reference conditions for determining the target region, the scale ratio is difficult to uniquely determine. When the scaling ratio is large, it is possible that results with a relatively high accuracy can be obtained with respect to the prototype, but the experiment cost is high, and the number of experiments is limited. Conversely, if the scaling ratio is small, the accuracy may be low depending on the target phenomenon. Therefore, based on the target phenomenon to be studied under the constraint conditions, ways to study the phenomenon using several scale ratios may be adopted.

2.3.1.3 Model Test Method and Design

The method of the model test is determined on the basis of the experiment's purpose and study of scaling ratio. Model tests can be conducted in various ways as described below.

Model tests are generally classified as pilot experiments or production experiments.[2.3.3] An experiment conducted to confirm whether the phenomenon observed in a model test

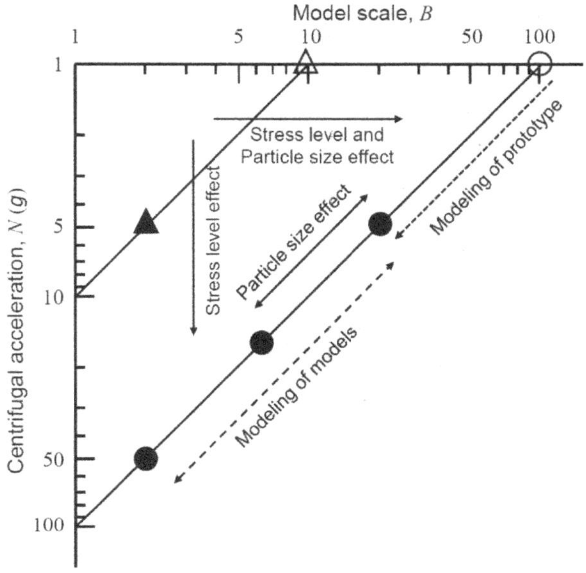

FIGURE 2.3.2
Conceptual diagram of modeling of models and modeling of prototypes in the centrifuge model test (adapted from note 2.3.4).

is similar to the actual phenomenon is called a pilot experiment. Moreover, production experiment is an experiment in which a model is used as a substitute for the prototype after the similarity is confirmed in a pilot experiment.

In geotechnical engineering, techniques known as modeling of models and modeling of prototypes[2.3.4] are available. As a specific example, Figure 2.3.2 shows a conceptual diagram related to a study on modeling of models and modeling of prototypes in a centrifuge model experiment. Refer to Section 2.3.2 for details of the centrifuge model experiment. Modeling is performed in centrifuge model experiments conducted on real objects of the same size at different scale ratios for comparison purposes. In this method, the actual phenomenon is predicted on the basis of the results of the model tests after confirming the validity of the experiments at different scale ratios. To confirm the validity of the model, it is important to compare by model test with a relatively wide scaling ratio. Conversely, modeling of prototypes is a method of comparing centrifuge model tests with the corresponding prototype. Okamura et al.[2.3.4] showed the importance of conducting studies using the modeling of models and modeling of prototypes approaches by citing the scale effect of soil particles and the scaling law regarding the time of the permeability phenomenon and vibration phenomenon. Please refer to the case shown in Section 2.3.5 as it contains relevant contents.

Once the model test method is fixed, an experiment plan is formulated. When conducting experiments, it is advisable to divide the experiments into preliminary experiment and actual experiment. For example, a preliminary experiment is carried out to confirm the validity of the experimental method and experimental data according to the drafted experimental plan. Additionally, we will also confirm the safety when conducting the experiment. Particularly, conducting an experiment may lead to a serious accident; thus, the experiment procedure, etc. should be carefully checked during the process so that all personnel engaged in the experiment can gain a common understanding of how to conduct the experiment. Based on the results of the preliminary test, the experimental plan will be revised as necessary. The actual experiment will be carried out according to the experimental method and experimental plan established in the preliminary test.

2.3.2 Classification and Target of Model Tests

As explained so far, a model test is carried out on the basis of constraints imposed by the area of study and the accuracy required for the target phenomenon, etc. Additionally, in experiments dealing with geomaterials, it is necessary to reproduce the same stress as the prototype to reproduce the mechanical behavior of the actual ground. The reason is that the shear strength exerted by the geomaterial differs depending on the constraint conditions (confining stress). Conversely, when applying geometric similarity to the geomaterial, if the scaling ratio is small, the properties of the material may differ between the prototype and model, and it may not be possible to obtain the same mechanical properties as the material of the actual ground even if the geomaterial is geometrically similar. For this reason, in model experiments conducted in the geotechnical engineering, besides experiments in a gravitational field, many experiments have been conducted in a centrifugal force field, thereby establishing the experimental techniques.

Nevertheless, being special and expensive, the equipment for model tests conducted in a centrifugal force field cannot be owned and used by everyone. Additionally, depending on the target phenomenon, it may not always be necessary to conduct the experiment in a centrifugal force field. Therefore, scaling laws have also been considered for experiments conducted in a gravitational force field, and many are still being carried out today.

Here, we will explain the advantages and disadvantages of experimental methods conducted in a gravitational force field or centrifugal force field while introducing specific examples of what kind of experiments have been carried out so far. Particularly, the principle will also be described in detail in the experiments conducted in a centrifugal force field.

2.3.2.1 Model Tests in Gravitational Force Field

In a model test conducted in a gravitational force field, it can be concluded that the biggest disadvantage is the inability to accurately reproduce the stress caused by dead weight of geomaterial. Particularly, when examining the stability of slopes and embankments and the bearing capacity of rubble mounds and foundation under large structures such as breakwaters and quays, the significance of a centrifuge model test is high because it is important to reproduce the stress condition of the ground. Nevertheless, in a centrifuge model test, a large and special apparatus (i.e., a centrifugal apparatus) is required, so it is not easy to take various measurements and apply an external force to the structure. Additionally, experiments in a centrifugal force field are conducted in a centrifugal force field of at least $10g$ to reduce the influence of gravity ($1g$). That is, when attempting to reproduce an object that is not so large in reality (for example, a thin sheet-like object) in a centrifuge model test, an extremely small model must be made. Additionally, not only is the interpretation of the results complicated, including the Coriolis force and changes in the centrifugal acceleration in the radial direction, but also the phenomenon is different from reality, and thus these effects may not be ignored. Such disadvantages of centrifuge model tests will be explained in detail in Section 2.3.2.2. For the above reasons, centrifuge model tests are not all-purpose, and gravitational force field model tests still play a big role and are being conducted in the field of geotechnical engineering even now. Regarding the two cases described below, we will show the significance of the experiment in a gravitational force field by showing that providing reproductions in a model test conducted in a centrifugal force field is difficult.

a. When reproducing a complicated construction process

   The stress history of the ground often affects the stability of the ground and the design of the structure. The deformation behavior of buried flexible pipes can be cited as an example. It is known that the mechanical behavior of buried flexible pipes is strongly influenced not only by pipe stiffness but also by soil stiffness. A centrifuge model test that can reproduce the stress caused by dead weight of the ground seems to be more appropriate. However, it is difficult to reproduce the stress history during surface compaction and other construction in a centrifuge model test.

   For example, in a field experiment for a FRP pipe with a diameter of 1,500 mm and a thickness of 15.5 mm, the pipe was vertically elongated because of backfilling up to the side of the pipe, and subsequently, the deformation tendency further became stronger because of backfilling up to the top of the pipe. It has been reported that the pipe remained vertically elongated even when backfilling was completed[2.3.5] (Figure 2.3.3). Although this example is a special case in which a pipe having a considerably lower stiffness than the usual flexible pipe is embedded, the deformation of the flexible structure is generally affected by soil compaction, etc. When reproducing the construction process in a centrifugal force field, it is necessary to spread the soil evenly and compact its surface in the centrifugal force field. If the centrifugal apparatus is stopped and an attempt is made to conduct these steps in a gravitational field, the stress will be released and the deformation of the pipe may not be reproduced. Although attempts have been made to reproduce the construction in a centrifugal field,

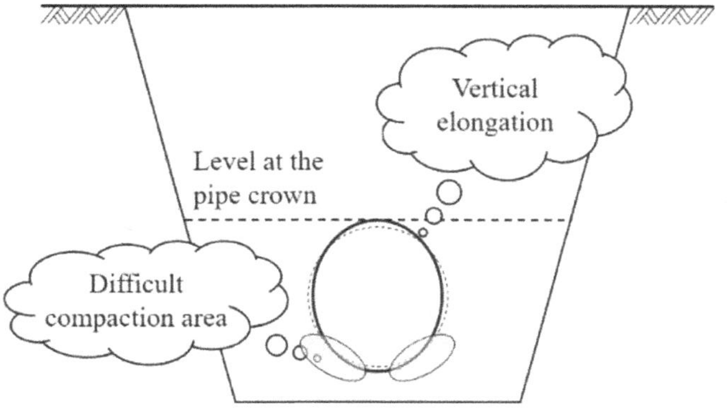

FIGURE 2.3.3
Deformation of a buried low stiffness pipe.

such as dynamic compaction by dropping a weight[2.3.6] and slope excavation,[2.3.7] no example in which process from soil spreading to compaction is conducted continuously in a centrifugal force field is found. Particularly, in the case of buried pipes, the area from the bottom to the side of the pipe cannot be compacted easily because the compacting machine cannot enter the area; hence, there have also been cases where careful compaction is performed neatly by human power (Figure 2.3.3). From the above, it can be said that an experiment in a gravitational force field is preferred when the effect of stress caused by the construction process cannot be ignored.

b. When reproducing a sheet having a complicated structure, etc.

It was mentioned earlier that the significance of centrifuge model tests to address stability problems in slopes is large, but when reproducing a sheet-like reinforcing material (for example, geotextile) in a centrifugal force field, the slip safety factor of the slope in the model and prototype must be matched.[2.3.8] For that purpose, it is necessary to set the tensile strength (note that the unit is in force per unit depth kN/m) and the tensile stiffness (in the same unit as the tensile strength) of the reinforcing material model to $1/N$ of the prototype. When the same material as the prototype is used for the model, the tensile strength and stiffness can be achieved by setting the thickness or mass of the reinforcing material to $1/N$ of the prototype.[2.3.8] Regarding the frictional properties at the interface between the reinforcing material and soil, if the frictional properties are characterized by a shear stress–shear strain relationship, the soil–reinforcement interface in the model may be the same as that in prototype, but if it is characterized by a shear stress–shear displacement relationship, greater care must be exercised, and the soil–reinforcement interface in the model must be made N times stiffer than the prototype. In this case, it will probably be difficult to satisfy the scaling laws pertaining to tensile strength and tensile stiffness at the same time. Modeling becomes more difficult when behaviors other than mechanical properties must also be reproduced. One example is geosynthetic clay liners (GCLs). GCLs are used in multilayer liners together with other synthetic liners to control fluid migration at solid waste landfills. The structure consists of powdered or granular bentonite sandwiched between a woven geotextile and a nonwoven geotextile or between two woven geotextiles, each of which is fixed by needle punching or stitch bonding.

Furthermore, various needle-punched GCL is available from different manufacturers, such as those with a thermal coating on the woven geotextile side (Figure 2.3.4). The tensile strength and tensile stiffness of GCLs are affected by those of the outer woven geotextile and nonwoven geotextile, and the permeability is affected by the fixing methods as needle punching.[2.3.9] Additionally, the resistance to internal shear between sheets is affected by the fixing method between sheets and whether there is any thermal coating,[2.3.10] although the shear strength of hydrated bentonite is extremely low. Moreover, the frictional properties at the interface between GCLs and other synthetic liners or soils show extremely complicated behavior, such as being affected by bentonite being squeezed out from the GCL woven geotextile surface.[2.3.11] As just described, GCLs are composed of several components, each of which exerts an impact on the mechanical and hydraulic properties. Although the thickness of

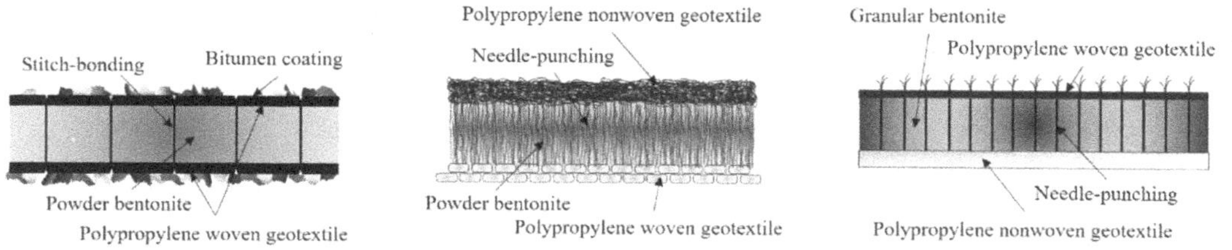

FIGURE 2.3.4
Types of reinforced GCL.

the actual hydrated GCL varies depending on the presence or absence of needle punches and the confining stress, it can be as thin as 4–12 mm.[2.3.9] Thus, producing a model that considers all the characteristics mentioned earlier seems to be extremely difficult. A centrifuge model test on a GCL has also been conducted, but the actual GCL has been used and the scaling law has not been investigated.[2.3.12] When reproducing multiple behavioral characteristics in a model test for a sheet having a complicated structure such as GCLs, it is considered realistic to use the actual object to make a model with the largest scale ratio possible and to carry out the model experiment in a gravitational force field.[2.3.13]

In the example given here, an experiment in a centrifugal force field may be more appropriate depending on the condition. Ultimately, it is important to choose an optimal experimental technique to reproduce the phenomenon to be captured after thoroughly considering what kind of phenomena govern the matter to be clarified.

2.3.2.2 Model Test in Centrifugal Force Field

In 1869, Edouard Philips[2.3.14] in France first proposed the possibility of model tests in a centrifugal force field, called geotechnical centrifuge modeling. Subsequently, a geotechnical centrifuge was developed in Russia by the 1930s, but it was not known to western European countries because of the iron curtain and was introduced by British researchers[2.3.15, 2.3.16]. At around the same time, it was reported that a centrifuge model test was conducted in the United States to simulate the collapse of the ceiling of a tunnel.[2.3.17–2.3.19] In 1964, for the first time in Japan, Prof. Mikasa of Osaka City University conducted an experiment on the problem of consolidation of the ground due to its own weight using a centrifugal loading device developed by himself.[2.3.20, 2.3.21] In the same year, Prof. Schofield in Cambridge University, UK, developed the one for testing the Cam Clay model behavior. Then, around 1966, experiments on slope stability and bearing capacity started at Cambridge University.[2.3.16, 2.3.18, 2.3.22]

Technical Committee 104 (TC104) "Physical Modelling in Geotechnics" of the International Society for Soil Mechanics and Foundation Engineering[2.3.23] has held international conferences focusing on centrifuge model tests once every four years since 1988, and a huge amount of research results on model tests are summarized in the conference report. Additionally, the results of ground model tests have been published continuously in the *International Journal of Physical Modeling in Geotechnics*, a scientific journal published by TC104.

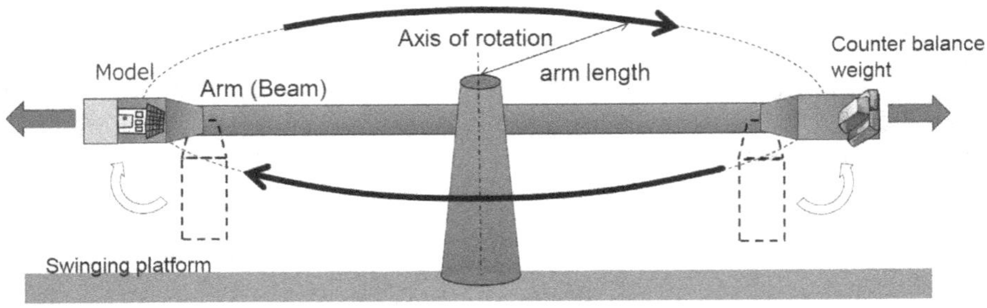

FIGURE 2.3.5
Beam-type geotechnical centrifuge.

Many research institutes began to operate the geotechnical centrifuge in the 1980s. There are two types of geotechnical centrifuge currently in use: the beam-type and drum-type. The former constitutes a majority because of the advantages described below. As shown in Figure 2.3.5, a beam-type centrifuge "beam" called an arm with a motor and has a swinging platform attached to the end of the beam where a model container is placed. One of the advantages of a beam-type device is the ability to build a model in a gravitational force field unlike the drum-type to be described later. The disadvantage is that the size of the model is restricted by the size of the swinging platform. Conversely, the drum-type device has a simple structure like the dehydrator of a washing machine. However, it is necessary to produce the model facing the side from the beginning, and advanced technology is required because the device for penetration test and other tests must be synchronized with the rotation of the drum.

One-way to achieve this is to blow sand while the drum is rotating to create the ground or install a frozen model ground at the prescribed location in the drum and let it melt during centrifugal loading before conducting the experiment. The advantage of the drum-type is that the distance between the side boundaries in the longitudinal direction can be longer than that of the soil container of a beam-type device. Also, depending on the experiment, a circular boundary may be assumed, which is believed to be suitable for experiments on hydraulic models.

In association with the field of hydraulics, Sassa and Sekiguchi[2.3.24] conducted a study on the interaction between waves and the ground using a beam-type device. Miyake et al.[2.3.25] and Imase et al.[2.3.26] examined the problem of breakwater stability due to tsunamis by making use of the advantages offered by drum-type devices. Recently, Takahashi et al.[2.3.27] conducted a study on the stability of a breakwater mound due to a tsunami.

The geotechnical centrifuges are used in model tests related to the ground because the strength characteristics of soil, which is a granular material, depend on the confining stress (soil becomes stronger as the confining stress increases). That is, by mounting a reduced model on the centrifugal loading device and applying a centrifugal force according to the scale in the gravitational direction, the confining stress (the pressure that surrounding soil exerts on soil in the ground) can be increased.

Consequently, it is possible to reproduce the mechanical properties, such as the stiffness and strength of the soil and stress–strain relationship which vary depending on the confining stress, closer to those of the actual ground. In this way, not only the behavior of the actual ground itself but also the behavior of the ground–structure system can be reasonably reproduced by a reduced model.

Next, the advantages and disadvantages of a centrifuge model test are listed.

a. Advantages

The centrifuge model tests offer three advantages. These advantages serve as a powerful motivation for conducting centrifuge model tests.

i. It is possible to reproduce the underground stress state closer to the mechanical properties of the actual ground.

As mentioned above, it can be said that this is the greatest advantage offered by this experimental method. The following advantage makes this possible.

ii. The scaling law is simple.

The scaling law described in detail in Chapter 3 for a centrifuge model test is given by the scaling factor of acceleration $\mu = a_m/a_p$. From this, simple scaling relationships that satisfy the geometrical, kinematic, and mechanical similarity relationships are derived.[2,3,28] In the centrifugal force field, the fact that the scaling factor of stress, strain, and velocity is one of the advantages that ought to be noted. It has also been shown that this scaling law satisfies Froude's law which is widely used in hydraulics. Besides these advantages, the following can be cited as practical advantages that have resulted in centrifuge model test being used widely.

iii. Since the model is compact, it is easy and economical to manufacture.

It is possible to conduct an experiment with a small amount of workforce, materials, and cost, allowing the experiment to be conducted several times under different conditions. Even with the world-class geotechnical centrifuge, the maximum soil container has a length of approximately 2 m in the longitudinal direction and a maximum height of approximately 1 m. This is considerably smaller compared with the one normally used in geotechnical model tests under 1g.

b. Disadvantages

Conversely, the following points can be listed as disadvantages of the geotechnical centrifuge model. Of these, (i)–(v) are disadvantages that cannot be avoided because of physical laws or equipment restrictions, whereas (vi)–(viii) are disadvantages due to equipment restrictions.

i. Centrifugal force changes in the radial direction

Since civil engineering deals with physical phenomena near the surface of the earth, the gravitational acceleration does not change in the depth direction and g can be regarded to be approximately 9.81 m/s^2. Therefore, the distribution of the underground vertical stress has a triangular shape similar to the hydrostatic pressure. Conversely, in a centrifugal force field, the centrifugal acceleration a is expressed as $a = r\omega^2$ using the radius r and the rotational angular velocity ω; so, the farther away it is from the central axis, the larger the acceleration will be. In other words, the vertical stress in the depth direction of the model ground changes according to the distance from the central axis. When the magnitude of this vertical stress is considered as a coordinate system z, assuming the ground surface as the origin and the downward (outward) direction as positive, the equation becomes a quadratic function of the depth z as shown below:

$$\sigma_v = \rho(r_m + z)\omega^2 z \tag{2.3.1}$$

Here, the density of the ground is ρ, and the distance (radius) from the center of rotation to the surface of the model ground is r_m.

By comparing the vertical stress distributions of the actual ground and the model ground schematically shown in Figure 2.3.6, a discrepancy in their magnitude is shown to exist.

Now, let us consider adjusting the centrifugal acceleration (changing ω in Equation (2.3.1)) to minimize this difference. When the centrifugal acceleration is varied, the shape of the vertical stress distribution of the model ground represented by Equation (2.3.1) changes, and the depth at which the magnitudes of both vertical stresses match (position (a) in Figure 2.3.6) also changes. At this time, the overestimated and underestimated values of the vertical stress shown in (b) and (c) also change in Figure 2.3.6, but the error of the vertical stress distribution is minimized when the absolute values of those evaluation values match. Assuming this as a constraint and H_L as the thickness of the model ground layer, it has been shown that both vertical stresses are equal at a depth of 1/3 H_L (position (a) in Figure 2.3.6) from the bottom of the soil container.[2.3.29] In other words, if $r_a = r_m + 2/3\ H_L$ is set as the effective radius and the desired centrifugal acceleration is set at this position (N times the gravitational acceleration g), the deviation of the vertical stress distribution will be minimized. Nevertheless, even when the centrifugal acceleration is set in this way, errors in the vertical stress cannot be avoided, and the stress–strain relationship of the soil will be affected by the difference. When this error is calculated assuming $N = 50$ in the trial, it

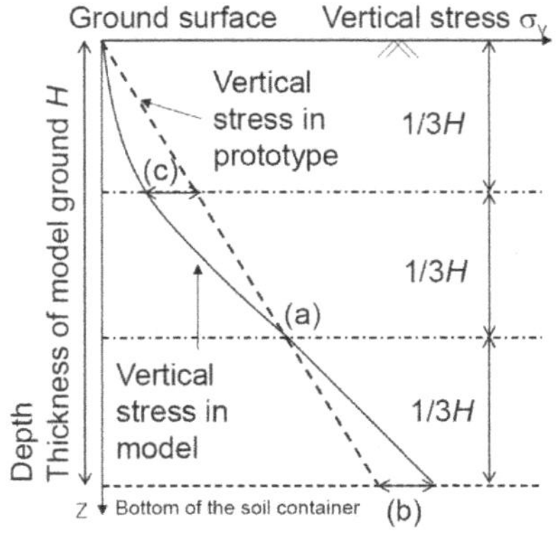

FIGURE 2.3.6
Schematic diagram in which the error in the vertical stress distribution between the actual ground and model ground is minimized (modified after Madabuhshi[2.3.29]: the depth at which both stresses match (a), the depth at which the stress is the most overestimated (b), and the depth at which the applied force is most underestimated (c). The error is exaggerated in the diagram.

is approximately 3.3% when the effective radius $r_a = 1.5$ m and the depth of the model ground $H_L = 0.3$ m. However, if the effective radius is increased to 2.5 and 5.0 m, the error will be reduced to approximately 2% and 1%, respectively. Therefore, a geotechnical centrifuge having a long arm length and a large effective radius is preferred. However, since the error itself is very small as calculated above, it is presumed that there will be no significant difference in the behavior of the ground unless a very precise model experiment is required.

ii. Gravitational force field in a model forms an arc

Although related to (i), in a centrifuge model test, the centrifugal acceleration acting on the surface of the flat model ground is different at each point in the direction of rotation (Figure 2.3.7). This can be easily understood by imagining the phenomenon in which the surface shape becomes an arc when only water is poured into the soil container. Particularly, when conducting experiments on the liquefaction of saturated sandy ground, the ground surface changes from a flat surface to an arc shape when it is completely liquefied. The smaller the radius of rotation, the more pronounced this effect will appear. For example, suppose that a centrifuge model test at a scale of 1/50 ($N = 50$) is performed with a geotechnical centrifuge with an effective radius of $r_a = 1.5$ m. Assuming a model ground with a layer thickness $H_L = 0.3$ m in a soil container with a width $W = 0.4$ m, when comparing the centrifugal acceleration at a position $r_1 = 1.32$ m on the ground surface at the edge of the container and at a position $r_m = 1.30$ m on the ground surface at the center of the container, the centrifugal acceleration at the edge of the container is approximately 1.16% larger. This error rapidly decreases to 0.38% when the effective radius is increased to $r_a = 2.5$ m and to 0.09% when increased to $r_a = 5.0$ m.

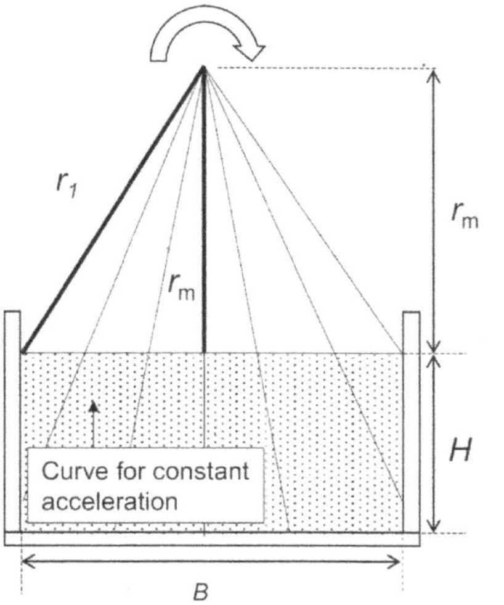

FIGURE 2.3.7
An example of constant acceleration curve that forms a circular arc in the model ground: curvature is exaggerated.

Since this effect becomes significant when the long side of the soil container coincides with the direction of rotation, many devices set the container's long side perpendicular to the rotating plane of the arm. Additionally, the ground surface on an arc may be constructed from the beginning so that the centrifugal acceleration acting on the ground surface becomes uniform. Tobita et al.[2.3.30] observed the difference in how the ground deforms when the ground surface is linear and when it is an arc by conducting dynamic experiments on the sloped ground with saturated sand.

iii. Coriolis force

Considering a system that rotates at a constant rotational angular velocity ω, the apparent acceleration acting on a body moving in it is not only the centrifugal acceleration $r\omega^2$ in the radial direction r but also the apparent acceleration $2\dot{r}\omega$ in the rotating direction. The force caused by this acceleration is called the Coriolis force. Since the Coriolis force increases in proportion to the velocity in the radial direction, care should be taken, for example, when creating a ground while letting sand rain down in a centrifuge model test, when conducting an experiment in which rainfall is reproduced, or when the vertical movement velocity of the body is fast due to excitation. As an example, as shown in Figure 2.3.8, let us consider injecting a ball into the ground in the radial direction in a centrifugal field. In this example, because of the influence of the Coriolis force, the orbit of the ball does not become perpendicular to the ground and shifts slightly backward from the assumed landing point. This deviation can be calculated by the formula shown below[2.3.29]:

$$D = \frac{1}{3} Ng\omega \left(\frac{2L_H}{Ng} \right)^{\frac{3}{2}} \quad (2.3.2)$$

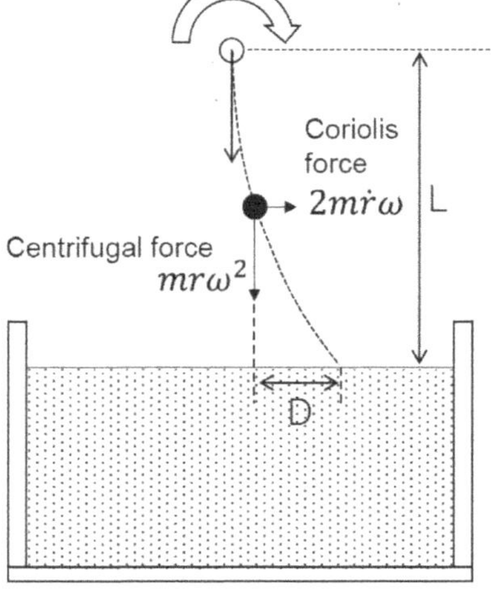

FIGURE 2.3.8
Apparent forces acting on centrifuge models (centrifugal and Coriolis forces).

Here, N is the similarity coefficient, ω is the rotational angular velocity, and L_H is the drop height.

As a trial, in the case of a device with an effective radius $r_a = 1.5$ m, assuming the centrifugal acceleration $Ng = 50g$ and a ball is dropped from a height $L_H = 0.3$ m at an initial velocity of 0 relative to the soil tank, the value will be as shown below:

$$\omega = \sqrt{\frac{Ng}{r_a}} = \sqrt{\frac{50 \times 9.81}{1.5}} = 18.10\frac{rad}{s} \tag{2.3.3}$$

Thus, the deviation of the orbit will be as follows:

$$D = \frac{1}{3} \times 50 \times 9.81 \times 18.10 \times \left(\frac{2 \times 0.3}{50 \times 9.8}\right)^{\frac{3}{2}} = 0.126m \tag{2.3.4}$$

Here, it is assumed that there is no change in acceleration in the radial direction, because the drop height is small relative to the effective radius of the device. Nevertheless, since magnitude of deviation D due to the Coriolis force is relatively large and cannot be ignored, error should be evaluated before the experiment in which an element moving at a high-speed in the vertical direction is included.

iv. Scale effect

A common question regarding centrifuge modeling is why the soil used is the same as the actual one and why the soil particles are not made smaller according to the scaling law. The simple answer to this is that reducing the size of soil particles, for example, from sand to silt, changes their mechanical properties. In other words, to reproduce the same stress–strain relationship as the actual one, it is necessary to use the same soil. However, another problem arises here. For example, consider the case in which a model is created with a scale of 1/100 for a pile with a diameter $D_p = 0.5$ m in prototype scale (prototype). In this instance, the diameter of the model pile will be 5 mm. Here, since the soil used in the centrifuge modeling is usually the same as the one in the actual ground, the number of soil particles in contact with the peripheral surface of the model pile is thus small. Assuming that mean grain size of the soil particles is $d_{50} = 0.2$ mm, the ratio of the diameter of the actual pile to the diameter of the soil particles is $(D_p/d_{50})_p = 2{,}500$, whereas it is $(D_p/d_{50})_p = 25$ in the model. Here, the subscripts p and m represent prototype and model, respectively. When a lot of soil particles are arranged around the pile, the behavior of the system can be regarded approximately as a continuum. Nevertheless, when the number of soil particles surrounding the pile decreases, it cannot be regarded as a continuum; hence, the interaction between the individual soil particles and the pile should be considered. Because of this effect, the deformation behavior of the model may be markedly different from the prototype.

The scale effect refers to the magnitude of shear forces acting on a model surface, restraining forces, and bearing capacity of foundation structures are expressed as functions of deformation, surface roughness of the model, and particle size of the soil. Klinkvort,[2.3.31] in a paper to introduce the latest technology, summarizes what should be considered when conducting model tests on monopile (single pile), which is often used as a foundation for wind turbines in offshore wind power generation. For example, it has been reported that the horizontal force acting on the model pile can be regarded as a continuum if D_p/d_{50} is 40–88 or more, but it has also been pointed out that the value varies depending on the roughness of the model pile surface. As an another reference, Tan and Scott[2.3.32] deal with scale effects related to the interaction between granules and interstitial fluids.

The constraint above is imposed by physical laws related to centrifuge modeling, but the error caused by it can be quantified as if it is calculated, so it can be examined in advance before the experiment.

v. Other restrictions in terms of the laws of physics

The strength characteristics of clay show strong dependency on strain. Here, in this textbook, hydraulic modeling is mainly dealt with. Therefore, the issue of strain dependency of clay may not be encountered and details can be found in some textbook of soil mechanics.

vi. Influence of the side boundary of a soil container

It is a fatal defect in conducting experiments on natural phenomena. As a solution, there is no better way than to use a large soil container, but if the size is a constraint, the deformation mode of the ground should be examined in advance by numerical analysis, etc., and the model cross section should be set so that the deformed area, slip surface, etc. are located sufficiently away from the sides. Alternatively, if the model scale cannot be reduced, experiments are often conducted as it is. Then, including a boundary effect, numerical analysis is conducted and validated.

vii. Interference between the model and the measuring device (sensor body and lead wire) buried in the soil

Since the model is small, the sensor itself may become a structure that is not originally there and interfere with the model. For this reason, technologies using sensors that are as small as possible or are wireless with no lead wires have been developed. However, in general, the effect is usually ignored.

viii. Phenomenon cannot be observed directly

There is a way to observe the behavior of the model from the side or top of the container using a video camera or a high-speed camera. Furthermore, recently, a method in which the deformation of the ground is visualized as a plane and observed as a video instead of a single point by applying the principle of photographic survey[2.3.33] has been developed.

The above disadvantages are those that cannot be avoided due to the physical laws related to centrifuge model tests or equipment constraints.

2.3.3 Target and Method of Measurement

A measuring instrument (sensor) is used to quantitatively evaluate the physical and mechanical phenomena during the experiment.

In recent years, the development of sensor technology has advanced significantly, with small and inexpensive sensors being developed one after another. At present, it has become possible to measure various quantities such as acceleration, velocity, position, angular velocity, angle, force, and pressure at high accuracy. An introduction to sensors mainly used in experiments related to geotechnical engineering is given here.

2.3.3.1 Pressure Gauge, Pore Pressure Gauge, and Earth Pressure Gauge

Devices to measure the pressure of liquids and gases are collectively known as pressure gauges. There are various types of pressure measurement depending on the measurement application. For example, devices that measure the wave pressure acting on a structure can be used to measure the pressure from the strain generated by a fluid acting on the pressure-receiving plate of a sensor.

A device to measure pore pressure in soil is called a pore pressure gauge (Figure 2.3.9). A stainless steel mesh, porous ceramic, is placed so that soil particles do not come into contact with the pressure-sensing surface of the pressure gauge, and the pressure of the pore water is measured. As a precaution, do not allow air bubbles to enter between the mesh or ceramic and surface subject to pressure. In an unlikely event that air bubbles are present, the pore pressure may not be measured accurately. For this reason, when installing a pore pressure gauge, it is necessary to saturate the inside of the mesh or ceramic with water or oil and to fill up the space between the mesh or ceramic and the pressure-sensing surface.

A device for measuring earth pressure is called an earth pressure gauge. A way to measure the earth pressure via a pressure-receiving plate like a pressure gauge is available, but if the size of the soil particles is large relative to the size of the pressure-receiving plate, values may not be measured accurately, so a measuring instrument with a size appropriate for the ground conditions must be used.

2.3.3.2 Load Cell

A device for measuring the force acting on a structure, etc. is called a load cell. Strain conversion methods for compression and tension purposes are available. The pressure gauge measures local change at a certain position, whereas the load cell measures the force, which is the integral value of the pressure over a certain area.

FIGURE 2.3.9
Micro pore pressure gauge. With permission from SSK Co. Ltd.

FIGURE 2.3.10
Soil moisture meter (Theta Probe ML3). With permission from Environment Measurement Japan Co. Ltd.

2.3.3.3 Soil Moisture Meter

A device to measure the amount of water contained in ground is called a soil moisture meter. Soil moisture meters based on the tensiometer method, time domain reflectometry (TDR) method, and amplitude domain reflectometry (ADR) method are available (Figure 2.3.10). The tensiometer method is used to measure the matrix potential (pF).[2.3.34] The TDR and ADR methods are used to obtain the volumetric water content from the dielectric constant of the soil. In the former method, the dielectric constant of the soil is determined from the propagation time of the interfering reflected wave, whereas in the latter method, the dielectric constant of the soil is found by measuring the impedance.[2.3.35]

2.3.3.4 Strain Gauge

A device to measure the strain of a material or a structure is called a strain gauge (Figure 2.3.11). There is a technique to measure the electrical resistance of the strain gauge caused by the deformation of the material by sticking a strain gauge to the material or structure to be measured. Besides measuring the strain, for example, the stress and load may be measured indirectly using the strain.

2.3.3.5 Displacement Meter

A device to measure the displacement of a structure, etc. is called a displacement meter. Displacement meters can be divided into two types. One of the device is a contact type

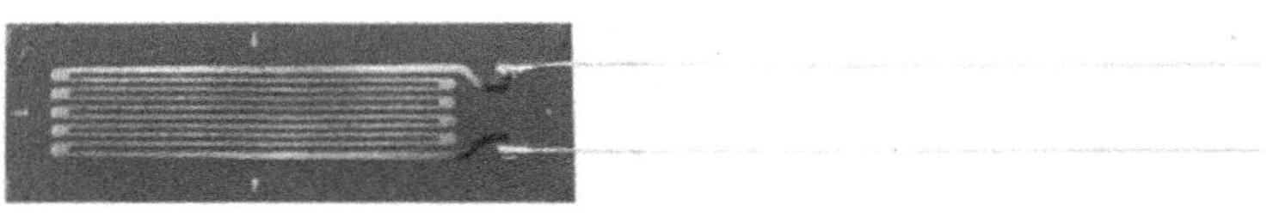

FIGURE 2.3.11
Strain gauge. With permission from Tokyo Measuring Instruments Laboratory Co. Ltd.

that measures the displacement of the measurement unit when the detector of the sensor touches the object, and the other one is a noncontact type that measures the displacement from the projection onto or reflection from the target object using laser light. Wire-type displacement meters, etc. that measure the displacement by pulling out a stainless steel wire are also available (Figure 2.3.12).

2.3.3.6 Accelerometer

A device to measure the acceleration and vibration of structures is called accelerometer. In recent years, it has been possible to measure the movement of an object with six degrees of freedom in combination with a gyro sensor (Figure 2.3.13). Furthermore, small and inexpensive accelerometers have been developed by applying microelectromechanical system (MEMS) technology. These have been introduced in devices used in daily life and are commonplace. Accelerometers are sometimes used as

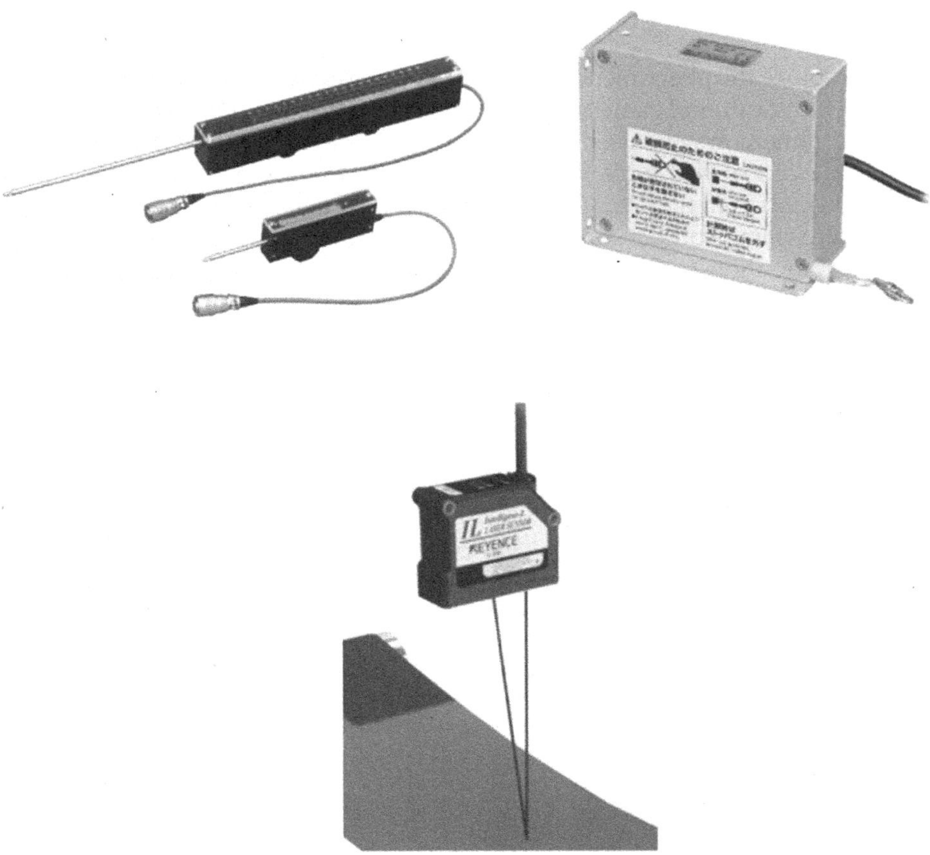

FIGURE 2.3.12
Displacement meter. With permission from (top left) Tokyo Measuring Instruments Laboratory Co. Ltd., (top right) Kyowa Electronic Instruments Co. Ltd., and (bottom) KEYENCE.

inclinometers in experiments. Additionally, trials are being carried out to understand the scouring condition due to changes in the acceleration response by burying accelerometers in the soil.[2.3.36]

2.3.3.7 Sand Level Gauge

A device used for measuring sand surface changes in movable bed experiments and observing local topography is called a sand level gauge. Sand level gauges can be divided into contact type and noncontact type. With the contact-type gauge, the detection unit is in direct contact with the measurement surface and measurements are taken on the basis of the distance from the reference position to the contact position (Figure 2.3.14a). Conversely, with the noncontact-type gauge, the shape of the sand surface is measured by emitting ultrasonic waves or laser to the measurement surface from a detection unit and measuring the reflection at the receiver (Figure 2.3.14b).

2.3.3.8 AV Equipment

A device to capture the phenomenon that occurred during an experiment is called an AV device. Generally, a digital camera or a video camera is used. Also, as a special device, a high-speed camera may be used for capturing images. Particularly, in an experiment conducted in a centrifugal force field, the phenomenon progresses faster according to the similitude ratio than the actual object; so, the phenomenon should be recorded using a high-speed camera. Additionally, when shooting with a high-speed camera, it is necessary to pay attention to the shooting speed, shutter speed, lighting, etc. to suit the target phenomenon.

2.3.3.9 Data Logger, Recorder

A device that reads measured values (voltage value, strain value, etc.) measured by various sensors, and saves them as digital data on a computer is called a data logger. In the past, it was necessary to have a data logger with specifications that correspond to the value measured by the sensor, but in recent years, multitype data logger units capable of taking measurements by combining different data loggers appropriate for the measurement value from each sensor regardless of what it may be are available.

FIGURE 2.3.13
Accelerometer and gyro sensor. With permission from Kyowa Electronic Instruments Co. Ltd.

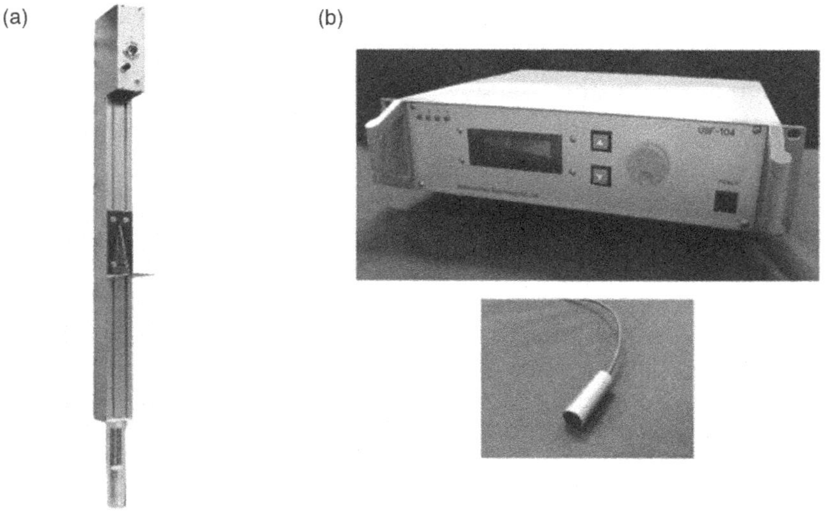

FIGURE 2.3.14
Sand level gauge. With permission from (a) KENEK Co. Ltd. and (b) Advanced New Technology.

2.3.4 Model Test Procedure

2.3.4.1 Setting of Boundary

As described above, the target area in the experiment is set based on the law of similarity and constraints such as the dimensions of the experimental equipment and experimental soil tank. In many model experiments, experiments are carried out using soil tanks, and boundary conditions are mainly assumed to be imposed by rigid walls, but it should be noted that the phenomenon may differ from the actual phenomenon when deformation is constrained. For example, to reduce the influence of boundaries, in a study targeting dynamic problems, an experimental soil tank called a laminar container (shear soil tank) is used.

In the model experiment, the area is specified so that the phenomenon to be examined is not affected by the boundary as much as possible; then, the scale ratio is determined or boundary measures are taken, and the influence of the boundary is identified in advance where necessary.

2.3.4.2 Method of Preparing Model Ground

a. Method of preparing sandy ground and cohesive soil ground

   Sand and cohesive soil have been used as ground material for the model, but the actual soil may not necessarily be composed of only sand or cohesive soil. Gravel, sand, silt, and clay are mixed together in several cases.

   However, since the behavior of the ground in which soil particles with different grain sizes are mixed is complicated, sand or cohesive soil was often used alone in the model test for simplification purposes. Therefore, methods for producing a ground model made of sand and cohesive soil will be described here.

The methods described below can be cited as the main ways of producing sandy ground that have been used in the past:

i. Freefall method using a sand hopper
ii. Simple stacking with a shovel
iii. Method in which water is poured in a sample container in advance and sand is lowered onto the surface of the water a little at a time to allow the sand to sink under water
iv. Mixing and stacking of sand and water

Figure 2.3.15 shows the sand hopper used in Method (i). In this method, the diameter of the discharge port at the bottom of the tank containing the sand is constricted to allow the sand to fall and deposit freely. In some cases, the sand is dropped through sieves so that the sand accumulates across a wide area.

Method (ii) is a method of simply stacking sand with a shovel, etc. In some cases, sand that has been weighed beforehand is sprinkled into the soil tank, and a tamper, rammer, etc. are used to finish the layer to a predetermined thickness in a large model, while a tamping rod, etc. are used for a small model.

In Method (iii), sand particles are deposited by placing them a little at a time on the surface of the water to ensure uniformity without any air bubbles and letting them sink down into the water. This method can be said to be more time-consuming as the sand is lowered a little bit at a time.

In Method (iv), the sand is moistened with water and then piled up by hand or with a shovel.

The uniformity and relative density of the resulting sandy ground differ greatly depending on the way these sandy grounds are made. Additionally, the reproducibility of each case is different. Of the four methods, the one that has the best uniformity, relative density control, reproducibility, etc. is probably the freefall method using a sand hopper as shown in (i). However, when preparing the sand hopper or when cranes

FIGURE 2.3.15
Sand hopper.

and other equipment are required because the sand hopper is larger in full-scale (large-scale) model tests, a preliminary study is required to adjust the relative density. Thus, this method is also somewhat laborious. It is necessary to adopt the optimum method for each model test, considering the equipment and labor required.

Cohesive soil is rarely used in gravitational field model tests. This is probably because a gravity model test using a relatively large model requires a large-scale loading device and a large amount of time for the consolidation process that is required to prepare the cohesive soil ground. Another possible cause is that in gravity model tests where the overburden pressure is small, reproducing the actual behavior is often difficult because the stiffness and strength of the cohesive soil ground are too small. For this reason, model tests on cohesive soil ground have been carried out extensively in a centrifugal force field where the model is relatively small, and the overburden pressure can be increased. One typical way of producing a cohesive soil ground in a centrifuge model test is to consolidate a hydrated slurry-like cohesive soil instead of using cohesive soil that is from the location that is to be reproduced. This is because, besides being able to produce a uniform ground, this method has a high reproducibility for each case, and the stiffness and strength of the ground are also easy to understand.

The soil collected at the site may be used or cohesive soil produced at a factory (e.g., kaolin) may be used as a material. The stiffness and strength of the ground are adjusted by the magnitude of consolidation stress and overburden pressure, and conducting a vane test of a miniature model that operates in a centrifugal force field is one-way of determining the strength.

b. Adjustment of relative density

When using sandy ground, the degree of clogging (relative density) of the sandy ground has a great influence on the deformation and fracture behavior of the ground, so it is very important to control the relative density. For example, regarding shear characteristic, shear resistance increases with strain in the loose sandy ground, but the dense sandy ground may soften after showing a peak value. If the density of the ground formed by fine sand under water is low, the pore pressure rises as the sandy ground gets compacted because of vibration, resulting in liquefaction of the ground. When the ground liquefies, the stiffness and strength are greatly reduced, so the deformation and failure behaviors of the ground are also greatly changed. It is important to consider the actual ground conditions to be reproduced and set the relative density of the ground appropriately.

As mentioned above, typical ways of preparing a sandy ground include (i) freefall method using a sand hopper, (ii) simple stacking with a shovel, (iii) method in which water is poured in a sample container in advance and sand is lowered onto the surface of the water a little at a time to allow the sand to sink under water, and (iv) mixing and stacking of sand and water. In Method (ii), not only a nonuniform ground is prepared, but also the relative density varies from case to case, and such value also varies depending on the stacking method.

However, as mentioned above, the density of the entire ground can also be controlled to some extent by sprinkling sand that has been weighed in advance into the soil tank and finishing the layer to a predetermined thickness by mechanical compaction, etc. In Methods (iii) and (iv), adjusting the relative density is difficult, and thus a single relatively loose sandy ground is formed. These methods have been used for liquefaction and other model tests because they can produce loose sandy ground. Conversely, in the freefall method of Method (i), the relative density can be adjusted. For this reason, this method is the most widely adopted. Specifically, the sand is dropped by narrowing the diameter and slit width of the discharge port below the tank containing the sand, but the relative density can be changed by changing the drop height, diameter, slit width, and other conditions such as whether or not to pass the sand through sieves when it drops.[2.3.37] Figure 2.3.16 summarizes the effects of each factor on the relative density. For example, if the drop height is increased, the relative density increases, allowing a dense sandy ground to be produced. Before making the sandy ground, the level of relative density of the sandy ground to be made shall be investigated using the sand and sand hopper that are actually used, and the conditions that allow the relative density required in the model test to be reproduced shall be clarified beforehand. Based on this, a ground having the prescribed relative density can be produced. Technology to automatically prepare a sandy ground by moving the sand hopper with a machine is also available. By using the machine, a ground with a relative density that is almost equal in each case can be produced.

c. Saturation control

Saturation control is also important when using sandy ground. This is because the behavior of sandy ground under water is completely different between saturated ground and ground containing air. In geotechnical engineering, the volume percent of the pore water to the pores is defined as the degree of saturation. The volume percent of completely dry soil is 0% and soil whose pores are completely filled with water is 100%. The difficulty with saturation lies in the high sensitivity of its effect on the behavior of the ground. The behavior does not change linearly with respect to the degree of saturation. For example, between soil with a 95% degree of saturation and soil with a 100% degree of saturation, the difference in the degree of saturation may be only 5%, but the hydraulic propagation characteristics vary greatly. This is because even

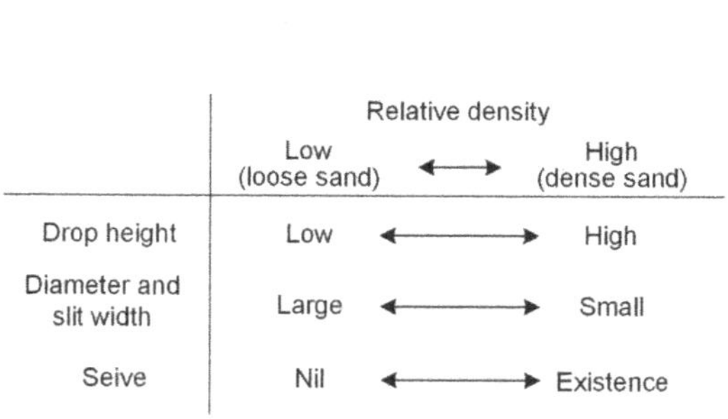

FIGURE 2.3.16
Effect of freefall conditions on relative density.

if an attempt is made to propagate the water pressure, when air is present, the air bubbles are compressed, making it difficult for the water pressure to propagate. The magnitude of the pore pressure has a great influence on the stiffness and strength of the ground, and consequently, it has a great influence on the results of the model test. It is important to understand and manage the degree of saturation of the model ground.

In element tests such as a triaxial compression test, tests are conducted on soil with the degree of saturation raised as high as possible, except for tests that examine the effects of the degree of saturation. In this method, carbon dioxide is poured into a dry sandy ground that is adjusted to a predetermined relative density to replace the air. Afterward, the air pressure is lowered by a deaeration pump. While maintaining the air pressure, water that has been deaerated in advance is poured from the lower part and the dry sandy ground is immersed in water. The reason for replacing air with carbon dioxide is that even if gas remains in the pores because of deaeration, it will dissolve in water. Furthermore, to reduce the volume of gas remaining in the soil as much as possible, loading is performed with the pore pressure raised to 200 kPa or other values. This pressure is known as backpressure. In a triaxial compression test, the response of the pore pressure to the confining stress (called the B value) is measured to confirm that the degree of saturation is increasing.

Even in model tests, when using a relatively small model such as a centrifuge model test, the system shown in Figure 2.3.17a is often used to replace carbon dioxide and pass water through a deaerated state to follow the element test technique. If the ground is saturated by this method, it will be close to the degree of saturation in the element test, making it easy to interpret the results of the model test. Takahashi et al.[2.3.38] investigated the effects that carbon dioxide substitution and deaeration have on saturation. Nevertheless, it takes time and effort to replace air with carbon dioxide, prepare a deaerated tank, etc. Additionally, the length of time water is passed through is also a problem. In tests using viscous fluids, it takes one to two days for the water to pass. Thus, recently, the number of cases in which water is passed through in a centrifugal force field as proposed by Okamura and Inoue[2.3.39] is increasing.

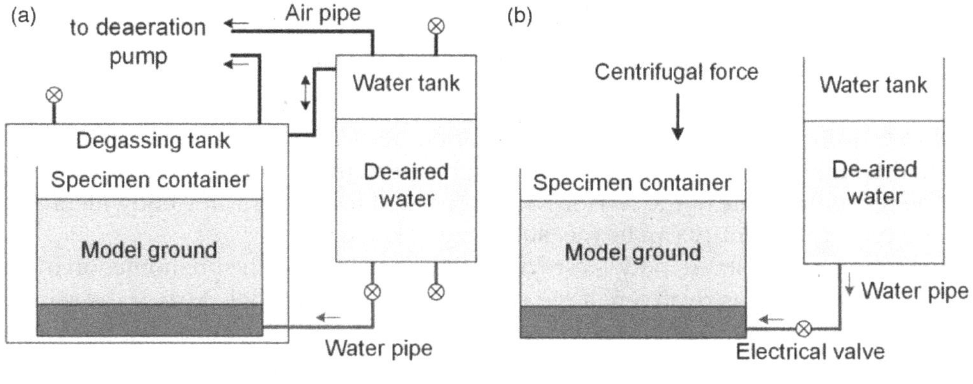

FIGURE 2.3.17
Control method of ground saturation.

In this method, a sandy ground is prepared in a dry state, and water is passed through with centrifugal force applied to the model (see Figure 2.3.17b). By applying centrifugal force to the ground, it is possible to reduce the amount of water sucked up due to suction during permeation, thereby significantly reducing the amount of air bubbles trapped in the pore water. Additionally, since the overburden pressure increases, water can be poured in at a large water pressure, allowing water to pass through in a short time (several hours even if it is a viscous fluid).

Common ways to prepare a model test with sandy ground containing water in a gravitational force field are to pour water into a sample container in advance and then letting sand fall onto the water surface and sink a little at a time, to stack a mixture of sand and water, and to prepare a dry ground and simply let water permeate through it. This is because relatively large models are often used in gravitational force field, so it is difficult to exhaust air or apply a centrifugal force to allow water to pass through. However, in the water pluviation method, sand is dropped in lumps of a certain size because it is difficult to drop sand one particle at a time. Thus, there is a high possibility that air bubbles will remain in the lumps. Additionally, air bubbles remain in the ground even when sand and water are mixed beforehand and when water is allowed to infiltrate without degassing.

Therefore, note that the ground is not completely saturated in a gravitational force field model test. A method to indirectly determine the degree of saturation by measuring the elastic wave velocity, suction, etc. has been considered.

2.3.4.3 Calibration of Measuring Instruments

To obtain physical values (acceleration, pressure, displacement, etc.) from the measuring instruments used in the experiment, calibration must be performed. Calibration is performed to investigate the relationship between measured values (voltage, strain, etc.) and physical values (calibration coefficient). It is also conducted before the experiment to confirm that the measuring instrument operates normally. Please refer to the detailed calibration method explained in Section 2.2.4.2.

2.3.4.4 Acquisition and Organization of Experimental Data

To quantitatively evaluate the phenomenon by experiment, a sensor is installed on the model to acquire data. Regarding the installation position of the sensor, the appropriate installation location will be studied through a preliminary experiment. At the same time, the response value of the sensor will be checked to see if a value equivalent to the expected physical quantity can be measured at the same time.

When conducting a small model test with a small-scale ratio, the phenomenon may be hindered if the sensor is relatively large with respect to the model. Also, if the sensor is wired, it may affect the phenomenon depending on the stiffness, thickness, and installation method of the cable; thus, measures have to be taken to minimize the effect. In the model test in a centrifugal force field, the wiring and securing method will be checked carefully to see if they are adequate. If not, the measurement may be disabled because of disconnection or a centrifugal force may be applied to the sensor, blowing it away and possibly leading to a serious accident depending on the situation.

The sensors used mainly in geotechnical engineering have already been dealt with in Section 2.3.3. The type, size, etc. of the sensor are selected according to the subject of the study. The sensor may be installed by either fixing it into a soil tank or burying it in the soil, for example. When burying sensors in soil, they are either laid directly in the soil or they are installed by making a fixture. In recent years, various efforts have been made, such as producing securing jigs using a 3D printer.

Data will be acquired by accelerometers, pressure gauges, soil moisture meters, strain gauges, etc. as point data. Although point data can capture local changes, multiple sensors are preferable to use for measurement to capture changes in physical quantities as a space (or surface) to determine if the measured data are a reasonable value or change in contrast to the phenomenon. Conversely, a value derived by integrating the pressure value and the strain value of a surface or space is obtained by a load cell or a displacement meter. Thus, what kind of sensor is appropriate will be considered on the basis of the target phenomenon and physical quantity to be acquired. For example, by combining a pressure gauge and load cell to take measurements, the validity of the measured value can be verified. Furthermore, it helps in understanding how to connect a point to a surface or space.[2.3.1]

The measured data are converted into physical quantity using the calibration coefficient obtained by calibration. The phenomenon will be evaluated qualitatively and quantitatively on the basis of the obtained physical quantity, but it is very important to judge whether the numerical value obtained is appropriate or not based on the phenomenon. Interpretation of the results obtained requires objective judgment from various perspectives.

Besides the sensor, the phenomenon is captured using an AV equipment to visually examine the phenomenon. In geotechnical engineering and other civil engineering model tests, several studies are conducted using two-dimensional cross sections assuming planar strain conditions due to the characteristics of the actual structures. In that case, a part of the side surface of the soil container is used as a visualization window to capture the phenomenon occurring in the cross section of the model. To satisfy the plane strain condition, the displacement in the direction orthogonal to the cross section due to the soil tank boundary and the shear stress in the direction parallel to the cross section should be set to zero. Additionally, to assume that the phenomenon that occurs in the model cross section occurs uniformly in the depth direction, it is necessary to devise ways to prevent friction from being generated between the visualization window and model cross section where possible.[2.3.1]

Phenomena are evaluated quantitatively by image analysis using captured images. One of the methods is particle image velocimetry (PIV). PIV is a measurement method to determine the amount of movement by tracking changes in the brightness of the measured images (or velocity by dividing the amount of movement by time interval) and has been used widely in recent years as a way to measure the flow velocity of a fluid. Additionally, in tracking the brightness, a tracer material is mixed in the fluid. For ground measurement, the deformation of the ground is quantified by using the ground material itself instead of a tracer material or embedding the target in the ground and analyzing changes in the target, for example. For more information on PIV, you may wish to refer to technical books.[2.3.40]

2.3.4.5 Implementation Order of Model Test

As for the order of conducting the experiments in a model test, as described above, a preliminary experiment is performed before the main experiment. During the preliminary experiment, the similarity, procedure, safety, etc. of the model test are verified according to the drafted experimental plan, which shall be positioned as a preliminary experiment for carrying out the main experiment. The experimental plan for the main experiment will be finalized on the basis of the procedures confirmed in the preliminary experiment and the experimental results obtained. The main experiment will be carried out according to the experimental method and experimental plan determined by the preliminary experiment.

2.3.5 Typical Model Test Example

2.3.5.1 Static Model Test for Bearing Capacity Problem in Sandy Ground

As a static problem, an example of a previous study on the problem of sand-bearing capacity is explained. Much research has been done so far on the problem concerning the bearing capacity of sandy ground (e.g., De Beer[2.3.41]). As stated by Okamura et al.,[2.3.42] if soil is assumed to be an isotropic and a completely rigid body, the bearing capacity theory related to the vertical loading conditions of a two-dimensional rigid and shallow foundation can be considered to be almost completed. Conversely, as the bearing capacity problem of sand ground is affected in a complicated manner by the dimensions and shape of the foundation, the loading conditions of the load, the boundary conditions, and the mechanical characteristics of the sand ground, further study is considered necessary for an accurate prediction. It has also been pointed out that the destruction progress must be considered,[2.3.43] and in considering the bearing capacity problem of coastal and marine structures due to ocean waves in hydraulic model tests, it is necessary to pay attention to these points in the study.

Tatsuoka et al.[2.3.44] conducted model tests in a gravitational force field and centrifugal force field on the bearing capacity problems of the strip foundation on top of sandy ground and conducted a modeling of the original using the same geomaterial in both cases. Figure 2.3.18 shows the relationship between the foundation width and bearing capacity factor converted to the prototype based on the results obtained in each experiment and the results obtained by finite element analysis. Tatsuoka et al.[2.3.45] defined the scale effect as the sum of the stress level effect and the grain size effect of the sand particles and foundation width. Looking at Figure 2.3.18, it is shown that a significant grain size effect is generated within a certain range.

Okamura et al.[2.3.42] investigated the scale effect (grain size effect) using a strip foundation and a circular foundation by a centrifuge model test. Figure 2.3.19 shows the relationship between the bearing capacity factor and foundation width according to the modeling of models. In the case of a strip foundation, the bearing capacity factor varies depending on the foundation width, whereas in the case of a circular foundation, the bearing capacity factor does not vary even if the foundation width changes. As a result of confirming the fracture mode using an X-ray film, it was observed that a clear shear zone was formed in the case of a strip foundation, resulting in a fracture

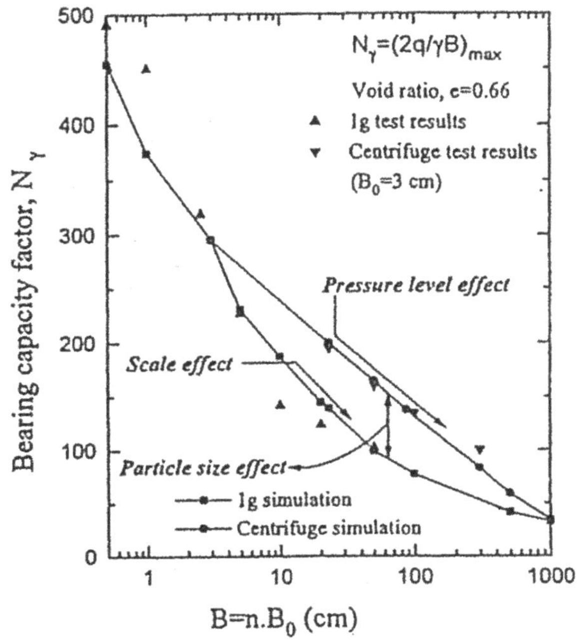

FIGURE 2.3.18
Soil particle effect for bearing capacity test of strip foundation.[2.3.44]

mode with a discontinuous surface. By contrast, in the case of a circular foundation, a fracture mode that is caused by continuous shear deformation occurs without a clear shear zone being formed.

From the above, the grain size effect may affect the formation of the shear zone, that is, the fracture mode, so the scale effect of the soil particles must be considered when studying the fracture mode in a two-dimensional wave tank model test.

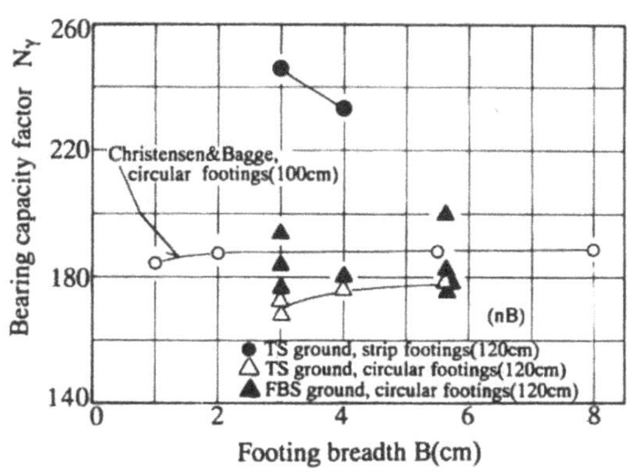

FIGURE 2.3.19
Relationship between bearing capacity factor and foundation width in modeling of models.[2.3.42]

2.3.5.2 Comparison of Gravity Model Test and Centrifuge Model Test in a Seismic Response

We will introduce a comparative study of a gravity model test and a centrifuge model test for a dynamic problem.

a. Study on soil structures

Hayashi et al.[2.3.46] conducted model tests in a gravitational force field and a centrifugal force field focusing on the liquefaction phenomenon of saturated sandy ground targeting soil structures (embankment) and compared the two. Table 2.3.1 shows the scaling law between a gravitational force field and a centrifugal force field. The scaling law derived by Iai[2.3.47] is applied to the gravitational force field. In this study, scales of 1/3, 1/5, and 1/10 are used in gravity model tests, whereas a scale of 1/20 is used in model tests in centrifugal force fields (Table 2.3.2).

As for the difference in the boundary conditions, shear soil tanks are used for all experiments except G-5. The input seismic motion is standardized as 60 sine waves with a maximum acceleration of 147 Gal. One of the features of this study is that we are analyzing the effect that a difference in the permeability of the ground exerts on the behavior of the model. The following results have been obtained through a series of experiments.

First of all, regarding the response acceleration in the ground, the damping and recovery show a similar tendency qualitatively, but they appear to be difficult to handle quantitatively.

Regarding the excess pore pressure, the increase in water pressure due to vibration was quantitatively consistent; however, the difference due to the viscosity of the pore water and the permeability of the soil ground has been verified for the dissipation process; thus, it has been shown that the behavior can be reproduced quantitatively by adjusting the viscosity of the pore water according to these measured values (Figure 2.3.20). Regarding the deformation at the top of the embankment, it has been shown that a difference appears in the input acceleration due to the boundary conditions and that the

TABLE 2.3.1

List of Similitude Ratio in Model Test by Hayashi et al.[2.3.46]

	Similitude	
Unit	**Gravitational Force Field**	**Centrifugal Force Field**
Length	$1/N$	$1/N$
Unit weight	1	$1/N$
Time	$1/N^{0.75}$	$1/N$
Stress	$1/N$	1
Pore water pressure	$1/N$	1
Displacement	$1/N^{1.5}$	$1/N$
Acceleration	1	N
Coefficient of permeability of soil	$1/N^{0.75}$	1

TABLE 2.3.2

Test Cases and Condition by Hayashi et al.[2.3.46]

Test Case	Prototype	G-3	G-5	G-10	C-20-1	C-20-2
Similitude ratio	1	1/3	1/5	1/10	1/20	1/20
Ground height (m)	4.5	1.5	0.9	0.45	0.225	0.225
Ground width (m)	–	2.8	1.8	1.0	0.60	0.60
Ground depth (m)	–	2.8	1.8	1.0	0.60	0.60
Test field Acceleration in vertical direction	Gravitational force 1g				Centrifugal force 20g	
Coefficient of permeability of soil (cm/s)	1.21×10^{-1}	4.3×10^{-2}	4.3×10^{-2}	2.2×10^{-2}	8.0×10^{-3}	5.6×10^{-3}

measured strain of the model and the scale of the strain are proportional to each other.

b. Study on complex structure–soil coupling problems

Kohama et al.[2.3.48] conducted model tests on the behavior of gravity-type quay wall during long period and long duration earthquake motion in a gravitational force field and centrifugal force field and made a comparison. Figure 2.3.21 shows a schematic diagram of each experimental model in a gravitational force field and a centrifugal force field. The scale is shown in Table 2.3.3. The scaling law derived by Iai[2.3.47] is applied to the gravitational force field. Model test has been conducted under conditions that do not cause liquefaction in gravitational and centrifugal force fields, and the following results were obtained.

Figure 2.3.22 shows a time history of the shaking table acceleration, excess pore pressure in landfilled ground, and caisson horizontal displacement in a gravitational force field and a centrifugal force field. The fluctuation in the response of excess pore pressure in landfill ground is small in both cases, indicating that stress changes in the ground are similar. Conversely, looking at the horizontal displacement of the caisson, the residual displacement in a centrifugal force field is less than half of that in a gravitational force field, but the maximum acceleration of the shaking table has been shown to be larger in the centrifuge model test. Via component analysis of the input seismic motion, the long period component has been confirmed to affect the residual displacement.

Kohama et al.[2.3.48] showed that useful information for clarifying the phenomenon can be obtained by extracting the factors influencing the phenomenon of interest and performing sensitivity analysis. In this way, to understand the phenomenon from the experimental results obtained, a data analysis method should be devised.

The example introduced above compares a gravity model test and a centrifuge model test for dynamic problems. The scaling law has been established in gravity model tests and centrifuge model tests, and the results obtained from both model tests show roughly similar phenomenon tendencies, which indicates the reproducibility of the phenomenon.

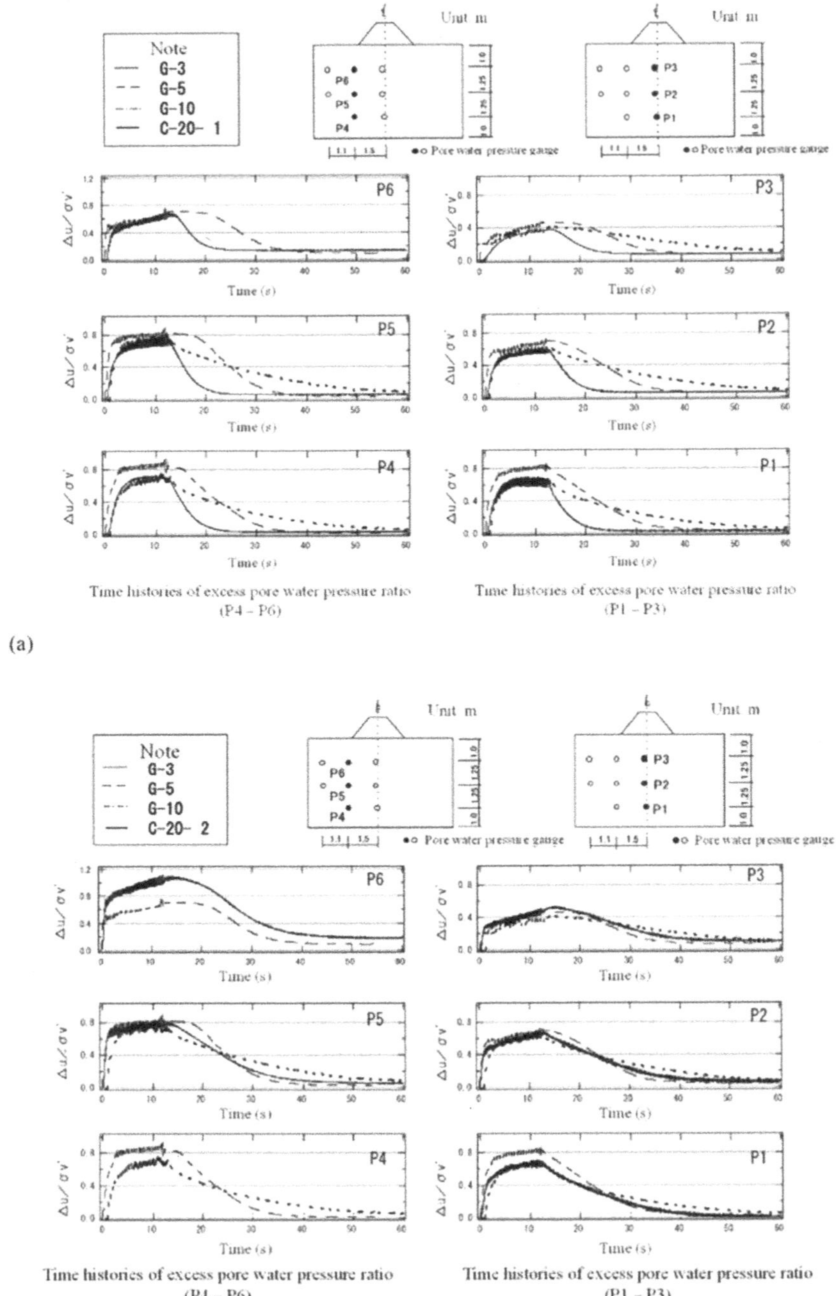

FIGURE 2.3.20
Generation and dissipation processes of excess pore water pressure (adapted from note 2.3.46). (a) Comparison between gravity model experiment and centrifuge model experiment (C-20-1), (b) Comparison between gravity model experiment and centrifuge model experiment (C-20-2).

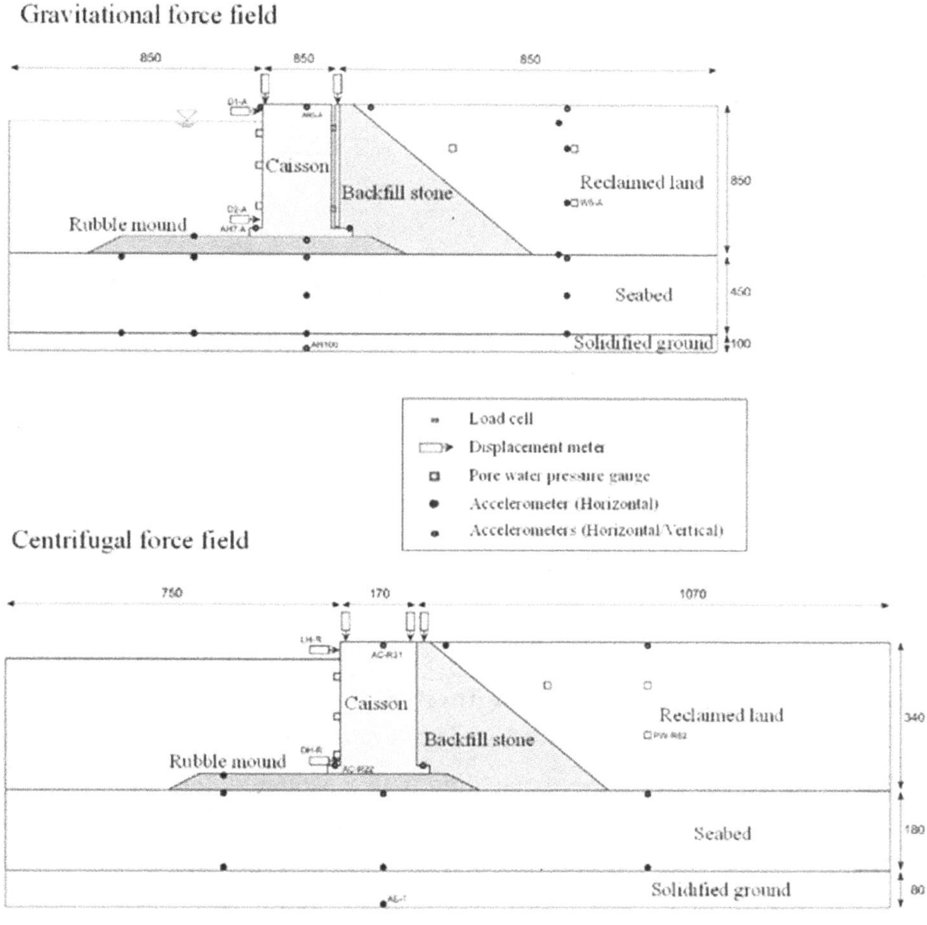

FIGURE 2.3.21
Schematic of model cross section of 1*g* model test and centrifuge model test (adapted from note 2.3.48).

TABLE 2.3.3
List of Similitude Ratio in Model Test by Kohama et al.[2.3.48]

	Prototype/Model	
Parameter	**Gravitational Force Field**	**Centrifugal Force Field**
Length	20	50
Density	1	1
Time	$20^{0.75}$	50
Stress	20	1
Acceleration	1	1/50
Velocity	$20^{0.75}$	1
Displacement	$20^{1.5}$	50

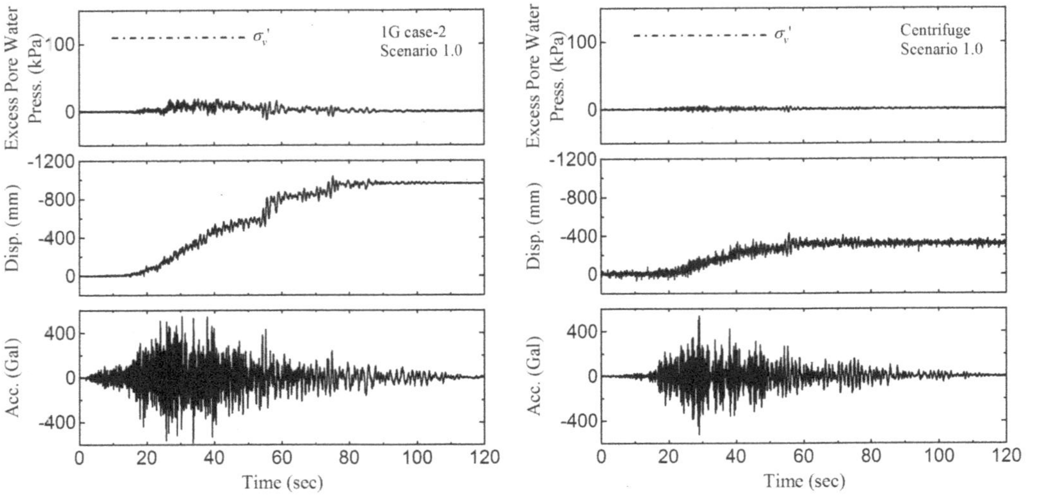

FIGURE 2.3.22
Time histories of shaking table acceleration, excess pore water pressure in reclaimed ground, and horizontal displacement of caisson in gravity model tests (*left*) and centrifuge model tests (*right*).[2.3.48]

2.4 Review on the Handling of Geomaterials and Organization of Issues in Japan (Published in Japanese)

2.4.1 Examples of Past Experiments Dealing with a Movable Bed in Hydraulic Model Tests

2.4.1.1 Model Test on Complex Wave Motion and Ground Problems

Examples of past experiments on (a) sediment movement on the seabed due to wave motion and (b) pore pressure response of the seabed due to wave motion are summarized from the *Journal of Japan Society of Civil Engineers*, Ser. B2 (Coastal Engineering), *Journal of Japan Society of Civil Engineers*, Ser. B3 (Ocean Engineering), and the predecessor of these journals in Japan (published in Japanese).

a. Examples of past experiments on sediment movement on the seabed due to ocean wave

   In a study on the movement of sediment on the seabed due to wave motion, gravity model test has been conducted to study the movement limits of the sand, the movement of suspended sand, the generation and movement of ripples, the movement of sand flowing like a sheet, the influence of mixed grain sizes on the movement, disappearance of ripples due to infiltration flow, and so on.

   Typical examples of studies on the movement of sediment on the seabed caused by wave motion are shown in Table 2.4.1, in which the geomaterials used in the experiments are specified. In the experiment, a wave tank and an oscillatory flow flume tank are used according to the subject of the study. In these experiments, model experiments can be carried out on the basis of Froude's law for the assumed wave-field, but detailed experimental conditions

TABLE 2.4.1

Previous Experiments on Wave-induced Sediment Movement on Seabed Soil in Japan (Published in Japanese)

Researchers	Focus Phenomenon	Test Field	Geomaterial
Iwagaki and Noda[2.4.1]	Scale effects in beach deformation experiments	Gravitational force	Sand: d_{50} = 0.36 mm, G_s = 2.70
Sato and Tanaka[2.4.2]	Sand movement limit ripple movement	Gravitational force	Sand: 1–2 mm, 0.35–0.71 mm and 0.125–0.25 mm
Honma et al.[2.4.3]	Movement of suspended sediment due to wave	Gravitational force	Fine sand: d_{50} = 0.178 mm and standard deviation is 0.032 mm
Matsunashi and Ohmi[2.4.4]	Generation mechanism of ripple	Gravitational force	Sand: d_{50} = 0.14 mm
Noda[2.4.5]	Suspension of sediment due to wave	Gravitational force	Vinyl chloride particle: d_{50} = 0.13 mm and G_s = 1.15
Nakamura et al.[2.4.6]	Movement of sand due to wave and current	Gravitational force	Sand: G_s = 2.64
Hino et al.[2.4.7]	Movement of sand due to wave	Gravitational force	Standard sand: d = 0.15 mm and G_s = 2.65
Yamashita et al.[2.4.8]	Oscillatory flow induced sheet flow sand movement	Gravitational force	Artificial pearls: d = 3 and 5 mm, and G_s = 1.58
Suzuki et al.[2.4.9]	Sand movement	Gravitational force	Fine sand: d_{50} = 0.18 mm and falling velocity of particle is 21 mm/s Coarse sand: d_{50} = 0.87 mm and falling velocity of particle is 100 mm/s
Goto et al.[2.4.10]	Sand movement	Gravitational force	Sand: d = 0.25 mm and G_s = 2.65
Mohammad et al.[2.4.11]	Sand moving layer thickness affecting mixed layer thickness	Gravitational force	Toyoura sand: d_{50} = 0.2 mm, G_s = 2.65 and falling velocity of particle is 23 mm/s Soma sand: d_{50} = 0.55 mm, G_s = 2.65 and falling velocity of particle is 70 mm/s Soma sand: d_{50} = 0.80 mm, G_s = 2.65 and falling velocity of particle is 100 mm/s
Sakai et al.[2.4.12]	Mixed grain size drifting sand migration process	Gravitational force	The cell of artificial pearls: d = 5.15 mm, 9.88 mm, and 15.60 mm (mixing); G_s = 1.318
Suzuki[2.4.13]	Extinction by seepage flow of the ripple	Gravitational force	Sand: d = 0.87 mm

such as the experimental scale were not described. Regarding the geomaterial used in the movable bed, although cases are showing that the same material as the actual geomaterial is used, in most cases the reasons for using it were not clearly described. Conversely, there have been cases in which geomaterials with a low specific density, such as artificial pearls, are used besides the actual geomaterials.

b. Examples of past experiments on stress changes in the seabed due to ocean wave

In a study on stress changes in the seabed due to ocean wave, model tests were conducted in a gravitational force field and centrifugal force field to study the pore pressure response in the ground, permeation flow, relationship of liquefaction and fluidization with densification, relationship between liquefaction and sand drift (sheet-flow), etc.

Table 2.4.2 shows a typical example of a study on the stress changes in the seabed due to wave motion, in which geomaterials used in the experiment are specified. Similar to (a) examples of past experiments on sediment movement in the seabed due to ocean wave, wave tanks and oscillatory flow flume tanks are used in studies on gravity model tests. Conversely, when targeting the change in stress of the seabed due to wave, a centrifuge model test is conducted. The reason is that in a centrifuge model test, the similarity factor of the stress, strain, and velocity is one as shown in Section 2.3.2.2, so it is possible to reproduce the same stress field as the prototype in the model.

TABLE 2.4.2

Previous Experiments on Wave-induced Stress Changes in Seabed Soil in Japan (Published in Japanese)

Researchers	Focus Phenomenon	Test Field	Geomaterial
Maeno et al.[2.4.14]	Liquefaction in sand layer surface	Gravitational force	Nabae sand: $d_m = 0.16$ mm; $d_{10} = 0.114$ mm; $U_c' = 1.53$; $G_s = 2.70$; $k = 2.3 \times 10^{-4}$ m/s
Zen and Yamazaki[2.4.15]	Seepage flow and liquefaction	Gravitational force	Toyoura sand: $d_{50} = 0.181$ mm; $U_c' = 1.79$; $G_s = 2.647$ Hasaki sand: $d_{50} = 0.16$-0.17 mm; $U_c' = 1.5$-1.6; $G_s = 2.689$
Izumiya et al.[2.4.16]	Response of porewater pressure	Gravitational force	Sand in Nigata port east area: $d_m = 0.18$ mm
Maeno and Tokutomi[2.4.17]	Variable of pore water pressure	Gravitational force	Same condition as Maeno et al.[2.4.14]
Maeno and Uchida[2.4.18]	Response of pore water pressure	Centrifugal force	Toyoura sand: $d_{50} = 0.208$ mm; $d_{10} = 0.170$ mm; $U_c' = 1.282$; $G_s = 2.637$
Sakai et al.[2.4.19]	Relationship between degree of saturation of ground and distribution of pore water pressure; response of pore water pressure and soil movement; drift sand	Gravitational force	Sand: $d_{50} = 0.25$ mm; $k = 3.9 \times 10^{-4}$ m/s
Nakano et al.[2.4.20]	Liquefaction of sediment and drift sand; sheet flow	Gravitational force	Light weight aggregate (Ishikawa light No.6): $d_{50} = 0.14$ mm; $G_s = 2.08$
Yamashita et al.[2.4.21]	Settlement, hardening, and liquefaction of loose sandy soil due to variable water pressure	Gravitational force	Silica sand: $d = 0.15$ mm
Miyamoto et al.[2.4.22]	Liquefaction of sand bed and its fluidization process	Centrifugal force	Silica sand No.7: $d_m = 0.14$ mm granulated marble: $d_m = 3$ mm; $G_s = 2.7$
Sakai et al.[2.4.23]	Drift sand of seabed soil	Gravitational force	Same conditions as Sakai et al.[2.4.19]

In some cases, the study may have been carried out on the basis of the actual wave conditions assumed. As for the geomaterial used for the movable bed, it is assumed that a material similar to the actual geomaterial is used. There are also cases where lightweight aggregates and granular marble stones are employed. In a centrifuge model test, a material equivalent to the prototype is used so that mechanical properties such as the stiffness and strength of the geomaterial and the stress–strain relationship match between the prototype and model. Additionally, a viscous fluid is used to satisfy the scaling law in the permeation time. See Chapter 3 for a detailed explanation of the reasons for using viscous fluids.

2.4.1.2 Model Test on Complex Wave–Structure–Soil Interactions Problem

Among examples of past experiments related to sand drift around a structure due to ocean wave, examples of model tests on complex wave motion–structure–soil interactions problems, which refer to model scales and geomaterials, are shown in Table 2.4.2.

As shown in Table 2.4.3, model tests are being conducted with various model scales due to constraints imposed by wave tanks, etc. However, the geomaterials used are determined in a way that allows actual phenomena such as erosion and suction to be reproduced at each model scale.

2.4.2 Extraction and Organization of Issues Related to the Handling of Geomaterials in Hydraulic Model Experiments

Key examples of past studies dealing with movable bed experiments in a hydraulic model test are shown in Section 2.4.1. Although experimental methods have been established in each field regarding hydraulic model tests and ground model tests that are related to wave motion. However, in hydraulic model tests to address complex water and soil problems shown in this section, the scaling laws necessary for interpreting the experimental results have not been examined adequately, and thus only independent reviews have been carried out in each field or by each institution. Thus, we believe that a sufficiently unified experimental technique has not been established.

Sediment movement phenomena caused by wave motion are problems involving a combination of water and soil and can be considered as multiphysics problems in which seepage force and other forces interact in a complex manner to change the bottom shear force (Shields number) that is the driving force for moving soil particles and the shearing force of the ground strength. The handling of geomaterials in hydraulic model tests is thus an important issue that must be fully discussed. Particularly, to resolve the problem, a multifield study based on knowledge of the coastal engineering field and geotechnical engineering field is considered indispensable.

Here, issues related to the handling of geomaterials in hydraulic model tests focusing on the phenomenon of sediment movement are identified.

- The geomaterial used for the movable bed is often a material with the same particle size as the assumed local geomaterial, but the conformity with the actual phenomenon is unclear.

TABLE 2.4.3

Previous Experiments on Wave–Structure–Soil Interaction in Japan (Published in Japanese)

Researchers	Object Structures	Experiment Scaling	Geomaterial	Determination of Geomaterials Rationale, Devices, etc.
Irie et al.[2.4.24]	Breakwater	1/75	Sand: $d_{50} = 0.14$ mm	Determined to be an L-type that results in deposition at the belly position
		1/50	Sand: $d_{50} = 0.07$ mm	Same as above
Umezawa et al.[2.4.25]	Wave-dissipating concrete blocks covering caisson breakwaters	1/75	Sand: $d_{50} = 0.16$ mm	Considering the similarity of the migration morphology of drifting sand, sand corresponding to a central grain size of 1 mm on a real scale was used
Takahashi et al.[2.4.26]	Caisson-type seawall	1/20	Sand: $d_{10} = 0.13$ mm; $n = 0.445$; $\rho_t = 1.90$ g/cm^3; $\rho_s = 2.70$ g/cm^3	Pointed out the possibility of small pressure attenuation appearing inside the sand layer
Kanatani et al.[2.4.27]	Upright breakwater	1/30	Sand: $d_{50} = 0.15$ m; $G_s = 2.66$; $U_c' = 1.88$; $e_{\text{max}} = 1.023$; $e_{\text{min}} = 0.595$	Confirmation that the similarity of the coefficient of permeability of the sandy soil is generally satisfactory
Sakakiyama and Kashima[2.4.28]	Caisson breakwaters covered with wave-dissipating sea wall	1/60	Sands: $d_{50} = 0.16$ mm and 0.2 mm	Examined the effect of model scale
Noguchi et al.[2.4.29]	Gentle slope-type coastal dike	1/2	Sand: $d_{50} = 0.32$ mm	Sand with d_{50} similar to that of the site was used. The model was simplified by omitting the broken stone in the upper backfill layer because erosion did not progress in the fully scaled model.
Suzuki et al.[2.4.30]	Wave-dissipating concrete blocks covering caisson breakwaters	1/16	Mixture of sand and silt:$d_{50} = 0.08$ mm	Use of small grain size material to prevent sand from moving easily
		1/4	Sand: $d_{50} = 0.2$ mm; $k = 0.05$ cm/s	Uses the same size sand as typical local
Yamamoto and Minami[2.4.31]	Coastal dike	1/30	Material mixed 2:1 with sand: $d_{50} = 0.20$ mm and $d_{50} = 0.66$ mm	Determination of geomaterials based on Ito and Tsuchiya similarity rule for bed sediments
Muraoka et al.[2.4.32]	Wave-dissipating concrete blocks covering caisson breakwaters	1/55.8	Sand: $d_{50} = 0.15$ mm	To avoid adhesive effects, sand with a grain size slightly larger than that determined by the similarity rule for settling velocity was used

- Although efforts have been made with a focus on the Shields number and materials with a lower specific gravity such as artificial pearls are used, the conformity with the actual phenomenon is not clear.
- Basically, only the median particle size (or mean particle size) is applied as a similarity parameter, and the characteristics of the ground material are not fully considered.

It should also be mentioned that the rate of advance associated with scouring, erosion, suction, and other actions and the similarity of the time until the topographic features reach an equilibrium state are unclear. Furthermore, the geotechnical engineering perspectives, such as the influence of bearing capacity failure for complex wave–structure–soil interactions problem targeted at stability issues in structures, must be considered.

The description above explains mainly the phenomenon which occurs when the ground is saturated and in water. However, when dealing with problems such as beach deformation and revetment stability, the study applies to unsaturated ground, and the impact due to the capillary phenomenon and suction must be considered.

Based on these problems, a scaling law is derived in Chapter 3, especially for sediment movement phenomena in which water and soil are combined, whereas the results of case studies adopted as an approach for resolving the issues described earlier are shown in Chapter 4 as we foster a deeper understanding of the law of similarity.

References

2.1 Purpose of model test

2.1.1 Japanese Geotechnical Society: Jiban kougaku ni okeru mokei zikken nyumon [Introduction to Model Tests in Geotechnical Engineering], Maruzen, 1994. (in Japanese)

2.1.2 Mikasa, M.: Model Testing in Soil Engineering, Soil mechanics and foundation engineering, 28(5), pp.1–2, https://dl.ndl.go.jp/info:ndljp/pid/10430785?tocOpened=1, 1980. (in Japanese)

2.1.3 Shibata, T. and Ohta, H.: Similitude in Soil-Model Test, Soil Mechanics and Foundation Engineering, 28(5), pp.9–14, https://dl.ndl.go.jp/info:ndljp/pid/10430788, 1980. (in Japanese)

2.1.4 Terzaghi, K.:Erdbaumechanik auf Bodenphysikalischer Grundlage, Deuticke, Wien, 1925.

2.1.5 Rocha, M.: The possibility of solving soil mechanics problems by the use of models, Proceedings of the 4th International Conference on Soil Mechanics and Foundation Engineering (ICSMFE), Vol.1, pp.183–188, 1957.

2.1.6 Phillips, E.: De l'equilibre des solides elastiques semblables, 68, C.R. Acad. Sci., Paris, pp.75–79, 1869.

2.2 Hydraulic model test

2.2.1 Sawaragi, T.: Kankyoken no atarashii kaigankogaku [New Coastal Engineering in Global Environment], Fuji Technosystem Co., Ltd, 1999. (in Japanese)

2.2.2 Hudson, R.Y.: Reliability of Rubble-mound Breakwater Stability Models, U.S. Army Engineer Waterways Experiment Station, 1975.

2.2.3 Hasegawa, I. and Arikawa, T.: Transition of hydraulic model experiment facilities in recent years, Journal of JSCE, Ser. B3 (Ocean Engineering), Vol.75, No.2, pp.I_349–I_354, 2019. (in Japanese)
2.2.4 Ikeya, T., Iwata, Y., Okuda, Y., Kikitsu, H., Ishihara, A., Hasegawa, I., Hashimoto, J., and Obata, D.: Influence of Experimental Channel Width on Tsunami Force Acting on Land Structure in Hydraulic Model Test, Journal of JSCE, Ser. B2 (Coastal Engineering), Vol.73, No.2, pp.I_901–I_906, 2017. (in Japanese)
2.2.5 Goda, Y., and Suzuki, Y.: Estimation of Incident and Reflected Waves in Random Wave Experiments, Proceedings of the 15th Coastal Engineering Conference, ASCE, Vol 1, pp.828–845, 1976.
2.2.6 Tanimoto, K., Tomida, E., and Muranaga, T.: Influence of Re-reflection from Wave Paddle on Irregular Incident Waves, Technical Note of the Port and Harbour Research Institute, No.467, 23pp., 1983. (in Japanese)
2.2.7 Frigaard, R. and Christensen, M.: An absorbing wave-maker based on digital filters, Proceedings of the 24th International Conference on Coastal Engineering, ASCE, pp.168–180, 1994.
2.2.8 Hirakuchi, H., Kashima, T., and Kawaguchi, T.: Suimenhakeiseigyohosiki ni yoru muhansyazohaki no fukisokuhajikken [Applicability of Nonreflected Wavemaker by Surface Waveform Control Method to Experiment of Random Wave], Proceedings of the Coastal Engineering, JSCE, Vol.35, pp.30–34, 1988. (in Japanese)
2.2.9 Goda, Y.: Random seas and design of maritime structure, Advanced Series on Ocean Engineering, Vol.33, World Scientific, 2010.
2.2.10 Homma, M. and Horikawa, K.: Kaigankankyokogaku kaigankatei no riron kansoku yosokuhoho [Coastal Environmental Engineering], University of Tokyo Press, pp.408, 1985. (in Japanese)
2.2.11 Tanimoto, K., Takahashi, S., Yoshimoto, Y.: Estimation of Shock Force from a Response in a Linear Damped Vibration System, Technical Note of the Port and Harbour Research Institute, No.474, 24pp., 1983. (in Japanese)
2.2.12 Takahashi, S., Tanimoto, K., and Shimosako, K.: Wave and block forces on a caisson covered with wave dissipating blocks, Report of the Port and Harbour Research Institute, Vol.29, No.1, 1990. (in Japanese)

2.3 Geotechnical model test

2.3.1 Japanese Geotechnical Society: Jiban kougaku ni okeru mokei zikken nyumon [Introduction to Model Tests in Geotechnical Engineering], Maruzen, 1994. (in Japanese)
2.3.2 Igarashi, T. and Sugiyama, H.: Ryutai kogaku to den-netsu kogaku no tameno zigen kaiseki katsuyou hou [Practical Application of Dimensional Analysis for Fluid and Thermal Engineering], pp.10, Kyoritsu Shuppan, 2013. (in Japanese)
2.3.3 Emori, I., Saito, K., and Sekimoto, K.: Mokei zikken no riron to ouyou [dai 3 han] [Theory and application of model tests], Gihodo Suppan, p.17, 2000. (in Japanese)
2.3.4 Okamura, M., Takemura, J., and Ueno, T.: Kouza-Enshin mokei zikken - Zikken gizyutsu to zitsumu heno tekiyou - 2 Enshin mokei no souzisoku, Zikken gizyutsu - Riten to genkai- [Lecture centrifuge model test: Experimental technics and application of business practice – 2 Scaling of centrifuge model, Experimental technics – advantages and limitations], Soil Mechanics and Foundation Engineering, 52(10), pp.37–44, 2004. (in Japanese)
2.3.5 Kawabata, T., Mohri, Y., and Kondo, T.: Behavior of buried large thin wall flexible pipe-field test and numerical analysis considered with stage of construction of buried flexible pipe, Transactions of the Japanese Society of Irrigation, Drainage and Reclamation Engineering, Issue167, pp.19–27, 1993. (in Japanese)
2.3.6 Loh, C.K., Tan, T.S., and Lee, F.H.: Three-dimensional excavation tests, *Proceedings of the International Conference Centrifuge 98*, pp.649–654, 1998.

2.3.7 Takada, N., Oshima, A., Ikeda, M., and Takeuchi, I.: Centrifuge model test on heavy tamping – ram penetration and ground deformation, Proceedings of JSCE, No.475/III-24, pp.89–97, 1993. (in Japanese)

2.3.8 Zornberg, J.G., Mitchell, J.K., and Sitar, N.: Testing of reinforced slopes in a geotechnical centrifuge, *Geotechnical Testing Journal*, Vol.20, No.4, pp.470–480,1997.

2.3.9 Petrov, R.J., Rowe, R.K., and Quigley, R.M.: Selected factors influencing GCL hydraulic conductivity, *Journal of Geotechnical and Geoenvironmental Engineering*, Vol.123, Issue 8, pp.683–695,1997.

2.3.10 Zornberg, J.G., McCartney, J.S., and Swan, R.H., Jr.: Analysis of a large database of GCL internal shear strength results, *Journal of Geotechnical and Geoenvironmental Engineering*, Vol.131, Issue 3, pp.367–380, 2005.

2.3.11 Vukelić, A., Szavits-Nossan, A., and Kvasnicka, P.: The influence of bentonite extrusion on shear strength of GCL/geomembrane interface, *Geotextiles and Geomembranes*, Vol.26, Issue 1, pp.82–90, 2008.

2.3.12 Viswanadham, B.V.S., Rajesh, S., and Bouazza, A.: Effect of differential settlements on the sealing efficiency of GCLs compared to CCLs: Centrifuge study, *Geotechnical Engineering Journal of the SEAGS & AGSSEA*, Vol.43, No.3, pp.55–61, 2012.

2.3.13 Sawada, Y., Nakazawa, H., Take, W.A., and Kawabat, T.: Effect of installation geometry on dynamic stability of small earth dams retrofitted with a geosynthetic clay liner, *Soils and Foundations*, Vol.59, No.6, pp.1830–1844, 2019.

2.3.14 Phillips, E.: De l'equilibre des solides elastiques semblables, 68, C.R. Acad. Sci., Paris, pp.75–79, 1869.

2.3.15 Pokorovski, G.I. and Fedorov, I.S.: Studies of soil pressures and soil deformations by means of a centrifuge, Proceedings of the 1st International Conference on Soil Mechanics and Foundation Engineering, 1, p.70, 1936.

2.3.16 Schofield, A.N.: Cambridge geotechnical centrifuge operations, Geotechnique, 30(3), pp.227–268, 1980.

2.3.17 Bucky, P.B.: Use of models for the study of mining problems, American Institute of Mining and Metallurgical Engineers, Tech. Pub., 425, pp.3–28, 1931.

2.3.18 Schofield, A.N., Dynamic and earthquake geotechnical centrifuge modelling, International Conferences on Recent Advances in Geotechnical Earthquake Engineering and Soil Dynamics, 2, 1981, https://scholarsmine.mst.edu/icrageesd/01icrageesd/session05/2

2.3.19 Kimura, T. et al.: Kouza-Enshin mokei zikken: 10 Zadankai [Lecture –centrifuge model test: 10 discussion meeting], Soil Mechanics and Foundation Engineering, 36(9), 1988, https://dl.ndl.go.jp/info:ndljp/pid/10434770?tocOpened=1. (in Japanese)

2.3.20 Mikasa, M. and Takada, N.: Enshin-ryoku wo riyou shita tsuchi-kouzoubutsu no mokei zikken [Centrifugal Model Testing of Soil Structures], Soil Mechanics and Foundation Engineering, 28(5), pp.15–23, 1980, https://ci.nii.ac.jp/naid/110003971000 (Viewed 2020.9.1). (in Japanese)

2.3.21 Geotechnical Engineering Laboratory, Osaka City University, http://geo.civil.eng.osaka-cu.ac.jp/~jibanken/ (accessed September 7, 2020)

2.3.22 Cambridge University: Geotechnical and Environmental Research Group HP, https://www-geo.eng.cam.ac.uk/directory/facilities/Physical (accessed September 7, 2020)

2.3.23 TC104: https://www.issmge.org/committees/technical-committees/fundamentals/physical-modelling, 2020.

2.3.24 Sassa, S. and Sekiguchi, H.: Wave-induced liquefaction of beds of sand in a centrifuge, Géotechnique, 49(5), pp.621–638, 1999.

2.3.25 Miyake, M., Sumida, H., Maeda, K., Sakai, H., and Imase, T.: Tsunami no enshinryoku-ba ni okeru zikken shuhou no kaihatsu to keison bouhatei heno tekiyou [Development of test method in centrifugal force field for tsunami and its application to caisson type breakwater], Proceedings of Civil Engineering in the Ocean, Vol.25, pp.87–92, 2009. (in Japanese)

2.3.26 Imase, T., Maeda, K., Miyake, M., Tsurugasaki, K., Sawada, Y., and Sumida, H.: Stability of coastal structure against tsunami wit account for seepage in coastal mound structure and seabed, Journal of Japan Society of Civil Engineers, Ser. A2 (Applied Mechanics (AM)), Vol.67, No.1, pp.133–144, 2011. (in Japanese)
2.3.27 Takahashi, H., Sassa, S., Morikawa, Y., Takano, D., and Maruyama, K.: Stability of caisson-type breakwater foundation under tsunami-induced seepage, Soils and Foundations, 54(4), pp.789–805, 2014.
2.3.28 Garnier, J., Gaudin, C., Springman, S.M., Culligan, P.J., Goodings, D., Konig, D., Kutter, B., Phillips, R., Randolph, M.F., and Thorel, L.: Catalogue of scaling laws and similitude questions in geotechnical centrifuge modelling, International Journal of Physical Modelling in Geotechnics, 7:01–23, 2007.
2.3.29 Adabhushi, G.: Centrifuge Modelling for Civil Engineers, CRC Press, Taylor & Francis Group, pp.292, 2015, ISBN:978-0-415-66824-8.
2.3.30 Tobita, T., Ashino, T., Ren, J., and Iai, S.: Kyoto University LEAP-GWU-2015 tests and the importance of curving the ground surface in centrifuge modelling, Soil Dynamics and Earthquake Engineering, 113, pp.650–662, 2018, http://dx.doi.org/10.1016/j.soildyn.2017.10.012.
2.3.31 Klinkvort, R.T., Black, J., Bayton, S., Haigh, S., Madabhushi, G., Blanc, M., Thorel, L., Zania, V., and Gaudin, C.: A review of modelling effects in centrifuge monopile testing in sand, 2018.
2.3.32 Tan, T.S. and R.F. Scott.: Centrifuge scaling considerations for fluid–particle systems, Geotechnique, 35, pp.461–70, 1985.
2.3.33 White D, Randolph M, and Thompson B.: An image-based deformation measurement system for the geotechnical centrifuge. International Journal of Physical Modelling in Geotechnics, 5(3), pp.1–12, 2005.
2.3.34 Nakamura, K.: Genchi hozyo ni okeru dozyo suibun kansoku ni tsuite omoukoto [Thoughts on Soil Moisture Observation in the Field], Journal of the Japanese Society of Soil Physics, Vol.138, pp.53–54, 2018. (in Japanese)
2.3.35 Nakashima, M., Inoue, M., Sawada, K., and Nicholl, C.: Measurement of soil water content by amplitude domain reflectometry method and its calibration, Journal of Groundwater Hydrology, Vol.40, Issue 4, pp.509–519, 1998. (in Japanese)
2.3.36 Kato, F., Suwa, Y., Hatogai, S., and Fujita, K.: A study on structures of coastal dike that persist disaster mitigation effects against tsunami overflow, Journal of Japan Society of Civil Engineers, Ser. B2 (Coastal Engineering), Vol.70, Issue 1, pp.31–49, 2014. (in Japanese)
2.3.37 Rad, N.S. and Tumay, M.T.: Factors affecting sand specimen preparation by raining, Geotechnical Testing Journal, Vol.10, Issue 1, pp.31–37, 1987.
2.3.38 Takahashi, H., Kitazume, M., Ishibashi, S., and Yamawaki, S.: Evaluating the saturation of model ground by P-wave velocity and modelling of models for a liquefaction study, International Journal of Physical Modelling in Geotechnics, Vol.6, Issue 1, pp.13–25, 2006.
2.3.39 Okamura, M. and Inoue, T.: Preparation of fully saturated models for liquefaction study, International Journal of Physical Modelling in Geotechnics, Vol.12, Issue 1, pp.39–46, 2012.
2.3.40 The Visualization Society of Japan: PIV hand bukku [PIV handbook], Morikita Publishing, 2002. (in Japanese)
2.3.41 De Beer, E.E.: Experimental determination of the shape factors and the bearing capacity factors of sand, Géotechnique, 20(4), pp.387–411,1970.
2.3.42 Okamura, M., Takemura, J., and Kimura, T.: A study on bearing capacities of shallow footings on sand, Doboku gakkai ronbunshu, 1993(463), pp.85–94, 1993. (in Japanese)
2.3.43 Kimura, T., Kusakabe, O., and Saitoh, K.: Geotechnical model tests of bearing capacity problems in a centrifuge, Géotechnique, 35(1), pp.33–45,1985.
2.3.44 Tatsuoka, F., Goto, S., Tanaka, T., Tani, K., and Kimura, Y.: Particle size effects on bearing capacity of footing on granular material, Proceedings of the IS-NAGOYA, pp.133–138, 1997.

2.3.45 Tatsuoka, F., Morimoto, T., Tani, K., Oshima, Y, Okahara, M., Takagi, S., Mori, H., and Tatsuta, M.: Relation among shear strength, experimental data and calculation formula for elucidating mechanism of bearing capacity of sand ground, 34th Doshitsu Symposium happyou kouenshu, pp.17–22, 1989. (in Japanese)

2.3.46 Hayashi, K., Fuji, N., Muramatsu, T., and Houjyou, K.: Direct comparison of gravity model and centrifuge model for the seismic problem, Doboku Gakkai Ronbunshu, No.582, III-41, pp.207–216, 1997. (in Japanese)

2.3.47 Iai, S.: Similitude for shaking table tests on soil–structure–fluid model in 1*g* gravitational field, Report of the Port and Harbour Research Institute, Vol.27, No.3, pp.3–24, 1988.

2.3.48 Kohama, E., Sugano, T., Miyata, M., and Noguchi, T.: Model shake table test on the dynamic behavior of gravity type quay wall during long-period and long-duration earthquake motion, Journal of Japan Association for Earthquake Engineering, Vol.29, pp.396–405, 2007. (in Japanese)

2.4 Review on the handling of geomaterials and organization of issues in Japan (published in Japanese)

2.4.1 Iwagaki, Y. and Noda, H.: Kaiganhenkei no jikkenn ni okeru syukusyakukouka no kennkyu [Scale effects in hydraulic model tests on beach profile changes], Proceedings of Coastal Engineering, JSCE, Vol.8, pp.139–143, 1961. (in Japanese)

2.4.2 Sato, S. and Tanaka, N.: Suiheisyo ni okeru nami ni yoru sunaidou ni tsuite [Sand Movement by Waves on Flat Seabed], Proceedings of Coastal Engineering, JSCE, Vol.9, pp.95–100, 1962. (in Japanese)

2.4.3 Homma, M., Horikawa, K., and Kashima, R.: Nami ni yoru fuyusa ni kansuru kenkyu [Study on Suspended Sediment by Waves] Proceedings of Coastal Engineering, JSCE, Vol.11, pp.159–168, 1964. (in Japanese)

2.4.4 Matsunashi, J. and Omi, K.: Nami ni yoru teishitu no henka ni tsuite [Study on Deformation of Sediment by Waves], Proceedings of Coastal Engineering, JSCE, Vol.11, pp.169–174, 1964. (in Japanese)

2.4.5 Noda, H.: Nami ni yoru teishitu no fuyu [Study on Suspension of Sediment by Waves], Proceedings of Coastal Engineering, JSCE, Vol.14, pp.306–314, 1967. (in Japanese)

2.4.6 Nakamura, M., Shiraishi, H., and Sasaki, Y.: Nami to nagare ni yoru suna no ido ni tsuite [Study on Sand Transportation caused by Waves in Current], Proceedings of Coastal Engineering, JSCE, Vol.14, pp.306–314, 1968. (in Japanese)

2.4.7 Hino, M. and Fukuoka, S.: Nami ni yoru saren no supekutoru ni tsuite no jikken [Experiments on Spectrum of Sand Ripples by Waves], Proceedings of Coastal Engineering, JSCE, Vol.15, pp.121–125, 1968. (in Japanese)

2.4.8 Yamashita, T., Sawamoto, M., Takeda, H., and Yokomori, G.: Idousyo no sindoryu kyokaiso to shi-tofuro-jyou sunaido ni kansuru kenkyu [Study on Oscillatory Boundary Layer and Sediment Transport under Sheet Flow Condition], Proceedings of Coastal Engineering, JSCE, Vol.32, pp.297–301, 1985. (in Japanese)

2.4.9 Suzuki, K., Watanabe, A., Isobe, M., and Dibajnia, M.: Shindoryuka ni okeru kongoryukeiteishitsu no idogensyo ni tsuite [Study on Transport Phenomena of Mixed Granular Bed under Oscillatory Flow Condition], Proceedings of Coastal Engineering, JSCE, Vol.41, pp.356–360, 1994. (in Japanese)

2.4.10 Gotoh, H., Sakai, T., Kashimura, M., Tanaka, H.: Hiatsukaiteijibannai no kangekisuiatsubunpu wo koryo shita teishitsu no idogenkai [Initiation of Sediment Motion Based on Distribution of Pore Water Pressure in Seabed Soil], Proceedings of Coastal Engineering, JSCE, Vol.42, pp.496–500, 1995. (in Japanese)

2.4.11 Dibajnia, M., Takazawa, D., and Watanabe, A.: Kongoryukeisa ni okeru idosoatsu to hyosaryo ni kansuru kenkyu [Study on transport layer and sediment transport rate of mixed granular sand], Proceedings of Coastal Engineering, JSCE, Vol.45, pp.481–485, 1998. (in Japanese)

2.4.12 Sakai, T., Gotoh, H., Oki, K., and Takahashi, T.: Kongoryukei shi-tofuro- hyosa no entyokubunkyukatei no kashika jikken [Vertical Sorting of Graded Sediment Transport Under Oscillatory Flow], Proceedings of the Coastal Engineering, JSCE, Vol.46, pp.516–520, 1999. (in Japanese)
2.4.13 Suzuki, K.: Saren no shintoryu ni yoru syometsu ni tsuite [Disappearing of sand ripples by seepage flow], Proceedings of the Coastal Engineering, JSCE, Vol.46, pp.516–520, 1999. (in Japanese)
2.4.14 Maeno, Y., Matsuoka, Y., Hayashida, H., and Mase, H.: Saren keijyou to kangekiatsu hattatsu tokusei to no kankei [Relation between sand ripple shape and pore pressure development characteristics], Proceedings of the Coastal Engineering, JSCE, Vol.36, pp.789–793, 1989. (in Japanese)
2.4.15 Zen, K., Yamazaki, H.: Harou ni okeru kaiteijiban tyu no shintouryu to ekijyouka genshou [Wave induced seepage flow and liquefaction in seabed], Proceedings of the Coastal Engineering, JSCE, Vol.37, pp.738–742, 1990. (in Japanese)
2.4.16 Izumiya, T., Komata, H., and Abe, K.: Kaitei jiban no atsumitsu keisu oyobi kangekiatu keisu no sokuteihou ni kansuru kenkyu [Study on the measurement method for coefficients of consolidation and pore pressure of seabed], Proceedings of the Coastal Engineering, JSCE, Vol.37, pp.743–747, 1990. (in Japanese)
2.4.17 Maeno, Y. and Tokutomi, K.: Harou ni yori hikiokosareru sasou nai kangekisuiatsu no isou tokusei [Phase characteristic of pore water pressure in the seabed induced wave], Proceedings of the Coastal Engineering, JSCE, Vol.37, pp.749–753, 1990. (in Japanese)
2.4.18 Maeno, Y. and Uchida, K.: Enshinsaikasouchi ni yoru harou wo ukeru kaiteijiban nai ouryokuba no saigen [Reproduction of stress field in seabed due to wave using geotechnical centrifuge], Proceedings of the Coastal Engineering, JSCE, Vol.37, pp.754–758, 1990. (in Japanese)
2.4.19 Sakai, T., Gotoh, H., Kawasaki J., and Takao K.: Shindouryu·suiatsu hendou kyouzon ka deno jiban nai kangeki suiatsu bunpu [Pore water pressure distribution in ground under combination of oscillatory flow and water pressure fluctuation], Proceedings of the Coastal Engineering, JSCE, Vol.43, pp.1006–1010, 1996. (in Japanese)
2.4.20 Nakano, S., Omura, S., Takahashi T., and Mitsui H.: Harou ni yoru teishitsu no ekijyouka ni oyobosu shintouryu no eikyo [Effect of seepage flow on wave induced liquefaction of seabed], Proceedings of the Coastal Engineering, JSCE, Vol.43, pp.536–540, 1996. (in Japanese)
2.4.21 Yamashita, T., Minamimura, T., and Akutsu T.: Harou ni yoru sunajiban no chinka·koudoka to ekijyouka [Settlement and hardening of sandy ground due to waves and liquefaction], Proceedings of the Coastal Engineering, JSCE, Vol.44, pp.911–915, 1997. (in Japanese)
2.4.22 Miyamoto, J., Sassa, S., Sekiguchi, H.: Harou ni yoru sashitsujiban no ekijyouka to ryuudou henkei katei [Flow deformation process of fluidization with liquefaction of seabed due to wave], Proceedings of the Coastal Engineering, JSCE, Vol.47, pp.921–925, 2000. (in Japanese)
2.4.23 Sakai, T., Gotoh, H., Harada E., Hama Y., and Inomoto Y.: Harou ni yoru kaitei jiban no ekijyouka ga hyousa ryou ni oyobosu eikyou [Effect for Rate of Littoral Transport with Liquefaction of Seabed due to Wave], Proceedings of Coastal Engineering, JSCE, Vol.48, pp.981–985, 2001. (in Japanese)
2.4.25 Umezawa, N., Endo, K., Yanase, T., and Ushijima, R.: Wave pressure exerted on sand mound breakwaters with steep slope, Proceedings of the Coastal Engineering, JSCE, Vol.36, pp.584–588, 1989. (in Japanese)
2.4.26 Takahashi, S., Suzuki, K., Tokubuchi, K., Shimosako, K., and Zen, K.: Hydraulic model experiment on the mechanism of settlement failures of caisson-type seawalls, Proceedings of the Coastal Engineering, JSCE, Vol.43, pp.666–670, 1996. (in Japanese)
2.4.27 Kanatani, M., Nishi, K., Sakakiyama, T., Yoshida, Y., and Ogasawara, M.: Stability of the foundation of caisson on permeable seabed under the action of breaking wave force, Proceedings of the Coastal Engineering, JSCE, Vol.43, pp.1046–1050, 1996. (in Japanese)

2.4.28 Sakakiyama, T. and Kashima, R.: Study on scour and wave conditions of wave dissipating works, Proceedings of the Coastal Engineering, JSCE, Vol.45, pp.886–890, 1998. (in Japanese)
2.4.29 Noguchi, K., Tanaka, S., Torii, K., and Sato, S.: Study on failure mechanism of gentle slope-type coastal dike covered with blocks by large-scale model experiment, Proceedings of the Coastal Engineering, JSCE, Vol.47, pp.756–760, 2000. (in Japanese)
2.4.30 Suzuki, K., Takahashi, S., Takano, T., and Shimosako, K.: Settlement failure of wave dissipating blocks in front of caisson type breakwater under the rubble mound: Field investigation and large scale experiment, Report of the Port and Airport Research Institute, Vol.41, No.1, pp.51–89, 2002. (in Japanese)
2.4.31 Yamamoto, Y. and Minami, N.: Experimental investigation on destruction mechanism of coastal dyke due to big waves, Journal of JSCE, Ser. B2 (Coastal Engineering), Vol.65, pp.901–905, 2009. (in Japanese)
2.4.32 Muraoka, H., Nakamura, T., Cho, Y., and Mizutani, N.: Experimental study on erosion of sand seabed under rubble mound of composite breakwater covered with wave dissipating blocks and effects of gabions, Journal of JSCE, Ser. B1 (Hydraulic Engineering), Vol.74, pp.I_595–I_600, 2018. (in Japanese)

3

Concept of the Law of Similarity in Model Tests

Hidenori Takahashi and Masaaki Ikeno

3.1 Role of Law of Similarity

3.1.1 Law of Similarity

Machines such as cars and airplanes and architectural and civil engineering structures such as buildings, bridges, and embankments can deform or be damaged by various forces acting on them. Objects must be designed and manufactured in such a way that harmful deformation and destruction do not occur. One may also want to predict the flow of air or water. In such a case, it is ideal to conduct a model test with the prototype to investigate its behavior. However, conducting model tests using actual structures is quite difficult, especially for large-scale civil engineering and building structures. For example, if you want to find out if a dike will be damaged by an earthquake, it would be almost impossible to build one and apply an earthquake. Thus, what is needed is a model test using a model that will substitute the prototype. If a model test is conducted using a model, the behavior can be observed in detail, and the time and cost will not be feasible. Currently, numerical analysis techniques have advanced together with improvements in the processing power of computers, but the "reality" (actual materials and actual movements that can be observed) offered by model tests is still appealing. This is because, in numerical analysis, a phenomenon that has already been elucidated is formulated to try reproducing its behavior, but there is no guarantee that it can reproduce an unexplained phenomenon. Particularly, when using "soil" whose behavior cannot be explained as a material in a unified manner, predicting the behavior of the structure only through numerical analysis is difficult.

The difficult and interesting point about model tests is that the same behavior cannot be reproduced when the actual structure is scaled down geometrically. The model is just a model and is different from the prototype. Model tests are useless if the "law of similarity," a law that expresses the relationship between phenomena that occur in models and prototypes, is not thoroughly examined. Especially in geotechnical model tests, it is impossible to satisfy the law of similarity for multiple phenomena at the same time in model tests conducted in Earth's gravity field where it is difficult to reproduce the stress of the soil; therefore, the model test must focus on the phenomenon to be reproduced and satisfy the law of similarity of such phenomenon. The so-called relaxation of the law of similarity is necessary. Even still, the observation is often affected by phenomena that is not properly scaled by the law of similarity, so it is essential to understand the

 DOI: 10.1201/9781003424758-3

law of similarity. Additionally, many constraints arise if multiple laws of similarity are not satisfied, so it is typical to expend efforts on maximizing the scale of the model. By contrast, in a centrifuge model test, the stress level of the soil in the prototype can be reproduced by applying a centrifugal force to the model. This test method can satisfy many laws of similarity at the same time even with a small model, which is useful for observing the behavior of the ground.

In the first half of this chapter, we will introduce the law of similarity when using a fluid such as water as a test material and the law of similarity when using soil. Many model tests using these materials have been carried out in the past. Since there is a law of similarity established from the governing equation, the generality and expandability are also high. Besides the derivation of the law of similarity and the similitude ratio in a model test conducted in Earth's gravity field, the law of similarity for a model test in centrifugal gravity field is summarized. The following three methods have been proposed for deriving the law of similarity.

- *Method 1:* All physical quantities related to a phenomenon of interest are enumerated, and dimensionless quantities are determined using Buckingham's π theorem to make the dimensionless quantities of the prototype and model equal.
- *Method 2:* Important forces related to the target phenomenon (inertial force, adhesive force, elastic force, frictional force, etc.) are listed, and dimensionless quantities are determined by their reciprocal ratios to make the dimensionless quantities of the prototype and model equal.
- *Method 3:* Dimensionless quantities are determined on the basis of a set of equations that governs the phenomenon, or the similitude ratio of a physical quantity is directly obtained.

In the law of similarity, the similitude ratio of each physical quantity is not determined directly but is often set via a dimensionless quantity. Also called π numbers, dimensionless quantities are defined as ratios between similar physical quantities that affect the phenomenon. If the dimensionless quantities of the prototype and model are equal, the ratios between the physical quantities are also the same, so the physical law (governing equation) related to the physical quantities of both is the same between the prototype and model, which allows the actual behavior to be understood with a model. For example, when considering the vibration of a cantilever, inertial and elastic forces are the main forces that determine the movement of the beam, and the ratio of these is a dimensionless quantity. If the ratio of the inertial force to the elastic force is equal between the prototype and model, the equation of motion of the prototype and model will be the same, and the movement of the prototype can be reproduced by the model. In this way, the similitude ratio of the physical quantity may be set so that the dimensionless quantity for the phenomenon to be reproduced is equal between the prototype and the model. Additionally, the phenomenon is often organized using dimensionless quantities. This means that if the law of similarity holds, the dimensionless quantities of the prototype and model are equal, so if you plot the relationship between the dimensionless quantities obtained in the model test, they lie on a curve regardless of whether the model test is conducted using the prototype or model. Particularly, many phenomena are organized by dimensionless quantities to understand the behavior of fluids in coastal engineering, etc. It is difficult to make all major dimensionless quantities equal in the prototype and model at the same time, so it is necessary to focus on dimensionless quantities for the physical phenomenon to be reproduced. For example, Froude and Reynolds numbers are famous dimensionless quantities in coastal engineering, but if you want to reproduce the phenomenon of sea waves, you must make the Froude number (ratio of inertial force and gravity) equal, conduct the model test within a scope in which the influence of water viscosity can be ignored, and allow changes in the Reynolds number. A model test under such conditions is called a "model test based on Froude's law of similarity." Under Froude's law of similarity, the similitude ratio of the flow velocity, gravitational acceleration, and length that compose the Froude number should be determined so that the Froude number, which is a dimensionless quantity, does not change. That is, since it is easy to understand the law of similarity in comparison with the phenomenon, Method 3 is adopted in this chapter to explain the respective law of similarity for the fluid and the ground.

In the latter half of this chapter, the law of similarity, which applies to a phenomenon such as drift sand in which a fluid is combined with the ground, is described. For this phenomenon, it is necessary to treat soil as particles and aggregates, making it all the more difficult to understand the phenomenon. Since the phenomenon is complicated and no comprehensive formulation has been established, it is difficult to theoretically derive a law of similarity from the governing equation based on Method 3 above. Thus, the law of similarity proposed so far is either based on Method 2, which focuses on the force acting on the particles, or Method 3, which determines the dimensionless quantity and similitude ratio from equations focusing on a part of the phenomenon. Furthermore, the equation is mainly based on field measurements and model tests and is not necessarily a theoretical equation; thus, the general applicability of the law of similarity is low currently. Research has just begun on the law of similarity in model tests in centrifugal gravity field. Since some of the contents introduced in this chapter have not been verified, it is expected that they will serve as a foundation for further research and studies on governing equations and laws of similarity in the future.

In this chapter, the similitude ratio is shown for various physical phenomena, and Figure 3.1.1 shows the relationship between them. The similitude ratios of the fluid and ground are shown first; then, the similitude ratio of the phenomenon with the fluid and ground combined is shown. Note that phenomena involving both a fluid and the ground are not limited to those in a narrow region near the ground surface where the fluid and the ground come into contact and mix with each other. This is also a problem over a wide area in which the entire ground becomes unstable because of waves and other hydraulic phenomena. It is also possible to apply conventional fluid–ground similitude ratios to this phenomenon. In this chapter, these are explained separately.

3.1.2 History of Law of Similarity

As for the history of model tests,[3.1.1] in the solid mechanics field, Cauchy[3.1.1] conducted tests on beams and plates that vibrate in model tests. In the field of fluid mechanics, Froude[3.1.2] created a ship model for model tests, while Reynolds[3.1.3] conducted a model test on the flow in a pipe. Various model tests have been conducted in Japan as well, and you can get a view of the background of model tests and laws of similarity in Japan

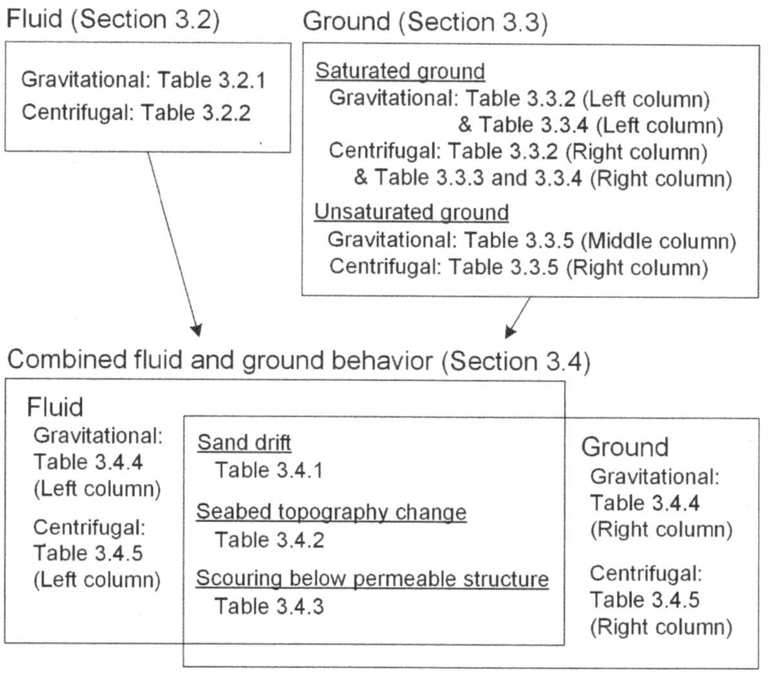

FIGURE 3.1.1
Relationships of similitude ratios for each phenomenon presented in this chapter.

from the model tests conducted on a fighter plane, which is introduced in a publication by Emori et al.,[3.1.1] for example. A feature of this model test is that a qualitative model test in which the shape and weight distribution are similar to the prototype and a quantitative (mechanical) model test in which the stress and the shape and weight are similar to the prototype are carried out. Although the cause of the problem may be investigated with the former model test, the problem could not be solved. Through trial and error, the stress similarity is applied to the problem known as "fracture" in the latter model test. In other words, model test theory has not been widely adopted during this period, and model tests were carried out at one's own discretion. Conversely, a technique to derive governing parameters by performing a dimensional analysis of the relevant quantities governing a certain physical phenomenon was proposed. As a historical backdrop, Rayleigh[3.1.4] came up with the original version of the law of similarity, and Buckingham[3.1.5] later proposed a technique that mathematically generalized Rayleigh's method. The technique is known as Buckingham's π theorem, which serves as the foundation of the mechanical law of similarity in model tests. It still plays an important role in considering the law of similarity for model tests.

For details on the history and development of hydraulic model tests, see Deguchi[3.1.6]. Here, the Chikugo River diversion model test in 1851 was probably the first hydraulic model test conducted in Japan. This is also described in Sato et al.[3.1.7] The first year of the Meiji era is 1868, and the idea that verification by model test is necessary was planted 15 years prior to that. According to Freeman,[3.1.8] it is said that hydraulic model tests conducted in Europe were observed and summarized, indicating that the origin of hydraulic model tests conducted in America are those conducted in Europe.

After World War II, studies on coastal engineering in Japan were stimulated. Deguchi[3.1.6] also carefully describes the concept of the law of similarity and the law of similarity for waves. Precautions and other notes on the surface tension due to restrictions on the size of the model are stated, so it can be seen that the model test was carried out with due attention paid to the law of similarity. Nagai and Tamai[3.1.9] state that Froude's law can be applied to the law of similarity for wave power tests, and they conducted a model test with dimensional ratios of 1/10 and 1/20 for comparison. As described in their paper, laws of similarity such as Froude's law were studied early in the United States, and the study of sand drift based on Froude's law was also more advanced. Institutions such as the Beach Erosion Board and Waterways Experiment Stations appear to have had doubts about the application of Froude's law. However, it is also stated that "the results obtained by model tests with a model scale of 1/20 to 1/24 can be applied directly to the site depending on the similarity rate," so even if there are any doubts, they did not seem to reach a resolution of the problem. Regarding studies on the law of similarity related to sand drift in Japan, Homma et al.[3.1.10] wrote in the conclusion, "We have not yet reached the stage of announcing any conclusions, so for the time being …" Subsequently, Noda and Iho,[3.1.11] Homma et al.,[3.1.12] and Matsunashi and Omi[3.1.13] compared the similarities between the model test and the actual site. It is shown that Ishihara et al.[3.1.14] thought that the dimensional scale ratio or Froude's law could be applied to the similarity regarding the transport mechanism of sand drift. There is also a study on tidal currents, and Hayami et al.[3.1.15] examined the similarity from the basic equations and comparison of the model test and local site.

In the field of geotechnical engineering,[3.1.16, 3.1.17] model tests have been conducted to clarify the mechanical properties of the soils and to elucidate the failure behavior of the soils and its interaction with structures. According to Mikasa,[3.1.16] to clarify the mechanical properties of the geo-material, model tests on earth pressure theory, bearing capacity theory, etc. have been conducted in the past by treating the geo-material as an elastic body. Since then, mechanical tests and in situ tests have been mainly conducted by Terzaghi[3.1.18] to demonstrate the importance of the physical and mechanical properties of soils. Nevertheless, as the behavior of the geo-material is complicated, the slip surface and crack generation mechanism, progressive failure,[3.1.16] behavior as a granular body, and so on must be elucidated, and model tests continue to be conducted even today as part of the technique used. In recent years, deriving optimal solutions by combining model tests, numerical model tests, and rigorous theories is considered important, so the role played by model tests as benchmark data for clarifying phenomena and studying the validity of numerical analysis is believed to be extremely important. In model tests conducted in the field of geotechnical engineering, the component factors governing the mechanical properties of soil are considered to derive the law of similarity. Particularly, besides $1g$ model tests, model tests conducted in centrifugal gravity field are unique in the geotechnical engineering field. Typical research examples are shown below.

Rocha[3.1.19] published a study on the development of the law of similarity for soil. This study examines the stress–strain similarity in static problems related to the bearing capacity of footing foundations in $1g$ model tests. Studies by Kagawa,[3.1.20] Kokusho and Iwatate,[3.1.21] and Iai[3.1.22] can be cited as studies on the laws of similarity for dynamic problems in a $1g$ model test. Kagawa proposed a method that utilizes the force interaction ratio to obtain an independent dimensionless quantity and then to equalize it between the prototype and the model, whereas Kokusho and Iwatate applied the Buckingham's

π theorem to the mechanical law of similarity. Additionally, Iai introduced a law of similarity applied to water–soil–structure coupling problems from a governing equation focusing on the equations of mutual balance and in/out balance. A law of similarity for centrifuge model tests was proposed by a document of Philips.[3.1.23, 3.1.24] The use of such model tests started in the Soviet Union and the United States around the 1930s (for example, Pokrobsky and Fedorov[3.1.25]), and, subsequently, full-scale centrifuge model tests (Mikasa et al.[3.1.26]; Avgherinos and Schofield[3.1.27]) were conducted in the 1960s.[3.1.28] Model tests on the stability problem of clay slopes as well as self-weight consolidation problem (Mikasa and Takada[3.1.29]) and bearing capacity problem (Ovesen[3.1.30]) of soft clay were cited as model tests that verified the law of similarity.[3.1.31]

3.2 Law of Similarity for Fluids

3.2.1 Derivation of Law of Similarity

When considering the behavior of a fluid, a phenomenon is often sorted out by a dimensionless quantity composed of physical quantities (inertial force, gravity, viscous force, etc.) that govern the phenomenon. Regarding the law of similarity of model tests, instead of setting the similitude ratio for each physical quantity, the physical phenomenon to be reproduced is focused on and similitude ratios of the physical quantities that make up the dimensionless quantity are specified in a way that will give a single value of the dimensionless quantity for that phenomenon in both the prototype and the model. In other words, the similitude ratio of the physical quantity is set so that the dimensionless quantity does not change. Thus, after deriving the dimensionless quantity, the similitude ratio of the specified physical quantity is described so that the dimensionless quantity does not change.

The most important dimensionless quantities with respect to fluid behavior in the field of coastal engineering are the Reynolds number R_e (ratio of inertial force and viscous force) and Froude number F_r (ratio of inertial force and gravity), both of which can be derived from the Navier–Stokes equation, which is the governing equation of fluids, as shown below.[3.2.1] Assuming that the fluid is uncompressed, the Navier–Stokes equation is expressed as follows:

$$\frac{\partial \boldsymbol{v}}{\partial t}+(\boldsymbol{v}\cdot\nabla)\boldsymbol{v}=-\frac{1}{\rho}\nabla p+\nu\nabla^2\boldsymbol{v}+g\boldsymbol{e}_g \tag{3.2.1}$$

Here, $\boldsymbol{v}$ is the flow velocity, t is the time, ρ is the fluid density, p is the pressure, g is the gravitational acceleration, ν is the kinematic viscosity coefficient, ∇ is the vector differential operator (spatial differentiation), and $\boldsymbol{e}_g$ is the unit vector facing the direction of gravity (vertically downward). Quantities in bold are assumed to be vectors, whereas those in the fine text are assumed to be scalars. The first term on the left-hand side is the time, the second term on the left-hand side is the advection (convection), the first term on the right-hand side is the pressure, the second term on the right-hand side is the viscosity (diffusion), and the third term on the right-hand side is the gravity.

Each physical quantity in the Navier–Stokes equation above shall be made dimensionless. The quantity can be made dimensionless by dividing the physical quantity by a representative value. For example, in the case of flow velocity, assuming v' is a dimensionless physical quantity and v is a representative value, $v' = v/\mathrm{v}$. From this equation, the flow velocity can be expressed as $v = \mathrm{v}v'$. Similarly, for each physical quantity, the product of the representative value and the dimensionless physical quantity is substituted for the physical quantity in Equation (3.2.1). If the representative value of each term is brought out as a coefficient, it will be as follows:

$$\frac{\mathrm{v}}{\mathrm{t}}\frac{\partial v'}{\partial t'} + \frac{\mathrm{v}^2}{L}(v' \cdot \nabla')v' = -\frac{\mathrm{p}}{\rho L}\nabla' p' + \frac{\nu \mathrm{v}}{L^2}\nabla'^2 v' + g e_g \tag{3.2.2}$$

Here, text appended with prime (′) refers to the dimensionless quantity of the physical quantity. v, t, L, and p are representative values. The equation becomes as shown below when L/v is substituted for t:

$$\frac{\mathrm{v}^2}{L}\left\{\frac{\partial v'}{\partial t'} + (v' \cdot \nabla')v'\right\} = -\frac{\mathrm{p}}{\rho L}\nabla' p' + \frac{\nu \mathrm{v}}{L^2}\nabla'^2 v' + g e_g \tag{3.2.3}$$

Dividing both sides by v^2/L results in the following equation:

$$\frac{\partial v'}{\partial t'} + (v' \cdot \nabla')v' = -\frac{\mathrm{p}}{\rho \mathrm{v}^2}\nabla' p' + \frac{\nu}{\mathrm{v}L}\nabla'^2 v' + \frac{gL}{\mathrm{v}^2} e_g \tag{3.2.4}$$

The equation becomes as shown below when $\rho\mathrm{v}^2$ is substituted for p.

$$\frac{\partial v'}{\partial t'} + (v' \cdot \nabla')v' = -\nabla' p' + \frac{\nu}{\mathrm{v}L}\nabla'^2 v' + \frac{gL}{\mathrm{v}^2} e_g \tag{3.2.5}$$

The reciprocal of the coefficients of the second and third terms on the right-hand side is represented by the following characters:

- Reynolds number (ratio of inertial force to viscous force): $R_e = \dfrac{\mathrm{v}L}{\nu}$
- Froude number (ratio of inertial force to gravity): $F_r = \dfrac{\mathrm{v}}{\sqrt{gL}}$

Using this, the equation becomes as shown in Equation (3.2.6):

$$\frac{\partial v'}{\partial t'} + (v' \cdot \nabla')v' = -\nabla' p' + \frac{1}{R_e}\nabla'^2 v' + \frac{1}{F_r^2} e_g \tag{3.2.6}$$

This is the dimensionless Navier–Stokes equation. For Equation (3.2.6) to be valid for both the model and the prototype, each term must be equal in the prototype and model, and R_e and F_r must also be equal in the model and the prototype.

According to Methods 1 and 2 explained in Section 3.1.1, other dimensionless quantities such as the Weber number W_e (surface tension) and Mach number M_a (elastic force) are also obtained, which are important dimensionless quantities related to the behavior of the fluid. Ideally, all dimensionless quantities such as R_e and F_r should be modeled so that they are equal between the prototype and the model. Nevertheless, it is difficult to ensure that these dimensionless quantities do not vary at the same time, so the model test is performed by limiting the dimensionless quantities only to those of the behavior of interest and setting the similitude ratio of each physical quantity to ensure that it does not vary.

3.2.2 Model Tests Involving Breaking Waves

In shallow waters where phenomena combining fluid and ground occur, breaking waves is an important physical phenomenon of fluids but may be complex and difficult to model. When the wave progresses to a shallow region, in association with shallow water deformation, the wave peak loses symmetry in the front and back direction and tilts forward, making it impossible for the wave to maintain its motion as a wave, thereby turning it into a breaking wave and shifts its form from a wave to a flow. When breaking waves occur, the waves form vortices and splashes, and air is also taken into the fluid as air bubbles, causing severe turbulence. Most of the problems of the combined phenomenon of fluid and ground are in shallow waters, so it is important to understand the characteristics and law of similarity of breaking waves. It is said that breaking waves form when the water particle velocity at the apex of the wave peak exceeds the wave velocity and the particles fly forward or when the asymmetry of the waveform becomes significant, and the front of the wave peak rises vertically. Hence, the water particle velocity, wave velocity, and waveform before breaking waves are generated following Froude's law of similarity and can be reproduced in a model test. The effects of R_e and W_e on the turbulence of the fluid in the surf zone should be considered, so it may be difficult to reproduce all of the wave breaking phenomena in a model test that adheres to Froude's law of similarity. For example, since the R_e of a model with a small dimension L and velocity v is smaller than the actual R_e, it is expected that the viscous force is relatively large and the degree of turbulence in the surf zone is smaller than the prototype.

Here, to study the law of similarity for breaking waves, we will take up the turbulence model proposed by researchers in the past. This is a method of time averaging and is widely used for turbulent flows. The Navier–Stokes equation expressed by Equation (3.2.1) is the governing equation of the fluid. To numerically model a breaking wave using this equation, it is necessary to make the time step shorter than the lifespan of the vortex generated by the breaking wave and to discretize it spatially smaller than the vortex. Since this is not realistic, it is thus common to rewrite it in the form of a Reynolds equation to solve it. The flow velocity and pressure at each moment are separated into time average and deviation values as shown in Equation (3.2.7):

$$v=\overline{v}+v',\quad p=\overline{p}+p' \tag{3.2.7}$$

By substituting this into the Navier–Stokes equation and using the continuity equation, the equation can be arranged into the form of the Reynolds equation as shown in Equation (3.2.8). The symbol –, which means the average time required for v and p, is omitted here:

$$\frac{\partial v}{\partial t}+(v\cdot\nabla)v=-\frac{1}{\rho}\nabla p+(\nu+\nu_t)\nabla^2 v+ge_g \tag{3.2.8}$$

Here, ν_t is the coefficient of kinematic eddy viscosity (turbulent diffusion coefficient),[3.2.2] which is a coefficient used to approximate the time average of the product of time-varying components as shown below (only the x direction is shown as an example). The time average of the product of time-varying components multiplied by the density is called the Reynolds stress, which has been virtually evaluated as a viscous force.

$$-\overline{u'u'} = \nu_t\left(\frac{\partial \overline{u}}{\partial x}+\frac{\partial \overline{u}}{\partial x}\right), -\overline{u'v'} = \nu_t\left(\frac{\partial \overline{u}}{\partial y}+\frac{\partial \overline{v}}{\partial x}\right), -\overline{u'w'} = \nu_t\left(\frac{\partial \overline{u}}{\partial z}+\frac{\partial \overline{w}}{\partial x}\right) \tag{3.2.9}$$

Here, u', v', and w' are deviations from the time-averaged flow velocity in the x, y, and z directions. By separating v and p into time average and deviation to study the time-averaged flow velocity and pressure, the effect of turbulence can be regarded as a fluid having a kinematic viscosity coefficient of ν_t as shown in Equation (3.2.9). Moreover, since ν_t is sufficiently larger than ν at the point breaking waves are generated, if the viscosity of water is ignored, then the Reynolds number due to the coefficient of kinematic eddy viscosity can be considered as a dimensionless quantity. Since ν_t has the dimension of velocity × length, it matches the dimension of the numerator of the Reynolds number and the Reynolds number is the same in the prototype and model. Based on this, there is a possibility that a phenomenon captured in a time step longer than the lifetime of the vortex can be reproduced by a model test based on Froude's law of similarity. The formula to estimate the wave height when breaking waves are generated has been confirmed to be consistent with the results of model tests based on Froude's law of similarity.[3.2.3]

3.2.3 Model Tests in Earth's Gravity Field

In wave and flow model tests, it is common to set the similitude ratio of physical quantities so as not to change the Froude number that affects many of its behaviors. Model tests based on the so-called Froude's law of similarity are often conducted. Hydraulic model tests of waves and water flows in Earth's gravity field are relatively large in scale, which can also be conducted within a range that is not affected by R_e, W_e, and M_a and can be conducted without causing F_r to change. Nevertheless, Froude's law of similarity is not universal, and the phenomenon of turbulence after breaking waves described in the previous section follows a different law of similarity. Thus, care must be taken when interpreting the phenomenon in the model test. Froude's law of similarity is expressed by Equation (3.2.10).

$$\frac{v_m}{\sqrt{g_m L_m}} = \frac{v_p}{\sqrt{g_p L_p}} \tag{3.2.10}$$

Here, the subscript p means the original, while the subscript m means the model. In 1g model tests, the gravitational force acting on both the model and prototype is the same. Thus, if the dimensional scale of the model is defined as $L_m/L_p = 1/N$, the velocity scale will be $v_m/v_p = \sqrt{L_m/L_p}$, which is proportional to the square root of $1/N$. From $t = L/v$, the timescale is proportional to the square root of $1/N$. If the mass is $M = \rho L^3$ and the density of water is the same in the model and prototype, the scale of the mass will be proportional to the cube of $1/N$. The scale of the force will be proportional to the cube of $1/N$. From this, the scale of the pressure will be proportional to $1/N$. These are summarized in Table 3.2.1 and allow the similitude ratio in a 1g model test to be obtained.

TABLE 3.2.1

Similitude Ratio between Prototype and Model by Froude's Similarity Law (Fluid, Earth's Gravity Field)

	Length	Time	Velocity	Acceleration	Mass	Pressure	Force
Prototype	1	1	1	1	1	1	1
Model	$1/N$	$1/\sqrt{N}$	$1/\sqrt{N}$	1	$(1/N)^3$	$1/N$	$(1/N)^3$

3.2.4 Model Tests in Centrifugal Gravity Field

Even in a centrifuge model test in which a wave or flow model test is conducted with a centrifugal force applied to the model, the model test is performed so that F_r does not change. Table 3.2.2 shows the similitude ratios that follow Froude's law of similarity. In centrifuge model tests, the model is often small (approximately 1/10–1/100) because of the restrictions of the dimensions of the equipment. In this case, since the wavelength becomes shorter, there is a concern that the influence of R_e (viscous force) and W_e (surface tension) cannot be ignored. However, as shown in Froude's law of similarity, since the flow velocity v does not decrease even in the model, R_e (viscous force) and W_e (surface tension) do not decrease as much as in Earth's gravity field. Also, M_a does not change. Thus, even a small model can be tested according to Froude's law.

3.2.5 Model Tests on Fluid–Structure Interaction

Although it may be a hydraulic model test, it is rare to conduct a model test using only fluid. In many cases, a model of structures such as a breakwater or a seawall is installed to investigate the behavior of the structure in a hydraulic environment. Regarding the concrete materials included in such structures and the wave-dissipating and covering blocks, Froude's law of similarity can often be satisfied by using a similarity model as well. Nevertheless, when reproducing the flow between crushed stones and blocks, it is difficult to reduce the viscosity of water. Thus, if the dimensions of the crushed stones are simply reduced according to the model dimensions, the Reynolds number R_e will become smaller, and the flow between the stones and blocks will approach a laminar flow. It has been pointed out that this causes differences in the following phenomena between the prototype and model.[3.2.4, 3.2.5]

- **Model scale effect of deformation amount of wave-dissipating blocks.**[3.2.4] In a 1*g* model test, the similitude ratio of the inertial resistance force and viscous resistance force of a fluid acting on a body such as a wave-dissipating block is different from the similitude ratio of the force according to Froude's law of

TABLE 3.2.2

Similitude Ratio between Prototype and Model by Froude's Similarity Law (Fluid, Centrifugal Gravity Field)

	Length	Time	Velocity	Acceleration	Mass	Pressure	Force
Prototype	1	1	1	1	1	1	1
Model	$1/N$	$1/N$	1	N	$(1/N)^3$	1	$(1/N)^2$

similarity, and the wave energy dissipation effect due to these forces will not follow Froude's law of similarity. Since the Reynolds number R_e is smaller and the drag is relatively larger in a small model, the degree of deformation of wave-dissipating blocks is larger in a small model than in a large model.

- **Model scale effect of wave overtopping rate.**[3.2.4] As described above, since the small model has a larger wave energy dissipation effect, the wave-dissipating effect is also larger. When reproducing the wave overtopping rate of a breakwater or revetment having wave-dissipating blocks in a model, the wave overtopping rate in a small model is smaller than that in a large model.
- **Model scale effect of reflectivity.**[3.2.4] As mentioned above, the wave-dissipating effect in the small model is relatively large. Thus, the reflectivity from the breakwater or revetment in the small model is smaller than that in the large model.
- **Model scale effect of transmitted wave height.**[3.2.5] When a nonbreaking wave passes through a permeable breakwater, the transmitted wave height is smaller in the small model than in the large model because the wave energy dissipation effect is larger in the small model. This tendency becomes more significant as the cycle becomes longer.
- **Model scale effect of seepage flow in crushed stone ground.** Since the Reynolds number R_e differs between the prototype and the model, it is considered difficult to reproduce the detailed flow between crushed stones and blocks with the model. Nevertheless, if the flow between crushed stones in the centrifugal force field is considered as turbulent flow, the pore pressure distribution is known to be reproduced macroscopically since the relationship between the hydraulic gradient and the average flow velocity in the interstices is maintained.[3.2.6] Furthermore, even in model tests conducted in Earth's gravity field, maintaining the relationship between the hydraulic gradient and the average flow rate in the interstices by changing the porosity is not impossible. Since the seepage force is generated by the difference in the pore pressure, the fact that the pore pressure distribution can be reproduced means that the similitude ratio of the seepage force matches the similitude ratio of the force according to Froude's law.

The model is just a model, so satisfying all laws of similarity is difficult. It is important to predict the phenomenon with the prototype while taking note of the differences above. Additionally, the crushed stone condition, method of producing the ground, fluid viscosity, gravitational acceleration (load of centrifugal force), etc. should be changed so that the degree of similarity between the model and prototype can be improved in terms of the phenomenon of interest.

3.3 Law of Similarity for Ground

3.3.1 Derivation of the Law of Similarity

This section considers the law of similarity for the ground. For the ground, instead of considering dimensionless quantities like fluids, the similitude ratio of each physical quantity has been directly obtained from the governing equations for use. This is possibly caused by the clarification of the ground behavior not being advanced enough to organize the phenomena with dimensionless quantities. First, for the sake of simplicity, consider a saturated state in which interstices of porous soil formed by soil particles are filled with water. In this case, the governing equation of the ground is as follows (Zienkiewicz et al.[3.3.1]). For stress and water pressure, compression is defined as being positive. Quantities in bold are tensor quantities and those in the fine text are scalar quantities. Also, $\boldsymbol{m}^T = (1,1,1,0,0,0)$, $\boldsymbol{L}$ is the strain displacement matrix (also called $\boldsymbol{B}$ matrix), ∇ is the vector differential operator (spatial differentiation), and Table 3.3.1 describes the meanings of the other characters.

- Definition of effective stress

$$\boldsymbol{\sigma} = \boldsymbol{\sigma}' + \boldsymbol{m}p \tag{3.3.1}$$

- Definition of strain

$$d\boldsymbol{\varepsilon} = -\boldsymbol{L}d\boldsymbol{u} \tag{3.3.2}$$

- Constitutive law of soil (ignoring strain accompanying creep deformation, temperature changes, etc.)

$$d\boldsymbol{\sigma}' = \boldsymbol{D}d\boldsymbol{\varepsilon} \tag{3.3.3}$$

- Equation of motion of soil (ignoring the relative acceleration of pore water with respect to the soil skeleton [u–p formulation])

$$-\boldsymbol{L}^T\boldsymbol{\sigma} + \rho\boldsymbol{g} = \rho\ddot{\boldsymbol{u}} \tag{3.3.4}$$

TABLE 3.3.1

Similitude Ratio of Geotechnical Model Test (Prototype/Model)

Physical Quantity	Description	Similitude Ratio
x	Length (dimension)	λ
ρ	Density of saturated soil	λ_ρ
ε	Strain of soil	λ_ε
g	Gravitational acceleration	λ_g
t	Time	$(\lambda\lambda_\varepsilon/\lambda_g)^{0.5}$
σ	Total stress of soil	$\lambda\lambda_\rho\lambda_g$
σ'	Effective stress of soil	$\lambda\lambda_\rho\lambda_g$
D	Tangent modulus of soil	$\lambda\lambda_\rho\lambda_g/\lambda_\varepsilon$
p	Pore water pressure	$\lambda\lambda_\rho\lambda_g$
k	Permeability coefficient	$(\lambda\lambda_\varepsilon\lambda_g)^{0.5}$
u	Displacement of soil	$\lambda\lambda_\varepsilon$
$\dot{u}$	Velocity of soil	$(\lambda\lambda_\varepsilon\lambda_g)^{0.5}$
$\ddot{u}$	Acceleration of soil	λ_g
w	Displacement of pore water relative to soil skeleton	$\lambda\lambda_\varepsilon$
$\dot{w}$	Rate of pore water flow	$(\lambda\lambda_\varepsilon\lambda_g)^{0.5}$
ρ_f	Density of pore water	λ_ρ

By substituting Equations (3.3.1)–(3.3.3) into Equation (3.3.4), the equation becomes a soil equation in which the displacement $\boldsymbol{u}$ and pore pressure p are unknown.

- Equation of motion of pore water

$$-\nabla p + \rho_f g = \frac{\rho_f g}{k} \dot{w} + \rho_f \ddot{u} \tag{3.3.5}$$

- Continuity equation of pore water (ignoring changes in the volume of water and soil particles themselves)

$$\nabla^T \dot{w} - \boldsymbol{m}^T \dot{\varepsilon} = 0 \tag{3.3.6}$$

By substituting Equations (3.3.2)–(3.3.5) into Equation (3.3.6), the equation becomes a pore water equation in which the displacement $\boldsymbol{u}$ and pore pressure p are unknown. The displacement $\boldsymbol{u}$ and pore pressure p can be obtained by solving the soil and pore water equations simultaneously.

In this way, Equations (3.3.1)–(3.3.6) serve as the governing equations for the ground. These equations are valid for both the prototype and model. For example, when considering Equation (3.3.1), Equations (3.3.7) and (3.3.8) both are valid. The subscripts p and m indicate the actual quantity and model quantity, respectively.

$$[\sigma]_p = [\sigma']_p + \boldsymbol{m}[p]_p \tag{3.3.7}$$

$$[\sigma]_m = [\sigma']_m + \boldsymbol{m}[p]_m \tag{3.3.8}$$

Assuming the ratio of the physical quantity in the prototype and model is $[\boldsymbol{\sigma}]_p/[\boldsymbol{\sigma}]_m = \lambda_\sigma$, $[\boldsymbol{\sigma}']_p/[\boldsymbol{\sigma}']_m = \lambda_{\sigma'}$, $[p]_p/[p]_m = \lambda_p$ and substituting it into Equation (3.3.7) gives Equation (3.3.7′).

$$\lambda_\sigma [\sigma]_m = \lambda_{\sigma'} [\sigma']_m + \boldsymbol{m}\lambda_p [p]_m \tag{3.3.7′}$$

For both Equation (3.3.8) and Equation (3.3.7′) to be valid, the coefficient of each term must be equal and the following relationship is obtained:

$$\lambda_\sigma = \lambda_{\sigma'} = \lambda_p \tag{3.3.1′}$$

Similar to Equation (3.3.1), the following relationship can be obtained by performing the same process for Equations (3.3.2)–(3.3.6). Here, λ is assumed as the dimensional ratio. From Equation (3.3.2),

$$\lambda_\varepsilon = \frac{\lambda_u}{\lambda} \tag{3.3.2′}$$

From Equation (3.3.3),

$$\lambda_{\sigma'} = \lambda_D \lambda_\varepsilon \tag{3.3.3′}$$

From Equation (3.3.4),

$$\frac{\lambda_\sigma}{\lambda} = \lambda_\rho \lambda_g = \lambda_\rho \frac{\lambda_u}{\lambda_t^2} \tag{3.3.4'}$$

From Equation (3.3.5),

$$\frac{\lambda_p}{\lambda} = \lambda_{\rho_f} \lambda_g = \lambda_{\rho_f} \lambda_g \frac{1}{\lambda_k} \frac{\lambda_w}{\lambda_t} = \lambda_{\rho_f} \frac{\lambda_u}{\lambda_t^2} \tag{3.3.5'}$$

From Equation (3.3.6),

$$\frac{\lambda_w}{\lambda \lambda_t} = \frac{\lambda_\varepsilon}{\lambda_t} \tag{3.3.6'}$$

There are 13 unknowns and 10 equations in Equation (3.3.1′) to Equation (3.3.6′). Nevertheless, there are only nine independent equations because Equations (3.3.4′) and (3.3.5′) contain one dependent equation. For this reason, assumptions about the four similitude ratios should be made since the number of equations is short by four. These four quantities are generally λ, λ_ρ, λ_ε, and λ_g (ratios of the dimensions, soil density, strain, and gravitational acceleration) in the law of similarity for geotechnical model tests. Given these similitude ratios, the similitude ratios of other physical quantities can be obtained from Equations (3.3.1′)–(3.3.6′) as shown in Table 3.3.1. These are the similitude ratios of the ground.

3.3.2 Model Tests in Earth's Gravitational Field

Kagawa[3.3.2] and Iai[3.3.3] provided details about the law of similarity of geotechnical model tests in Earth's gravity field. Here, $\lambda_\rho = 1$, $\lambda_\varepsilon = \lambda^{0.5}$, $\lambda_g = 1$, assuming the dimensional ratio is λ. Greatly changing the density of soil is not easy, so the similitude ratio of the density of saturated soil is set to 1. Moreover, since it is a model test in Earth's gravity field, the similitude ratio of gravitational acceleration is 1. The strain similitude ratio is the square root of the dimensional ratio. Assuming the similitude ratio of strain in this way, the similitude ratio of tangential stiffness is $\lambda_D = \lambda^{0.5}$, and the similitude ratio of effective stress is $\lambda_\sigma = \lambda$ to the power of 0.5. Since the effective confining stress σ'_m of the sandy ground and the deformation coefficient G have a relationship of approximately G–$\sigma'^{0.5}_m$, it is convenient because the similitude ratio can be automatically satisfied even if the same sand as the actual one is used in the model. Using these similitude ratios, the similitude ratios in a 1g model test are as shown in Table 3.3.2.

The relationship between stress and strain shown in the constitutive law of soil in Equation (3.3.3) is complicated, and various problems occur in 1g model tests where the actual stress cannot be reproduced. The similitude ratio of the stress, tangential stiffness, and strain is expressed as a constant such as $[\boldsymbol{\sigma}]_p/[\boldsymbol{\sigma}]_m = \lambda_\sigma$, $[\boldsymbol{D}]_p/[\boldsymbol{D}]_m = \lambda_D$, $[\boldsymbol{\varepsilon}]_p/[\boldsymbol{\varepsilon}]_m = \lambda_\varepsilon$. This means that the shapes of the stress–strain curve of the prototype and the model match, as shown in Figure 3.3.1 (Rocha's assumption[3.3.4]). In reality, it is also known that the tangential stiffness varies depending on the historical stress and strain, and the shape of the stress–strain curve does not match in the extreme state when the

TABLE 3.3.2

Similitude Ratio of Geotechnical Model Test (Prototype/Model) (Gravitational and Centrifugal Force Fields)

Physical Quantity	Description	Similitude Ratio	
		Gravitational	Centrifugal
x	Length (dimension)	λ	λ
ρ	Density of saturated soil	1	1
ε	Strain of soil	$\lambda^{0.5}$	1
g	Gravitational acceleration	1	$1/\lambda$
t	Time	$\lambda^{0.75}$	λ
σ	Total stress of soil	λ	1
σ'	Effective stress of soil	λ	1
D	Tangent modulus of soil	$\lambda^{0.5}$	1
p	Pore water pressure	λ	1
k	Permeability coefficient	$\lambda^{0.75}$	1
u	Displacement of soil	$\lambda^{1.5}$	λ
$\dot{u}$	Velocity of soil	$\lambda^{0.75}$	1
$\ddot{u}$	Acceleration of soil	1	$1/\lambda$
w	Displacement of pore water relative to soil skeleton	$\lambda^{1.5}$	λ
$\dot{w}$	Rate of pore water flow	$\lambda^{0.75}$	1
ρ_f	Density of pore water	1	1

failure is imminent. Thus, note that the similitude ratio can be used when considering the response during an earthquake to the extent in which it will not lead to failure in the ground.

Moreover, when the similitude ratio of the gravitational field shown in Table 3.3.2 is used, the similitude ratio of the displacement to the similitude ratio λ of the length (dimension) is $\lambda_\varepsilon = \lambda^{1.5}$; thus, there is also the issue of the displacement getting smaller relative to the dimensions in the model. For example, in a 1/10 scale model test, the displacement of the ground is not 1/10 but approximately $1/10^{1.5} = 1/32$. For this reason, when using a model to reproduce a phenomenon in which large deformation occurs,

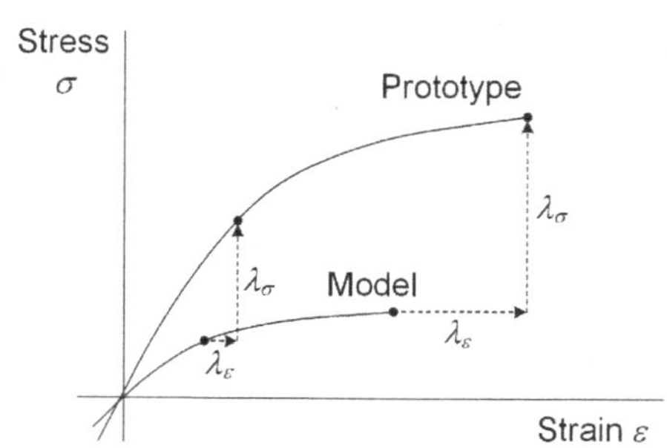

FIGURE 3.3.1
Stress–strain curve between prototype and model.

the results of the prototype and the model will be different. Furthermore, although the similitude ratio of tangential stiffness is $\lambda_D = \lambda^{0.5}$, the value is not necessarily valid even for sand, and in the case of cohesive soil, the value is not valid at all. In the model, it is also necessary to use a soil material that matches the similitude ratio of tangential stiffness that is different from the prototype.

3.3.3 Model Tests in Centrifugal Gravity Field

A major constraint of the law of similarity in 1g model tests is the similitude ratio of the gravitational acceleration: $\lambda_g = 1$. Since the model test is conducted in the same gravitational field for both the prototype and the model, λ_g is 1, and the stress of the prototype and the model could not be equalized unless a material with a high density was used. Since the similitude ratio of stress does not become 1, various assumptions were required in the constitutive laws of soil. A test method that can solve this problem is the centrifuge model test (see Section 2.3). Since the gravitational acceleration can be virtually increased by rotating the model to apply a centrifugal force, the similitude ratio of stress can be set to 1 by setting the gravitational acceleration to match the dimensional similitude ratio. Even if the same soil is used in the prototype and the model, the similitude ratio of the tangential stiffness and strain of the soil is 1, and the constitutive law of soil of the prototype directly matches that of the model (Equation (3.3.3)). Consequently, various assumptions are not necessary for the constitutive law, removing constraints posed by a 1g model test. Assuming the dimensional ratio is λ and the other similitude ratios are $\lambda_\rho = 1$, $\lambda_\varepsilon = 1$, and $\lambda_g = 1/\lambda$, the similitude ratios in the centrifuge model test are as shown in Table 3.3.2. As shown in the table, the similitude ratio of the centrifugal force field is simpler than that of Earth's gravity field. Note that centrifuge model tests are not universal. Centrifugal force is applied by rotating the model, but as the centrifugal force increases in proportion to the radius of rotation, the centrifugal force in the model ground is not uniform. To reduce this effect, a large centrifuge machine has been devised to increase the turning radius, but it cannot be completely eliminated. Additionally, if there is any movement in the model, Coriolis force may be generated, and its influence cannot be ignored in some cases. Nevertheless, given these shortcomings, no one can deny the superiority of centrifuge model tests as they can solve many of the problems of the law of similarity in 1g model tests.

When conducting a model test of a ground that liquefies in a centrifuge model test, a viscous fluid is often used instead of water. The reason for this will be explained from the law of similarity. As shown in Table 3.3.2, the similitude ratio of the permeability coefficient is 1 (it must be assumed as 1). The permeability coefficient k is the ratio of the average flow velocity $\bar{v}$ of water in the pores to the hydraulic gradient i and is a value indicating how much flow velocity is generated when a certain head gradient is imparted to the soil. For example, if the particle size of soil is large, the permeability coefficient is also large, and if the viscosity of water is large, the permeability coefficient is small. The following Dupuit–Forchheimer equation is used to find the factors that determine the permeability coefficient:

$$i = \alpha_0 \frac{\nu}{g} \frac{(1-n)^3}{n^2 d_{15}^{\,2}} \bar{v} + \beta_0 \frac{1}{g} \frac{1-n}{n^3 d_{15}} \bar{v}^2 \tag{3.3.9}$$

The first term on the right-hand side indicates the viscous resistance and the second term indicates the inertial resistance. As the flow of water in fine pores like soil is laminar, the first term is thus predominant. Thus, the reciprocal of the first term on the right-hand side of Equation (3.3.9) becomes permeability coefficient k as shown in:

$$i = \alpha_0 \frac{\nu}{g} \frac{(1-n)^3}{n^2 d_{15}^{\ 2}} \overline{v} = \frac{1}{k} \overline{v} \tag{3.3.10}$$

According to this equation, the permeability coefficient depends on the viscosity coefficient, gravitational acceleration, porosity, and 15% particle size. When the same soil is used for the prototype and model, the porosity and the 15% particle size do not vary, so as the gravitational acceleration increases, the permeability coefficient increases proportionally. Hence, the law of similarity can be satisfied by increasing the viscosity coefficient as the gravitational acceleration increases and not changing the permeability coefficient. Takahashi et al.[3.3.5] explained in detail the law of similarity when the particle size is changed or when the second term is predominant like the flow in a gravel or stone ground.

As can be seen from the derivation process of the equation, the similitude ratio of the permeability coefficient is derived from the equation of motion of pore water (3.3.5). When the similitude ratio of the permeability coefficient is 1, the similitude ratio of the time in the equation of motion for pore water is λ. The time is also included in the equation of motion of soil (3.3.4), and this similitude ratio is λ. Consequently, the similitude ratio of both times matches. For example, it is possible to carry out a model test of liquefying ground in which the time that a dynamic phenomenon occurs during an earthquake matches the time that excess pore pressure generated in the pore water dissipates. When a viscous fluid is not used, excess pore pressure dissipates faster than a dynamic phenomenon because of the large permeability coefficient, and even if a liquefaction model test is conducted using the same soil as the prototype, it cannot be said that the behavior of the prototype is reproduced correctly.

In the case of static model tests to investigate the consolidation phenomenon of the ground for which dynamic phenomenon is not an issue, water is generally used instead of a viscous fluid. Let's think about the similitude ratio in this case. When dynamic phenomena are not being dealt with, the acceleration terms in the equation of motion of soil (3.3.4) and the equation of motion of pore water (3.3.5) can be ignored, resulting in eight independent equations against 13 unknowns in Equations (3.3.1′)–(3.3.6′), which allows five similitude ratios to be determined. Thus, besides the four ratios λ, λ_ρ, λ_ε, and λ_g (dimension, soil density, strain, and gravitational acceleration ratios, respectively), the similitude ratio of the permeability coefficient λ_k may be assumed as a given condition. As mentioned above, when water is used, the permeability coefficient increases in proportion to the gravitational acceleration; thus, it is convenient to assume a similitude ratio of the permeability coefficient of $1/\lambda$. In this case, the similitude ratio of time is λ^2, which is convenient because consolidation can proceed much faster than the prototype. Table 3.3.3 shows an excerpt of only the parts that differ from the similitude ratio described above.

3.3.4 Model Tests on Soil–Structure Interaction

In a geotechnical model test, the model test is often conducted with structures such as those made of concrete or iron. Here, let us consider the law of similarity for structures.

TABLE 3.3.3
Similitude Ratio of Geotechnical Model Test (Prototype/Model) (Centrifugal Force Field, Pore Fluid: Water)

Physical Quantity	Description	Similitude Ratio
k	Permeability coefficient	$1/\lambda$
t	Time	λ^2
$\dot{w}$	Rate of pore water flow	$1/\lambda$

Since the governing equation of the structure is equal to the governing equation of soil which ignores pore water, it can be expressed by Equations (3.3.2)–(3.3.4). Therefore, the similitude ratio can be calculated in the same way and summarized as shown in Table 3.3.4. In the case of a 1g model test, it can be seen that the ratio of the displacement to the dimensional ratio decreases and the inclination angle also becomes smaller than the prototype. Thus, when reproducing a phenomenon with a large displacement of the structure in a model, the results of the prototype and model will be different. Furthermore, when the deformation of the structure itself becomes a problem, it is necessary to use a material having an elastic modulus different from that of the prototype. Conversely, in the case of a centrifuge model test, the similitude ratio of the dimension and displacement is the same, so the inclination angle is also the same in the prototype and model. Moreover, since the similitude ratio of the elastic modulus is 1, the model can thus be made of the same material as the prototype.

3.3.5 Law of Similarity for Unsaturated Ground

Up to the previous section, we have considered the similitude ratio of a saturated ground where the pores between soil particles are filled with water. In the actual ground, air is often mixed in the pores between soil particles and the ground is often in an unsaturated state; hence, we will consider the law of similarity for unsaturated ground here. The behavior of saturated ground is complicated, but the behavior of unsaturated ground is even more complicated. This is not simply because water and air are mixed in the pore but because the water pressure in the pore becomes a negative pressure lower than the atmospheric pressure. Water itself is highly agglutinative, forming a concave surface called a meniscus in constricted parts such as the contact

TABLE 3.3.4
Similitude Ratio of Structure (Prototype/Model)

		Similitude Ratio		
Physical Quantity	**Description**	**General**	**Gravitational**	**Centrifugal**
F	Force	$\lambda^3 \lambda_\rho \lambda_g$	λ^3	λ^2
u	Displacement	$\lambda\lambda_\varepsilon$	$\lambda^{1.5}$	λ
M	Moment	$\lambda^4 \lambda_\rho \lambda_g$	λ^4	λ^3
ϑ	Inclination angle	λ_ε	$\lambda^{0.5}$	1
E	Modulus of elasticity	$\lambda\lambda_\rho \lambda_g/\lambda_\varepsilon$	$\lambda^{0.5}$	1

point of soil particles. The water pressure in this portion is lower than the pressure of the interstitial air by an amount that is just equal to that contributed by the surface tension of water. This negative pressure is called suction, but since suction attracts soil particles and increases effective stress, the effect of suction when considering the behavior of the ground should be taken into account. For details on unsaturated ground, see the reference document.[3.3.6] Note that the law of similarity shown here is one-way of thinking because the characteristics of unsaturated ground are still in the research stage. Note that a different law of similarity may also be proposed in the future.

Let us consider how the governing Equations (3.3.1)–(3.3.6) of saturated ground change as the ground becomes unsaturated. Three equations must be changed: the equation that defines effective stress (3.3.1), the equation of motion for pore water (3.3.5), and the continuous equation for pore water (3.3.6). The respective equations are as follows:

- Definition of effective stress

$$\sigma' = (\sigma - mu_a) + \chi m(u_a - p) \tag{3.3.1''}$$

- Equation of motion of pore water

$$-\nabla p + \rho_f g = \frac{\rho_f g}{K}\dot{w} + \rho_f \ddot{u} \tag{3.3.5''}$$

- Continuity equation of pore water (ignoring changes in the volume of water and soil particles themselves)

$$\nabla^T \dot{w} - S_r m^T \dot{\varepsilon} + n\dot{S}_r = 0 \tag{3.3.6''}$$

Here, u_a is the pore air pressure, χ is the physical quantity representing the contribution level of suction to the effective stress, K is the unsaturated permeability coefficient, S_r is the degree of saturation, and n is the porosity. Equation (3.3.1″) was proposed by Bishop[3.3.7, 3.3.8] and is still the most common effective stress equation even now. In this equation, χ is 1 for saturated soil, 0 for dry soil, and between 0 and 1 for unsaturated soil. The higher the degree of saturation, the higher the contribution of suction to the effective stress. Various formulas have been proposed to express χ such as the following formula by Vanapalli et al.[3.3.9]:

$$\chi = S_e = \frac{\theta - \theta_r}{\theta_s - \theta_r} \tag{3.3.11}$$

$$\theta = nS_r \tag{3.3.12}$$

Here, S_e is the effective saturation, θ is the volumetric water content, and θ_s and θ_r are the saturation and residual volumetric water content, respectively. The only difference between Equation (3.3.5″) and Equation (3.3.5) is the permeability coefficient, and K varies greatly depending on the water content and the distribution condition of pore water.

When the water content decreases, K decreases exponentially, but this characteristic also changes greatly depending on the soil. There is no established formula for K. However, for example, the following van Genuchten–Mualem model[3.3.10–3.3.12] is available:

$$K = kS_e^{\xi}\left\{1-\left(1-S_e^{1/m}\right)^m\right\}^2 \tag{3.3.13}$$

Here, k is the saturated permeability coefficient, whereas ξ and m are dimensionless parameters. Equation (3.3.6″) is a continuous equation of pore water. In saturated ground, water movement is converted into soil strain, but in unsaturated ground, water movement is also converted into the degree of saturation. Hence, Equation (3.3.6) is turned into Equation (3.3.6″). Here, in the case of a model test that considers the interaction between waves and the ground, the movement of water due to changes in the degree of saturation is considered to be greater than the movement of water because of soil strain, and the second term on the left-hand side of Equation (3.3.6″) is ignored. Besides the above equation, it is necessary to consider the equation (soil moisture characteristic curve) governing the relationship between volumetric water content and suction, which can be said to be the most important in unsaturated ground. Besides changing greatly depending on the soil, this relational expression also has hysteresis characteristics in which different paths are followed in the water absorption process and drainage process. For example, the following van Genuchten equation[3.3.11, 3.3.12] has been proposed:

$$S_e = \left\{1+\left(\alpha p\right)^n\right\}^{-m} \tag{3.3.14}$$

Here, α and n are dimensionless parameters. In Equations (3.3.13) and (3.3.14), the combination assuming $m = 1 - 1/n$ is the most widely used unsaturated infiltration characteristic model. If these are combined, the displacement $\boldsymbol{u}$ and pore pressure p can be solved. Since the unsaturated ground is composed of three phases, namely, soil particles, pore water, and interstitial air, the interstitial air pressure u_a can be obtained by solving the equation related to interstitial air simultaneously besides the equation above. Nevertheless, for the sake of simplicity in explaining the law of similarity, u_a will be ignored here.

As in the case of deriving the similitude ratio of saturated ground, let us consider the law of similarity by assuming that the above governing equation is valid for both the prototype and the model. In the governing Equations (3.3.1)–(3.3.6) of saturated ground, there were 9 independent equations and 13 unknowns. Thus, by assuming four similitude ratios, other similitude ratios could be set. In the group of governing equations for unsaturated ground, the number of physical quantities increases by six (χ, K, S_r, n, S_e, and θ), but the number of equations increases by five (Equation (3.3.11) (two equations), Equation (3.3.12), Equations (3.3.13), and Equation (3.3.14)). Hence, it will be sufficient to assume five similitude ratios, which is one more than for saturated ground. However, since Equations (3.3.13) and (3.3.14) are complicated exponential functions and these equations are not theoretical equations, the same similitude ratio for both equations is difficult to find. Therefore, it is probably more practical to make a separate assumption in setting the similitude ratio and determine the type of soil used in the model test so that physical quantities such as p, K, and S_e satisfy Equations (3.3.13) and (3.3.14). Here, we decided not to find the similitude ratio of the saturated permeability coefficient k and to find the similitude ratio using the assumption that the similitude ratio of the soil and water density is equal ($\lambda_\rho = \lambda_{\rho f}$). Besides the four similitude ratios assumed in saturated ground, the similitude ratio of the degree of saturation was assumed to be $\lambda_{Sr} = 1$. Consequently, the similitude ratios of the model tests in the gravitational field and centrifugal force field are as shown in Table 3.3.5. Similitude ratios different from those of saturated ground are enclosed in a box.

Looking at the similitude ratio of the gravitational field, the similitude ratio regarding permeability differs between saturated and unsaturated ground; therefore, care must be taken when reproducing the seepage flow in a saturated and unsaturated state at the same time in a model test. The similitude ratio of the porosity and volumetric water content is 1, but the similitude ratio of the pore pressure is λ and the unsaturated permeability coefficient is $\lambda^{0.25}$; thus, a soil whose similitude ratios satisfy Equations (3.3.13) and (3.3.14) should be used in the model. In other words, it is necessary to use soil that has the same water content in the model as the prototype but has a small suction and unsaturated permeability coefficient. To reduce the suction, while it may be sufficient to

TABLE 3.3.5

Similitude Ratio of Unsaturated Ground (Prototype/Model) (Gravitational and Centrifugal Force Fields)

Physical Quantity	Description	Similitude Ratio		
		General	Gravitational	Centrifugal
x	Length (dimension)	λ	λ	λ
ρ	Density of soil	λ_ρ	1	1
ρ_f	Density of pore water	λ_ρ	1	1
ε	Strain of soil	λ_ε	$\lambda^{0.5}$	1
g	Gravitational acceleration	λ_g	1	$1/\lambda$
t	Time	$(\lambda\lambda_\varepsilon/\lambda_g)^{0.5}$	$\lambda^{0.75}$	λ
σ	Total stress	$\lambda\lambda_\rho\lambda_g$	λ	1
σ'	Effective stress of soil	$\lambda\lambda_\rho\lambda_g$	λ	1
D	Tangent modulus of soil	$\lambda\lambda_\rho\lambda_g/\lambda_\varepsilon$	$\lambda^{0.5}$	1
p	Pore water pressure	$\lambda\lambda_\rho\lambda_g$	λ	1
K	Unsaturated permeability coefficient	$\boxed{(\lambda\lambda_g/\lambda_\varepsilon)^{0.5}}$	$\boxed{\lambda^{0.25}}$	1
u	Displacement of soil	$\lambda\lambda_\varepsilon$	$\lambda^{1.5}$	λ
$\dot{u}$	Velocity of soil	$(\lambda\lambda_\varepsilon\lambda_g)^{0.5}$	$\lambda^{0.75}$	1
$\ddot{u}$	Acceleration of soil	λ_g	1	$1/\lambda$
w	Displacement of pore water relative to soil skeleton	$\boxed{\lambda}$	$\boxed{\lambda}$	λ
$\dot{w}$	Rate of pore water flow	$\boxed{(\lambda\lambda_g/\lambda_\varepsilon)^{0.5}}$	$\boxed{\lambda^{0.25}}$	1
χ	Suction contribution ratio	1	1	1
n	Porosity	1	1	1
S_r, S_e	Saturation degree, effective degree of saturation	1	1	1
θ	Volume water content	1	1	1

just increase the pore, permeability will increase. A trial and error method is required, such as slowing the permeability using an appropriate viscous fluid. Conversely, looking at the similitude ratio of the centrifugal force field, it can be seen that the similitude ratio of the saturated ground and the unsaturated ground is the same and therefore suitable for model tests in which both are combined. Additionally, the similitude ratio of p, K, and S_e is 1, and if the same soil as the prototype is used in the model, Equations (3.3.13) and (3.3.14) are automatically satisfied. Furthermore, since $\lambda_\varepsilon = 1$, there is no need to ignore the second term on the left side of Equation (3.3.6″), so a centrifuge model test is better in satisfying the law of similarity.

3.4 Law of Similarity in Combined Fluid and Ground Model Tests

3.4.1 Sand Drift

3.4.1.1 Derivation of Dimensionless Quantity

The first phenomenon that comes to mind as a complex phenomenon involving fluid and ground is sand drift. This is a phenomenon in which soil near the ground surface drift due to the flow of water, resulting in large scouring and sedimentation. Although research on sand drift has been conducted extensively in the past, a unified governing equation has not yet been established. This is probably because soil cannot be treated as a continuum, and thus, it is necessary to treat it as particles and aggregates, making it even more difficult to understand the phenomenon. Most of the equations that have been proposed so far to understand the phenomenon are based on experiences such as on-site measurements and model tests. Here, we will introduce laws of similarity that have been studied in the field of hydraulics, but we must add that since these are not necessarily theoretical laws of similarity, different laws of similarity may be proposed depending on future studies.

First, the dimensionless quantity[3.4.1] related to sand drift is derived. Since sand drift is a phenomenon in which soil particles on the ground surface move due to the flow of water, assuming we focus on one soil particle on the ground surface and an inertial force acting on the particle is F_i, underwater gravity is F_b, and shear force is F_τ, each one of them can be expressed as shown below:

$$F_i = \rho_w v^2 \left(\frac{1}{4} \pi d^2 \right) \tag{3.4.1}$$

$$F_b = \left(\rho_s - \rho_w \right) g \left(\frac{1}{6} \pi d^3 \right) \tag{3.4.2}$$

$$F_\tau = \rho_w u_*^2 \left(\frac{1}{4} \pi d^2 \right) \tag{3.4.3}$$

Here, ρ_w is the density of water, v is the horizontal flow velocity, d is the particle size of sand, ρ_s is the density of sand, g is the gravitational acceleration, and u is the frictional velocity. Assuming the ratio of the shear force F_τ to the underwater gravity F_b is a dimensionless quantity π_1, the equation becomes as shown below:

$$\pi_1 = \frac{F_\tau}{F_b} = \frac{\rho_w u_*^2\left(\frac{1}{4}\pi d^2\right)}{(\rho_s - \rho_w)g\left(\frac{1}{6}\pi d^3\right)} = \frac{\left(\frac{3}{2}\right)u_*^2}{\frac{(\rho_s - \rho_w)}{\rho_w}gd} = \left(\frac{3}{2}\right)\frac{u_*^2}{sgd} = \left(\frac{3}{2}\right)\tau_* \tag{3.4.4}$$

Here, s $(= (\rho_s - \rho_w)/\rho_w)$ is the specific gravity of sand under water, and τ_* $(= u_*^2/sgd)$ is the dimensionless tractive force (Shields number). Next, assuming that the ratio of the inertial force F_i to the shear force F_τ is a dimensionless quantity π_2, the equation becomes as follows:

$$\pi_2 = \frac{F_i}{F_\tau} = \frac{\rho_w v^2\left(\frac{1}{4}\pi d^2\right)}{\rho_w u_*^2\left(\frac{1}{4}\pi d^2\right)} = \left(\frac{v}{u_*}\right)^2 \tag{3.4.5}$$

To make the ratio of the shear force and underwater gravity of sand similar in the prototype and model, it is necessary to make τ_* equal in the prototype and model. To make the ratio of the inertial force to the shear force similar, $(v/u_*)^2$ shall be made equal. Conversely, the relationship with the final sedimentation velocity w_s when the underwater gravity of sand is in equilibrium with the underwater resistance is given by the following equation[3.4.2]:

$$(\rho_s - \rho_w)g\left(\frac{1}{6}\pi d^3\right) = \frac{1}{2}\rho_w C_D w_s^2\left(\frac{1}{4}\pi d^2\right) \tag{3.4.6}$$

Here, C_D is the drag coefficient of the particle. Equation (3.4.6) can be transformed as follows.

$$sg = \frac{3C_D w_s^2}{4d} \tag{3.4.7}$$

The following equation is derived by substituting Equation (3.4.7) into Equation (3.4.4):

$$\pi_1 = \frac{F_\tau}{F_b} = \left(\frac{2}{3}\right)\frac{u_*^2}{sgd} = \frac{2u_*^2}{C_D w_s^2} = \left(\frac{2}{C_D}\right)\left(\frac{u_*}{w_s}\right)^2 \tag{3.4.8}$$

C_D is the function of the particle Reynolds number R_{e*2} $(= dw_s/v)$, and it is known that as R_{e*2} increases, C_D becomes smaller.[3.4.2] When the particles settle quietly without disturbing the dispersion medium and the inertia is very small relative to the viscosity ($R_{e*2} \leq 0.5$), C_D is expressed by the following equation according to Stokes's law:

$$C_D = \frac{24}{R_{e*2}} \tag{3.4.9}$$

Also, when the inertial force is dominant ($10^3 \leq R_{e*2} \leq 3 \times 10^5$), C_D is constant at 0.44. Making the ratio of the shear force to the underwater gravity of sand similar between the prototype and model is also the same as making the ratio $(u_*/w_s)^2/C_D$ similar. Finally, let us consider the ratio of the inertial force. F_i to the underwater gravity F_b as a dimensionless quantity. When Equation (3.4.7) is substituted and transformed in the same way as Equation (3.4.8), the following equation is derived:

$$\pi_3 = \frac{F_i}{F_b} = \frac{\left(\frac{3}{2}\right)v^2}{\frac{(\rho_s - \rho_w)}{\rho_w}gd} = \left(\frac{3}{2}\right)\frac{v^2}{sgd} = \frac{2v^2}{C_D w_s^2} = \left(\frac{2}{C_D}\right)\left(\frac{v}{w_s}\right)^2 \quad (3.4.10)$$

To make the ratio of the inertial force to the underwater gravity of sand similar between the prototype and the model is also the same as making $(v/w_s)^2/C_D$ equal between the prototype and the model. However, among the three dimensionless quantities above, two are independent dimensionless quantities. If these two dimensionless quantities are equal in the prototype and model, the remaining one dimensionless quantity will be equal automatically.[3.4.1] Empirically, the dimensionless quantities τ_*, v/u_*, u_*/w_s, and v/w_s obtained as described above have been the focus in understanding the phenomenon of sand drift as explained below.

3.4.1.2 Law of Similarity in Each Phenomenon

Sand drift is a phenomenon in which water near the bottom moves due to waves and currents, causing soil at the bottom of the sea to move accordingly. There are various types of sand drift such as bed load transport in which soil particles move by rolling or hopping and suspended load transport in which soil particles move by being incorporated in a flow of water in the vertical direction or a vortex. Various phenomena may occur in a complex manner and mutually influence one another. Here, we introduce the law of similarity[3.4.3–3.4.5] for these phenomena. The law of similarity for seabed topography changes because of sand drift is described in the next section.

a. Law of similarity of sand drift due to traction

The flow of water exerts a shear force to the seabed called a tractive force. Based on the law of action and reaction, the tractive force is also the shearing force exerted to the flow from the seabed. In other words, the force is a bottom frictional force for a fluid. This bottom friction force, τ_0, is considered to be proportional to the velocity energy $v^2/2$ and is expressed by the following equation:

$$\tau_0 = \rho f' \frac{v^2}{2} \quad (3.4.11)$$

Here, ρ is the density of water and f' is the coefficient of friction. No unified formula for f' has been proposed, and various formulas have been applied to each phenomenon. For example, Equation (3.4.12) has been proposed for considering a uniform flow in a waterway and Equation (3.4.13) for considering a flow fluctuating like a short-wavelength wavefield.[3.4.3, 3.4.4, 3.4.6]

$$f' = \frac{2gn^2}{R^{1/3}} = \frac{2g}{R^{1/3}}\left(\frac{k_s^{1/6}}{7.66\sqrt{g}}\right)^2 \cong 0.0341\left(\frac{k_s}{R}\right)^{1/3} \tag{3.4.12}$$

$$f' = 0.47\left(\frac{k_s}{a_\delta}\right)^{3/4} \tag{3.4.13}$$

Here, R is the hydraulic radius, n is the Manning roughness coefficient, k_s is the roughness of the seabed, and a_δ is the amplitude of water particle fluctuation due to waves on the seabed. Normally, k_s is replaced by the particle size d of the soil particles that make up the seabed. As can be seen from Equations (3.4.12) and (3.4.13), f' is considered to be a function of the particle size and length (hydraulic radius and amplitude) of the soil particles, although the coefficients and exponents are different.

The flow of water exerts a shear force on the seabed in the direction of the flow. This shear force is called tractive force. When it becomes large, sand drift occurs, so the tractive force is an important physical quantity that determines whether sand drift will occur or not. The tractive force is expressed by Equation (3.4.11). This tractive force τ_0 is divided by the density ρ of water, and the square root is then taken to obtain a dimensional velocity called the frictional velocity u_*, which is used as a representative value of the velocity in the flow near the bottom surface. This can be expressed by the following equation:

$$u_* = \sqrt{\frac{\tau_0}{\rho}} \tag{3.4.14}$$

The dimensionless tractive force (Shields number) τ_* is obtained by dividing the tractive force by the density ρ of water, the specific gravity s of the soil particles in water, the gravitational acceleration g, and the particle size d of the soil particle and making it dimensionless. It can be expressed by the following equation:

$$\tau_* = \frac{\tau_0}{\rho s g d} = \frac{u_*^2}{sgd} \tag{3.4.15}$$

Furthermore, the dimensionless tractive force when soil particles start to move is called the dimensionless critical tractive force τ_{*C}, and soil particles are deemed to move when τ_* is larger than τ_{*C}.

Considering Equation (3.4.15), the similitude ratio for bed load transport is as shown in Table 3.4.1. For the physical quantities from L to v in the table, the similitude ratio is assumed as a given condition, the similitude ratio of the particle size of soil particles is assumed to be $1/M$, and the similitude ratio of the flow velocity is assumed to follow Froude's law. Note that the similitude ratios from f' to τ_* in the table are derived, but the similitude ratios τ_0 and τ_* differ depending on the concept of the frictional coefficient f'. When conducting a model test, it is important to consider the phenomenon to be reproduced and

TABLE 3.4.1

Similitude Ratio of Drift Sand (Model/Prototype)

Physical Quantity	Description	Similitude Ratio	
		Gravitational	Centrifugal
L	Length (dimension)	$1/N$	$1/N$
d	Diameter of soil particle	$1/M$	$1/M$
ρ	Density of water	1	1
s	Specific gravity of soil particles in water	1	1
g	Gravitational acceleration	1	N
R	Hydraulic mean depth	$1/N$	$1/N$
a_δ	Amplitude of water particle fluctuations	$1/N$	$1/N$
v	Flow velocity	$(1/N)^{1/2}$	1
f'	Friction coefficient	$(N/M)^{1/3}$ $(N/M)^{3/4}$	$(N/M)^{1/3}$ $(N/M)^{3/4}$
τ_0	Tractive force	$(N/M)^{1/3}\cdot 1/N$ $(N/M)^{3/4}\cdot 1/N$	$(N/M)^{1/3}$ $(N/M)^{3/4}$
u^*	Frictional velocity	$(N/M)^{1/6}(1/N)^{1/2}$ $(N/M)^{3/8}(1/N)^{1/2}$	$(N/M)^{1/6}$ $(N/M)^{3/8}$
τ_*	Dimensionless tractive force	$(M/N)^{2/3}$ $(M/N)^{1/4}$	$(M/N)^{2/3}$ $(M/N)^{1/4}$

* Upper section: calculated from Equation (3.4.12); lower section: calculated from Equation (3.4.13).

use a similitude ratio suitable for the phenomenon. If the similitude ratios of the length and particle size of the soil particles can be made equal ($1/N = 1/M$), the similitude ratio of the dimensionless tractive force will be 1 in both the gravitational field and the centrifugal force field, thereby satisfying the law of similarity. However, since the particle size is considerably smaller than that of the prototype, significant changes in the physical properties of the ground may be an issue, so in practice, it is often difficult to use sand with a particle size of $1/N$ in model tests. Thus, the phenomenon in the prototype should be understood by considering also the similitude ratio of the phenomenon you wish to know by model test (e.g., changes in the seabed topography as a result of sand drift) or conversely to set another similitude ratio by assuming the similitude ratio of the phenomenon. Furthermore, although the law of similarity shown in the table assumes that the flow velocity follows Froude's law, the flow velocity does not follow Froude's law when the flow condition changes significantly because of the frictional force at the bottom. To reproduce such a phenomenon, it is necessary to consider the deviation from Froude's law and take measures such as multiplying the similitude ratio by a correction coefficient.

b. Law of similarity of sand drift due to suspension

When the seabed topography changes, it is necessary to consider suspended load besides the bed load. In suspended load transport, soil sinks and accumulates on the seabed after floating up due to the flow of water. At the same time, the soil moves in the downstream direction due to the horizontal flow. Therefore, it is necessary to make the ratio of the sedimentation velocity w_s of the soil particles and the horizontal flow velocity v equal in the prototype and model. In Equation (3.4.16), the subscript p represents the physical quantity of the prototype, and the subscript m represents the physical quantity of the model:

$$\frac{w_{sp}}{v_p} = \frac{w_{sm}}{v_m} \tag{3.4.16}$$

If Equation (3.4.16) is rewritten and Froude's law of similarity is applied, the equation becomes as follows:

$$\frac{w_{sm}}{w_{sp}} = \frac{v_m}{v_p} = \left(\frac{NgL_m}{gL_p}\right)^{1/2} = \left(\frac{NL_m}{L_p}\right)^{1/2} \tag{3.4.17}$$

Here, the gravitational acceleration of the model is considered to be N times that of the prototype. For 1g model tests, $N = 1$. The ratio of the sedimentation rate to the dimensional ratio of the model considering N is set, the type of soil used in the model is determined. Additionally, according to the Airy wave theory,[3.4.7] the similitude ratio of the horizontal flow velocity v of the water particles can be expressed by the similitude ratio $(2\pi/T)(H/2)$; thus, Equation (3.4.17) can also be transformed into the following equation:

$$\left(\frac{H}{w_s T}\right)_p = \left(\frac{H}{w_s T}\right)_m \tag{3.4.18}$$

Here, H is the wave height and T is the period. Examples of model tests using the law of similarity of the sedimentation velocity include model tests by Suzuki et al.,[3.4.8] Matsuda et al.,[3.4.9] Arikawa et al.,[3.4.10] and Araki et al.[3.4.11]

As mentioned above, when conducting a model test of suspended load transport, bed load transport also occurs at the same time; thus, the ratio of the frictional velocity to sedimentation velocity that characterizes them must be made equal as shown in the following equation[3.4.4, 3.4.5]:

$$\frac{u_{*p}}{w_{sp}} = \frac{u_{*m}}{w_{sm}} \tag{3.4.19}$$

The sedimentation velocity in the model can also be determined from this equation. Sand in an open channel is known to move in a form that changes as follows, assuming the ratio of the frictional velocity to sedimentation velocity as a parameter.[3.4.4, 3.4.5]

$\frac{u_*}{w_s} < 1$: Traction is superior

$\frac{u_*}{w_s} > 1.7$: Suspension is superior

$1 < \frac{u_*}{w_s} < 1.7$: Traction and suspension coexist

3.4.2 Seabed Topography Changes

Seabed topography changes when sand drift occurs. Here, let us consider the law of similarity for changes in the seabed topography due to sand drift. From past model test studies, the dimensionless sand drift quantity Q_T is expressed as a function of the dimensionless tractive force, and a relationship as expressed by the following equation has been proposed[3.4.3, 3.4.5, 3.4.12–3.4.14]:

$$\frac{Q_T}{B\sqrt{sgd^3}} = a \cdot \tau_*^b \tag{3.4.20}$$

Here, B is the sand drift width, s is the underwater specific gravity of the soil particles, g is the gravitational acceleration, d is the particle size of the soil particles, a and b are coefficients obtained from the model test, and τ_* is the dimensionless tractive force. The equation expresses a relationship in which the amount of sand drift increases as the dimensionless tractive force increases.

Focusing on a certain area, the seabed topography will change if the amount of sand drifting in and out of the area is different. This relationship can be expressed in an equation as shown below:

$$\frac{\partial Z_b}{\partial t} + \frac{1}{1-\lambda}\left\{\frac{\partial}{\partial x}\left(\frac{Q_b}{B}\right) - P + w_s c_b\right\} = 0 \tag{3.4.21}$$

$$\frac{\partial}{\partial x}\left(\frac{Q_s}{B}\right) = -P + w_s c_b \tag{3.4.22}$$

Here, Z_b is the height of the surface of the seabed topography, λ is the porosity of the sandy ground, Q_b is the amount of bed load transport, Q_s is the amount of suspended load transport, B is the sand drift width, P is the amount of suspended sand stirred up from the seabed, $w_s c_b$ is the amount of sedimentation, w_s is the sedimentation velocity, and c_b is the concentration of suspended sand near the seabed. It is known that the amount of suspended sand stirred up is a function of the dimensionless tractive force (Shield number), similar to the amount of bed load transport.[3.4.15] Simplifying as $Q_T = Q_b + Q_s$, the equation can be transformed as follows:

$$\frac{\partial Z_b}{\partial t} = -\frac{1}{1-\lambda}\left\{\frac{\partial}{\partial x}\left(\frac{Q_T}{B}\right)\right\} \tag{3.4.23}$$

Substituting Equation (3.4.20) into this equation and modifying the partial derivative to a differential form, the following relationship is obtained:

$$\Delta Z_b = -\frac{\sqrt{sgd^3} \cdot a \cdot \Delta\tau_*^b}{(1-\lambda)\Delta x}\Delta t \tag{3.4.24}$$

Considering Equations (3.4.20) and (3.4.24), the similitude ratio related to the amount of sand drift is as shown in Table 3.4.2. It is assumed that the similitude ratio of length is $1/N$, the similitude ratio of the soil particle size is $1/M$, the similitude ratio of the specific gravity of soil particles is 1, and that the similitude ratio of time follows Froude's law. Also, you may wish to note that the following similitude ratios are examples because similitude ratios such as τ_* differ depending on the concept of the frictional coefficient f'. When conducting a model test, it is necessary to use the similitude ratio of τ_* in the

TABLE 3.4.2

Similitude Ratio of Seabed Topography Change (Model/Prototype)

Physical Quantity	Description	Similitude Ratio	
		Gravitational	Centrifugal
L	Length (dimension)	$1/N$	$1/N$
d	Diameter of soil particle	$1/M$	$1/M$
ρ	Density of water	1	1
s	Specific gravity of soil particles in water	1	1
g	Gravitational acceleration	1	N
λ	Porosity of sandy ground	1	1
a,b	Coefficient	1	1
t	Time	$(1/N)^{1/2}$	$1/N$
τ_*	Dimensionless tractive force	$(M/N)^{2/3}$ $(M/N)^{1/4}$	$(M/N)^{2/3}$ $(M/N)^{1/4}$
Q_T	Sediment transport rate	$(N^2M^3)^{-1/2}\cdot(M/N)^{2b/3}$ $(N^2M^3)^{-1/2}\cdot(M/N)^{b/4}$	$(NM^3)^{-1/2}\cdot(M/N)^{2b/3}$ $(NM^3)^{-1/2}\cdot(M/N)^{b/4}$
Z_b	Seabed topography	$(N/M^3)^{1/2}\cdot(M/N)^{2b/3}$ $(N/M^3)^{1/2}\cdot(M/N)^{b/4}$	$(N/M^3)^{1/2}\cdot(M/N)^{2b/3}$ $(N/M^3)^{1/2}\cdot(M/N)^{b/4}$

* Upper section: calculated from Equation (3.4.12); lower section: calculated from Equation (3.4.13).

phenomenon to be reproduced. If the similitude ratio between the length and particle size of the soil particles can be made equal ($1/N = 1/M$), the similitude ratio of the dimensionless tractive force will be 1 in both the gravitational field and the centrifugal force field as described above. Additionally, regardless of whether it is Earth's gravity field or centrifugal force field, the similitude ratio of the height Z_b of the seabed topography surface is $1/N$, which is consistent with the similitude ratio of the model dimensions. Nevertheless, since the particle size is considerably smaller than that of the prototype, significant changes in the physical properties of the ground may be an issue; therefore, in practice, it is often difficult to use sand with a particle size of $1/N$ in model tests. Thus, the phenomenon in the prototype must be estimated by considering the similitude ratio of the phenomenon, or conversely, another similitude ratio should be set by assuming the similitude ratio of the phenomenon. Additionally, as described in the similitude ratio of the tractive force, the law of similarity shown in the table is the similitude ratio when Froude's law is assumed to be valid. If it deviates from Froude's law, corrections must be made by multiplying the similitude ratio by a correction coefficient.

3.4.3 Scouring Phenomenon below Permeable Structure

When waves act on a structure in which stones or blocks are set on a sandy ground, sand will float and flow outside the stones and blocks into the sea, which results in the so-called scouring phenomenon occurring below a permeable structure. Subsidence of breakwaters, detached breakwaters, and submerged breakwaters covered with wave-dissipating blocks are often caused by this scouring phenomenon. This phenomenon occurs below the structure; thus, investigating it by conducting an underwater survey out in the field is difficult. Therefore, observing the phenomenon in a model test will be useful for clarification purposes (see the case study in Section 4.2). Nevertheless, the law of similarity of this phenomenon is complicated. Although Froude's law of similarity can be applied as a law of similarity of fluids because the flow is generated by waves, the change in the Reynolds number in the flow between stones and blocks also cannot be ignored. Additionally, when studying the ground in combination, the movement of sand is caused not only by traction but also by suspension, and the flow comprises a mixture of fluid and sand particles. Thus, it is necessary to consider the law of similarity for multiple phenomena at the same time. Nevertheless, it is important to prioritize satisfying the law of similarity for predominant phenomena given that it is difficult to satisfy all the laws of similarity at the same time.

Sand drift phenomena can be roughly divided into two types depending on whether the sand is transported by traction or suspension, but the importance of reproducing suspension in a scouring phenomenon in which sand is stirred up at the bottom of a permeating structure has been pointed out. The suspension of sand is determined by the balance between the force acting on the sand particles and gravity when it is dragged by the flow. Assuming the phenomenon can be reproduced by balancing the flow velocity and sedimentation velocity, each of these should be made similar according to Froude's law (Dean[3.4.16, 3.4.17]). Among these, various w_s have been proposed for the sedimentation velocity. For example, according to Jimenez and Madsen,[3.4.18] it is expressed by the following equation:

$$w_s = \sqrt{sgd_N}\left(0.954 + \frac{5.12}{S_*}\right)^{-1} \tag{3.4.25}$$

$$S_* = \frac{d_N}{4\nu}\sqrt{sgd_N} \tag{3.4.26}$$

Here, s is the specific gravity of soil particles in water, g is the gravitational acceleration, d_N is the representative diameter of the soil particles (assuming it is equal to $d/0.9$), d is the soil particle size or size of a spherical particle equal in volume to the crushed stone, and ν is the kinematic viscosity coefficient of water. Based on this equation, the similitude ratio is as shown in Table 3.4.3. The length similitude ratio is assumed to be

TABLE 3.4.3

Similitude Ratio of Scouring Phenomenon below Permeable Structure (Model/Prototype)

		Similitude Ratio	
Physical Quantity	**Description**	**Gravitational**	**Centrifugal**
L	Length (dimension)	$1/N$	$1/N$
d	Diameter of soil particle	$1/M$	$1/M$
ρ	Density of water	1	1
ν	Kinematic viscosity of water	1	1
g	Gravitational acceleration	1	N
H	Wave height	$1/N$	$1/N$
T	Wave period	$(1/N)^{1/2}$	$1/N$
v	Flow velocity	$(1/N)^{1/2}$	1
w_s	Sedimentation velocity	w_{sm}/w_{sp}	w_{sm}/w_{sp}

$1/N$, and the particle size similitude ratio of the soil particles is assumed to be $1/M$. The similitude ratio w_{sm}/w_{sp} of the sedimentation velocity in the table cannot be expressed in a simple form, but it can be calculated from the following equations using Equations (3.4.25) and (3.4.26):

$$\frac{w_{sm}}{w_{sp}} = \left(\sqrt{sgd_N} \left(0.954 + \frac{5.12}{S_*} \right)^{-1} \right)_m \Bigg/ \left(\sqrt{sgd_N} \left(0.954 + \frac{5.12}{S_*} \right)^{-1} \right)_p \qquad (3.4.27)$$

In Equations (3.4.27), the subscript p represents the physical quantity of the prototype, while the subscript m represents the physical quantity of the model. Additionally, Froude's law is applied to the similitude ratio of the wave period and flow velocity. As mentioned above, when considering sand drift due to suspension, it is necessary to follow the ratio of the sedimentation velocity w_s of sand in the following equation, assuming Froude's law and the Airy wave theory:

$$\left(\frac{H}{w_s T} \right)_p = \left(\frac{H}{w_s T} \right)_m \qquad (3.4.28)$$

Here, H is the wave height, T is the period, and Equation (3.4.28) is called the Dean number. That is, the ratio of the force acting on the particles described earlier (horizontal wave power) and gravity (vertical force) is matched in the prototype and model. Using the similitude ratios of H and T shown in Table 3.4.3, the similitude ratio w_{sm}/w_{sp} of the sedimentation velocity should be assumed as $(1/N)^{1/2}$ in Earth's gravity field and 1 in centrifugal gravity field. The particle size, etc. of the soil particles may be adjusted using Equation (3.4.27) so that this similitude ratio is obtained.

3.4.4 Combined Wave and Ground Model Test

In the previous sections, sand drift and scouring were treated as phenomena in which the fluid and ground are combined, but these are phenomena that occur in a narrow area near the ground surface where the fluid and ground come into contact and mix with each other. When the target area is expanded, besides sand drift and scouring, other phenomena in which the fluid and ground are combined may also occur. Ground instability due to waves is also one phenomenon that may occur. For example, liquefaction of the seabed due to traveling waves is a problem. This is because the water pressure acting on the seabed fluctuates from place to place when waves pass, so a shear force acts repeatedly on the ground, and liquefaction occurs in loosely deposited sandy ground (Sekiguchi et al.[3.4.19]; Sassa & Sekiguchi[3.4.20]). Additionally, when the ground is not completely saturated, the propagation of water pressure fluctuations on the seabed to the inside of the ground is delayed, and an upward seepage flow is generated when the valley of the wave passes, thereby destabilizing the ground (Zen[3.4.21]). It is necessary to consider the ground stress and pore pressure characteristic within a range of several meters from the ground surface and not only the area near the ground surface in these phenomena. When the ground becomes unstable or liquefies, the stiffness and strength of the ground are reduced significantly, which may cause displacement of superstructures and underground structures. For example,

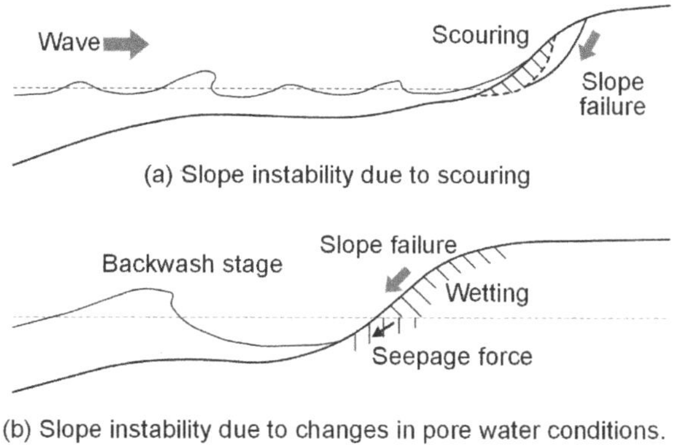

FIGURE 3.4.1
Slope instability problem due to waves. (a) Scouring and (b) Changes in pore water conditions.

Miyamoto et al.[3.4.22] studied the floating of submarine pipes in ground where liquefaction caused by waves occurred.

Besides the problem of liquefaction caused by waves, there is also the problem of slope instability caused by waves, as shown in Figure 3.4.1. When waves act on a slope formed by soil, the slope is directly scoured or the area around the foot of the slope is scoured, and the slope becomes unstable and collapses. Moreover, when high waves act on the slope, water permeates and moistens the ground, changing the water pressure conditions on the ground surface; thus, the degree of saturation and pore pressure in the ground change. In this case, the ground becomes unstable because of the downward seepage force generated in the backwash (Takahashi & Morikawa[3.4.23]).

Studies have shown that the instability of the ground due to the increase in pore pressure caused by waves affect the amount of sand drift and scouring. It was also pointed out above that waves may cause scouring and destabilize the slope. When it is necessary to reproduce such sand drift and scouring in a model, the law of similarity described in the previous sections should be considered. Let us consider the law of similarity regarding the problem of ground instability caused by waves other than these. In this problem, the ground can be thought of as a continuum rather than at the particle level, so the respective law of similarity explained in Sections 3.2 and 3.3 may be used for the fluid and ground. However, regarding the interaction between the two, the law of similarity of the force at the contact surface between the fluid and ground should be considered. The forces acting on the interaction between fluid and ground include the force in the direction perpendicular to the plane and the force in the direction parallel to the plane. The force in the perpendicular direction is due to the water pressure, and the force in the parallel direction is due to the tractive force. The tractive force of water acting on the ground cannot be ignored when considering the behavior of soil at the particle level, but it is not large enough to have a significant effect on the stability of the entire ground. Hence, the similitude ratio of the water pressure, which is mainly a perpendicular force, can be almost the same between the fluid and ground. Considering this, the similitude ratio between fluid and ground is summarized in Tables 3.4.4 and 3.4.5. For fluids, Froude's law of similarity is applied to reproduce the waves in a model. Moreover, since it is rare to use a viscous fluid in a 1g model test, the similitude ratio when water

TABLE 3.4.4

Similitude Ratio of Combined Wave and Ground Model Test (Model/Prototype) (Earth's Gravity Field, Pore Fluid: Water)

Physical Quantity	Description	Similitude Ratio of Fluid	Similitude Ratio of Ground
x	Length (dimension)	$1/N$	$1/N$
ρ	Density	1	1
ε	Strain of soil	–	$(1/N)^{0.5}$
g	Gravitational acceleration	1	1
ν	Kinematic viscosity of water	1	1
t	Time (dynamic)	$(1/N)^{0.5}$	$(1/N)^{0.75}$
t	Time (seepage)	–	$(1/N)^{1.5}$
σ	Stress	–	$1/N$
D	Tangent modulus of soil	–	$(1/N)^{0.5}$
p	Water pressure, pore water pressure	$1/N$	$1/N$
k	Permeability coefficient	–	1
u	Displacement	$1/N$	$(1/N)^{1.5}$
$\dot{u}$	Velocity	$(1/N)^{0.5}$	$(1/N)^{0.75}$
$\ddot{u}$	Acceleration	1	1

TABLE 3.4.5

Similitude Ratio of Combined Wave and Ground Model Test (Model/Prototype) (Centrifugal Gravity Field)

Physical Quantity	Description	Similitude Ratio of Fluid	Similitude Ratio of Ground
(a) When using viscous fluid			
x	Length (dimension)	$1/N$	$1/N$
ρ	Density	1	1
ε	Strain of soil	–	1
g	Gravitational acceleration	N	N
ν	Kinematic viscosity of water	N	N
t	Time (dynamic)	$1/N$	$1/N$
t	Time (seepage)	–	$1/N$
σ	Stress	–	1
D	Tangent modulus of soil	–	1
p	Water pressure, pore water pressure	1	1
k	Permeability coefficient	–	1
u	Displacement	$1/N$	$1/N$
$\dot{u}$	Velocity	1	1
$\ddot{u}$	Acceleration	N	N
(b) When using water (Other similitude ratios are the same as (a).)			
ν	Kinematic viscosity of Water	1	1
t	Time (seepage)	–	$(1/N)^2$
k	Permeability coefficient	–	N

is used is shown for the 1g model test. In a centrifuge model test, a viscous fluid is often used to match the similitude ratio of the seepage time of the ground with the similitude time ratio of the fluid, thereby demonstrating the respective similitude ratios when using the viscous fluid and water. Additionally, the similitude ratio of the ground shows that the ground is saturated. In a 1g model test, the problem of the similitude ratio of the ground is as explained in Section 3.3. The other problem is that the similitude time ratio between the ground and fluid does not match. By using a viscous fluid in 1g model tests, the similitude time ratio of the fluid can be matched to the similitude time ratio of seepage in the ground, but when dealing with dynamic problems where the acceleration of the ground cannot be ignored, note that the similitude time ratios are not matched. Accordingly the conditions are ideal for model tests in centrifugal gravity field since all the similitude time ratios are the same.

For the combined wave and ground model test, the merits offered by a model test in centrifugal gravity field are great. This is because the stress in the ground can be reproduced, making it suitable for evaluating the stability of the ground. As mentioned above, the similitude time ratio can also be matched. For this reason, many past model tests were also conducted in centrifugal gravity field. For example, Takahashi et al.[3.4.24] examined the effectiveness of combined wave and ground model tests in centrifuge model tests. The modeling of models technique is used to demonstrate the effectiveness of centrifugal model tests. In this method, model tests are conducted with various combinations of model scale and centrifugal acceleration to verify that the actual conversion values overlap and to verify the reproducibility of the behavior on a prototype scale. It has been confirmed that the pore pressure responses are equal even at different centrifugal accelerations.

References

3.1 Role of Law of Similarity

3.1.1 Emori, I., Saito, K., and Sekimoto, K. (2000): Mokei jikken no riron to ouyou [Theory and Application of Model Experiments –Third Edition–], Gihodo Publishing. (in Japanese)
3.1.2 Froude, W. (1868): Observations and suggestions on the subject of determining by experiment the resistance of ships: Correspondence with the British Admiralty. In: A. D. Duckworth, The papers of William Froude, The Institution of Naval Architects: London, UK, pp.120–128, 1955.
3.1.3 Reynolds, O. (1883): An experimental investigation of the circumstances which determine whether the motion of water shall be direct or sinuous, and of the law of resistance in parallel channels, Philosophical Transactions, The Royal Society, Vol.174, pp.935–982.
3.1.4 Rayleigh, L. (1892): On the question of the stability of the flow of liquids, Philosophical Magazine, Vol.34, No.206, pp.59–70.
3.1.5 Buckingham, E. (1914): On physically similar systems; illustrations of the use of dimensional equations, Physical Review, Vol.4, No.4, pp.345–376.
3.1.6 Deguchi, T. (1962): Suiri mokei zikken (Dai 1 kou) [Lecture "Hydraulic model experiment (1st lecture)" General remarks], Journal of the Agricultural Engineering Society, Vol.30, No.1, pp.25–29. (in Japanese)
3.1.7 Satoh, K., Igarashi, H., Dohgaki, E., and Nakaoka, R. (1984): A tentative report of chronological tables on civil engineering history of Japan before Meiji era, Papers of the Research Meeting on the Civil Engineering History in Japan, Vol.4, pp.191–197. (in Japanese)

3.1.8 Freeman, J.R. (1929): Hydraulic laboratory practice, American Society of Mechanical Engineers (ASME): New York.

3.1.9 Nagai, S. and Tamai, S. (1962): Kousei bouhatei chokuritubu no katsudou jikken –1/10 mokei ni okeru jikken– [Sliding experiment of upright part of mixed breakwater – experiment in 1/10 model–], Proceedings of Coastal Engineering Research Presentation, Vol.9, pp.127–132. (in Japanese)

3.1.10 Homma, H., Horikawa, K., and Yu, S. (1960): Saiha oyobi kaihinhenkei no hendou ni tsuite [On breaking waves and beach topography changes], Proceedings of the Coastal Engineering Research Presentation, Vol.7, pp.91–99. (in Japanese)

3.1.11 Noda, H. and Ibo, T. (1964): Nami ni yoru kaiteisa no idou genkai to saren no hassei [Limitation of movement of seabed sand due to waves and generation of sand ripple], Proceedings of Coastal Engineering Research Presentation, Vol.11, pp.153–158. (in Japanese)

3.1.12 Homma, H., Horikawa, K., and Kashima, R. (1964): Nami ni yoru fuyuusa ni kansuru kenkyuu [Research on suspended sand caused by waves], Proceedings of Coastal Engineering Research Presentation, Vol.11, pp.159–168. (in Japanese)

3.1.13 Matsunashi, J. and Omi, K. (1964): Nami ni yoru teishitsu no henkei ni tsuite [On deformation of sediment sand caused by waves], Proceedings of the Coastal Engineering Research Presentation, Vol.11, pp.169–174. (in Japanese)

3.1.14 Ishihara, T., Sawaragi, T., and Amano, T. (1958): Hyousa no undoukikou ni kansuru kisoteki kenkyuu (dai 1 pou) [Basic research on the mechanism of sediment transport (1st report)], Proceedings of the Coastal Engineering Research Presentation, Vol.5, pp.65–71. (in Japanese)

3.1.15 Hayami, S., Higuchi, A., and Yoshida, K. (1958): Chouryuu wo fukumu suiri mokei jikken no soujisei ni tsuite [Similarity of hydraulic model experiments including tidal currents], Proceedings of Coastal Engineering Research Presentation, Vol.5, pp.169–173. (in Japanese)

3.1.16 Mikasa, M. (1980): Model testing in soil engineering, Soil Mechanics and Foundation Engineering, Vol.25, No.5, pp.1–2. (in Japanese)

3.1.17 Shibata, T. and Ohta, H. (1980): Model testing in soil engineering, Soil Mechanics and Foundation Engineering, Vol.25, No.5, pp.9–14. (in Japanese)

3.1.18 Terzaghi, K. (1925): Erdbaumechanik auf Bodenphysikalischer Grundlage, Deuticke: Wien.

3.1.19 Rocha, M. (1957): The possibility of solving soil mechanics problems by the use of models, Proceedings of the 4th International Conference on Soil Mechanics and Foundation Engineering (ICSMFE), Vol.1, pp.183–188.

3.1.20 Kagawa, T. (1978): On the similitude in model vibration tests of earth-structures, Proceedings of the Japan Society of Civil Engineers, Vol.275, pp.69–77. (in Japanese)

3.1.21 Kokusho, T. and Iwatate, T. (1979): Scaled model tests and numerical analyses on nonlinear dynamic response of soft grounds, Proceedings of the Japan Society of Civil Engineers, Vol.285, pp.56–67. (in Japanese)

3.1.22 Iai, S. (1989): Similitude for shaking table tests on soil-structure-fluid model in 1*g* gravitational field, Soils and Foundations, Vol.29, No.1, pp.105–118.

3.1.23 Philips, E. (1869): De l'equilibre des solides elastiques semblables, C.R. Acad. Sci.: Paris, Vol.68, pp.75–79.

3.1.24 Taylor, R.N. (1995): Geotechnical centrifuge technology, CRC press.

3.1.25 Pokrovsky, G.I. and Fedorov, I.S. (1936): Studies of soil pressures and soil deformations by means of a centrifuge, Proceedings of the 1st International Conference on Soil Mechanics and Foundation Engineering (ICSMFE), Vol.1, pp.70–71.

3.1.26 Mikasa, M., Takada, N., and Kishimoto, Y. (1965): Enshinryoku souchi niyoru jijyu atsumitsu jikken (Dai 1 pou) [Self-weight consolidation tests using centrifuge facility (1st Report)], Proceedings of the 20th JSCE Annual Meeting, III-25. (in Japanese)

3.1.27 Avgherinos, P.J. and Schofield, A.N. (1969): Drawdown failures of centrifuged models, Proceedings of the 7th International Conference on Soil Mechanics and Foundation Engineering (ICSMFE), Vol.2, pp.497–505.

3.1.28 Mikasa, M., Takada, N., and Mochizuki, A. (1980): Centrifugal model testing of soil structures, Tsuchi to Kiso, Vol.28, No.5, pp.15–23. (in Japanese)
3.1.29 Mikasa, M. and Takada, N. (1966): Enshinryoku souchi niyoru jijyu atsumitsu jikken (Dai 3 pou) [Self- weight consolidation tests using centrifuge facility (3rd Report)], Proceedings of the 21st JSCE Annual Meeting, III, pp.46/1–2. (in Japanese)
3.1.30 Ovesen, N.K. (1979): The scaling law relationship: panel discussion, Proceedings of the 7th European Conference on Soil Mechanics and Foundation Engineering, Vol.4, pp.319–323.
3.1.31 Takada, N. and Kusakabe, O. (1987): Kouza – Enshin mokei jikken – 3. Genri [Lecture course – Centrifuge model test – 3. Principle], Tsuchi to Kiso, Vol.35, No.12, pp.89–94. (in Japanese)

3.2 Law of Similarity for Fluids

3.2.1 Deguchi, T. (1962): Suiri mokei zikken (Dai ikkou) [Lecture "Hydraulic model experiment (1st lecture)" General remarks], Journal of the Agricultural Engineering Society, Vol.30, No.1, pp.25–29. (in Japanese)
3.2.2 Fujima, K. (2003): Development policy and theory of numerical wave channel (CADMAS-SURF), Lectures of the 39th Summer Seminar on Hydraulic Engineering, pp.B.3.1–B.3.16. (in Japanese)
3.2.3 Goda, Y. (1975): Deformation of irregular waves due to depth-controlled wave breaking, Report of Port and Harbour Research Institute, Vol.14, No.3, pp.59–106. (in Japanese)
3.2.4 Kashima, R., Sakakiyama, T., Matsuyama, M., Sekimoto, T., and Kyooya, O. (1992): Antei genkai wo koeru harou ni taisuru shouhakou no henkei to bouha kinou no henka nitsuite [Deformation of wave-dissipating work and change of wave-proof function for waves exceeding the stability limit], Journal of Coastal Engineering, Vol.39, pp.671–675. (in Japanese)
3.2.5 Sakakiyama, T. and Ogasawara, M. (1993): Sentei niyoru shougeki saiharyoku no Teigen to jikken suke-ru kouka [Reduction of impact breaking force by diving dike and experimental scale effect], Proceedings of Coastal Engineering, Vol.40, pp.746–750. (in Japanese)
3.2.6 Takahashi, H., Sassa, S., Morikawa, Y., Takano, D., and Maruyama, K. (2014): Stability of caisson-type breakwater foundation under tsunami-induced seepage, Soils and Foundations, Vol.54, Issue 4, pp.789–805.

3.3 Law of Similarity for Ground

3.3.1 Zienkiewicz, O.C., Chang, C.T., and Bettess, P. (1980): Drained, undrained, consolidating and dynamic behaviour assumptions in soils, Geotechnique, Vol.30, No.4, pp.385–395.
3.3.2 Kagawa, T. (1978): On the similitude in model vibration tests of earth-structures, Proceedings of the Japan Society of Civil Engineers, Vol.275, pp.69–77. (in Japanese)
3.3.3 Iai, S. (1989): Similitude for shaking table tests on soil-structure-fluid model in 1*g* gravitational field, Soils and Foundations, Vol.29, No.1, pp.105–118.
3.3.4 Rocha, M. (1957): The possibility of solving soil mechanics problems by the use of models, Proceedings of the 4th International Conference on Soil Mechanics and Foundation Engineering (ICSMFE), Vol.1, pp.183–188.
3.3.5 Takahashi, H., Sassa, S., Morikawa, Y., Takano, D., and Maruyama, K. (2014): Stability of caisson-type breakwater foundation under tsunami-induced seepage, Soils and Foundations, Vol.54, Issue 4, pp.789–805.
3.3.6 Japanese Geotechnical Society (2004): Fuhouwa jiban no kyodou to hyouka [Behaviour and assessment of unsaturated soils], Japanese Geotechnical Society, p.217 (in Japanese)
3.3.7 Bishop, A.W. (1960): The principle of effective stress, Norwegian Geotechnical Institute, No.32, pp.1–5.

3.3.8 Bishop, A.W. (1961): The measurement of pore pressure in the triaxial test, Pore Pressure and Suction in Soils, Butterworth: London, pp.38–46.
3.3.9 Vanapalli, S.K., Fredlund, D.G., Pufahl, M.D., and Clifton, A.W. (1996): Model for prediction of shear strength with respect to soil suction, Canadian Geotechnical Journal, Vol.33, No.3, pp.379–392.
3.3.10 Mualem, Y. (1976): A new model for predicting the hydraulic conductivity of unsaturated porous media, Water Resources Research., Vol.12, pp.513–522.
3.3.11 van Genuchten, R. (1978): Calculating the unsaturated hydraulic conductivity with a new closed-form analytical model, Research Report, 78-WR-08, Princeton University: Princeton.
3.3.12 van Genuchten, M.Th. (1980): A closed-form equation for predicting the hydraulic conductivity of unsaturated soils, Soil Science Society American Journal, Vol.44, pp.892–898.

3.4 Law of Similarity in Combined Fluid and Ground Model Tests

3.4.1 Emori, I., Saito, K., and Sekimoto, K. (2000): Mokei jikken no riron to ouyou [Theory and Application of Model Experiments –Third Edition–], Gihodo Publishing. (in Japanese)
3.4.2 National Institute of Advanced Industrial Science and Technology, Noda A.: Chinkouhou niyoru ryuushikei sokutei [Particle size measurement by sedimentation method] (in Japanese) https://staff.aist.go.jp/a.noda/memo/settle/settle/settle.html (viewed May 16, 2020)
3.4.3 Suga, K. (1974): Kasen mokei jikken no riron to ouyou [Recent Advances in River Hydraulic Model Experiments], Lectures of the 10th Summer Seminar on Hydraulic Engineering, pp.A.9.1–A.9.22. (in Japanese)
3.4.4 Tsubaki, T. (1974): Suirigaku II [Hydraulics II], Chapter 14 Basic Civil Engineering Encyclopedia 7, Morikita Publishing. (in Japanese)
3.4.5 Fukuoka, S. and Hayashi, M. (1988): Idousyou suiri jikken no soujisoku no teian – hijuu no chiisai kasyou zairyou wo motiita baai – [Proposal of similarity rules for moving bed model experiments – when using riverbed materials with low specific density–], Civil Engineering Note, Vol.30, No.9, pp.469–476. (in Japanese)
3.4.6 Steven A. Hughes (1993): Physical models and laboratory techniques in coastal engineering, World Scientific, 588pp.
3.4.7 Goda, Y. (1990): Kouwan kouzoubutu no taiha sekkei – haroukougaku heno josetsu – [Wave-resistant design of harbor structures – introduction to wave engineering –], Kajima Institute Publishing. (in Japanese)
3.4.8 Suzuki, K. et al. (2002): Settlement failure of wave dissipating blocks in front of caisson type breakwater due to scouring under the rubble mound – field investigation and large scale experiment, Report of Port and Airport Research Institute, Vol.41, No.1, pp.51–89. (in Japanese)
3.4.9 Matsuda, T., Miura, K., Sato, J., Iwayama, K., Sawada, Y. (2017): Water pressure change in movable bed regarding similarity law for geomaterial by Dean Number on wave flume experiment, Journal of Japan Society of Civil Engineers, Ser. B2 (Coastal Engineering), Vol.73, No.2, pp.I_1117–I_1122. (in Japanese)
3.4.10 Arikawa, T., Ikeda, G., and Kubota, K. (2014): Experimental study on scour behind seawall due to tsunami overflow, Journal of Japan Society of Civil Engineers, Ser. B2 (Coastal Engineering), Vol.70, No.2, pp.I_936–I_930. (in Japanese)
3.4.11 Araki, S., Sawada, Y., Miyamoto, J., Ushiyama, H., Tanaka, Y., and Kotake, Y. (2018): Examination of sand scour under rubble mound of a composite breakwater with dissipating blocks in a drum centrifuge, Journal of Japan Society of Civil Engineers, Ser. B2 (Coastal Engineering), Vol.74, No.2, pp.I_1093–I_1098. (in Japanese)
3.4.12 Shimizu, T. (1988): Similarity of model experiment on onshore–offshore beach profile change, Report of Central Research Institute of Electric Power Industry, U87059, 41pp. (in Japanese)

3.4.13 Ito, M. (1990): Time scale for modeling beach change, Journal of Japan Society of Civil Engineers, Vol.423, No. II-14, pp.151–160. (in Japanese)
3.4.14 Ito, T., Nagayama, T., Kaizuka, K., Hayakawa, T., and Watanabe, Y. (2012): Preliminary experimental methods for bed variations using artificial light and heavy materials, Journal of Japan Society of Erosion Control, Vol.64, No.5, pp.3–13. (in Japanese)
3.4.15 Ikeno, M. et al. (2009): Experiments on sediment transport rate due to tsunami and proposal of pickup rate formula of suspended sand, Report of Central Research Institute of Electric Power Industry, V08064, 43pp. (in Japanese)
3.4.16 Dean, R.G. (1973): Heuristic models of sand transport in the surf zone, Proceedings of the Conference on Engineering Dynamics in the Surf Zone, Sydney.
3.4.17 Dean, R.G. (1985): Physical modelling of littoral processes, Physical Modelling in Coastal Engineering, CRC Press, pp.119–139.
3.4.18 Jimenez, J.A. and Madsen, O.S (2003): A simple formula to estimate setting velocity of natural sediments, Journal of Waterway, Port, Coastal, and Ocean Engineering, Vol.129, No.2, pp.70–78.
3.4.19 Sekiguchi, H., Kita, K., and Okamoto, O. (1995): Response of poro-elastoplastic beds to standing waves, Soils and Foundations, Vol.35, No.3, pp.31–42.
3.4.20 Sassa, S. and Sekiguchi, H. (1999): Wave-induced liquefaction of beds of sand in a centrifuge, Géotechnique, Vol.49, Issue 5, pp.621–638.
3.4.21 Zen, K. (1993): Study on the wave-induced liquefaction in seabed, Technical Note of the Port and Harbour Research Institute, No.755, 112pp. (in Japanese)
3.4.22 Miyamoto, J., Sassa, S., Tsurugasaki, K., and Sumida, H. (2020): Wave-induced liquefaction and floatation of a pipeline in a drum centrifuge, Journal of Waterway, Port, Coastal, and Ocean Engineering, ASCE, Vol.146, Issue 2, pp.04019039-1–04019039-12.
3.4.23 Takahashi, H.and Morikawa, Y. (2017): Centrifuge model tests examining stability of seawalls subjected to high waves, Proceedings of the 19th International Conference on Soil Mechanics and Geotechnical Engineering, pp.971–974.
3.4.24 Takahashi, H., Morikawa, Y., and Kashima, H. (2019): Centrifuge modelling of breaking waves and seashore ground, International Journal of Physical Modelling in Geotechnics, Vol.19, Issue 3, pp.115–127.

4

Case Study of Model Experiments

Susumu Araki, Kojiro Suzuki, Junji Miyamoto, Takenori Shimozono, Fuminori Kato, Takayuki Suzuki, and Kenichi Maeda

4.1 Viewpoint Concerning the Case Study

From Chapters 1-3, we described the method for examining the issues regarding fluid and ground using model experiments, challenges in examining issues that compound fluid and ground by model experiments, and the law of similarity required to interpret the experimental result. This chapter is aimed at improving the understanding of the law of similarity through specific study cases in which examination regarding the combined issue of fluid and ground is made on the basis of the law of similarity. Although you may find it difficult to understand the law of similarity, to understand the actual phenomenon using the result of model experiments, the law of similarity should be fully understood since it presents the relationship between the phenomenon measured in the experiment and the actual phenomenon.

This chapter describes two case studies regarding complex phenomena of fluid and ground that are the most significant phenomena covered in this book. The first phenomenon is the subsidence of wave-dissipating work due to scouring under the permeable structure composed of rubble stones and wave-dissipating blocks; the second phenomenon is the formation of beach scarp as a result of erosion of sand for beach nourishment in the form of banking.

The former case is related to the scouring phenomenon observed below the permeable structure described in Section 3.4.3. Its main subject is suspended sediment transport that causes topographical change by the movement of sand below the permeable structure rolled up by wave motion and current in the water. The conditions necessary to reproduce the subsidence of the wave-dissipating work associated with this by experiments in different scales are examined. In this examination case, the status of reproduction of the scouring phenomenon is summarized in terms of the shape and progress rate of scouring by the law of similarity of fall speed using Dean number (Dean's law). Additionally, we will focus on the turbulence within the rubble mound as a factor in reproducing the scouring phenomenon. The latter case is related to the combined experiment regarding waves and ground discussed in Section 3.4.4. Its subject is the beach scarp formed as a result of erosion of sand for beach nourishment in the form of banking located in the proximity of the shoreline caused by run-up waves. Bed load sediment transport in which sand particles move on the seabed surface in rolling or jumping motion is examined as the main cause of the topographical change. Regarding

DOI: 10.1201/9781003424758-4

examples that require the application of the law of similarity that compounds fluid and ground, note that in beach nourishment filling above the run-up height of the waves, the law of similarity has to consider the characteristics of unsaturated ground.

In the former example, not only gravity model tests but also centrifuge model tests are conducted. Regarding the centrifuge model test, it may be difficult to imagine its experimental approach and the possible examination it brings or how to interpret its results, but since the general experimental approach of centrifuge model test is described in Chapter 2, along with this, we expect that the presented examination cases will help advance the understanding. For the interpretation of the results of the centrifuge model test, as Froude's law of similarity for centrifuge model test described in the previous chapter is also applied, further understanding of the law of similarity regarding centrifugal force field is expected as well.

4.2 Case (1) Scouring Phenomenon under Wave-dissipating Blocks

Numberless wave-dissipating blocks are installed throughout the world as a means to cover breakwaters and revetments. Generally, the size (required mass) of wave-dissipating blocks is determined by wave heights using formulas such as the Hudson formula. Nevertheless, a specific design technique regarding scouring of sandy ground that supports wave-dissipating blocks is not established. There are cases where waves smaller than the height determined by the stability evaluation of caisson and wave-dissipating blocks cause the wave-dissipating blocks to sink, and as a result, blocks that were above the water sink beneath sea level. It has become apparent that this is caused by wave-induced scouring of sandy ground at the seabed on which blocks are installed, resulting in block subsidence. Here, we will discuss the applicability of the law of similarity to the sandy ground in movable bed experiment using an experiment to reproduce this phenomenon.

4.2.1 Subsidence Phenomena of Wave-dissipating Blocks

An example of subsidence damage of blocks of breakwater covered with wave-dissipating blocks is shown in Figure 4.2.1. It was damaged by waves 6.3 m high and a wave period of 10.4 s, which are below the design wave (wave height of 11 m and wave period of 16 s).[4.2.1] This is an example in which scouring of the sandy ground under the wave-dissipating block is considered to be the main cause of subsidence of blocks. When waves act on structures in which stones and blocks are installed on sandy ground, sand will drift from the gaps between the stones and blocks into the sea and a so-called scouring phenomenon under the permeable structure occurs. Subsidence of breakwaters, detached breakwaters, and submerged breakwaters covered with wave-dissipating blocks are often caused by this scouring phenomenon.

An example of subsidence of wave-dissipating blocks of a detached breakwater at Niigata Nishikaigan is shown in Figure 4.2.2. This is an example considered to be due to sandy ground under blocks being worn away because of scouring, and the measurements were taken using cross-hole seismic survey method.[4.2.2] After construction, new wave-dissipating blocks had been added as they sank, and consequently, several blocks

FIGURE 4.2.1
Subsidence failure of blocks of a breakwater covered with wave-dissipating blocks.[4.2.1]

sank into the sandy ground. You can see that the number of wave-dissipating blocks that have sunk into the sandy ground is far greater than those above the sandy ground. Figure 4.2.3 is an example of subsidence of blocks of breakwater covered with wave-dissipating blocks at Kochi Port, and it shows the subsidence condition of the mound. This was found when blocks were removed in association with the extension of the breakwater. The mound has sunk because the sand under the mound got scoured, and consequently, the wave-dissipating blocks installed on the mound have sunk.[4.2.1]

Such subsidence of wave-dissipating blocks not only impairs the function of wave-dissipating blocks in reducing wave force but can also produce impulsive breaking wave force that can possibly damage the whole breakwater. Thus, it is necessary to take measures such as adding blocks after subsidence. Nevertheless, the amount of damage is huge: breakwaters covered with wave-dissipating blocks cost tens of millions to hundreds of millions of yen each. If you take an extensive view of our country, there are revetments, detached breakwaters, submerged breakwaters, etc. covered with wave-dissipating blocks, and one can imagine that it requires considerable costs.

Nevertheless, the cause of these subsidence phenomena had not been elucidated for many years. This was because the scouring phenomenon of sandy ground under permeable structure (stones and blocks) as shown in Figures 4.2.2 and 4.2.3 was difficult to confirm even with a submersible survey performed by divers. It is a contrast to scouring phenomena that occur as a result of flow acting directly on sandy ground, such as L-type scouring in front of breakwaters, whose condition can easily be confirmed with submersible surveys by divers or bathymetric surveys.

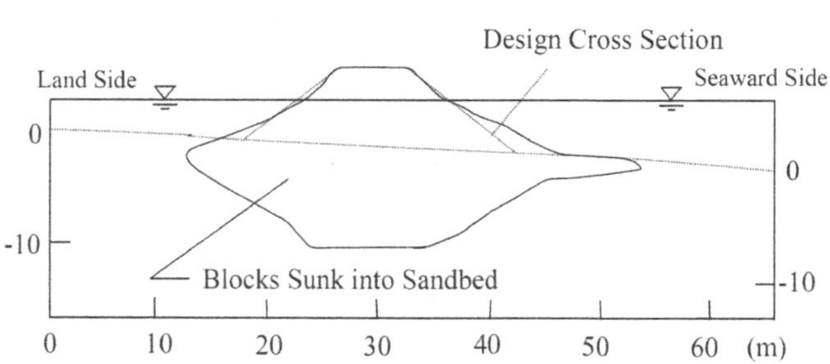

FIGURE 4.2.2
Subsidence of blocks due to scouring of the sandbed under detached breakwaters.[4.2.2]

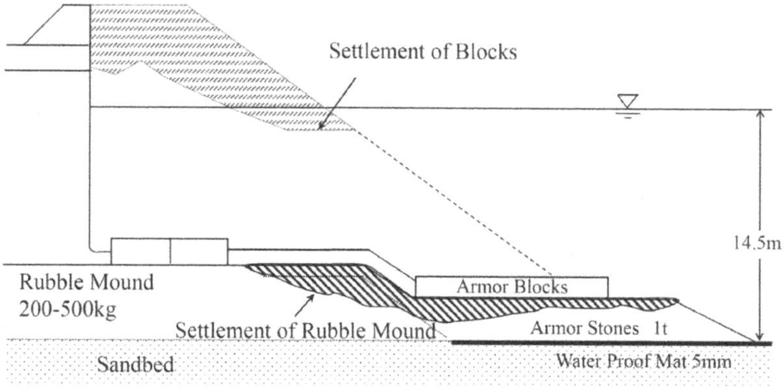

FIGURE 4.2.3
Subsidence of rubble mound and wave-dissipating blocks observed when blocks were temporarily removed.[4.2.1]

The scouring phenomena under permeable structure may be confirmed via a movable bed experiment. For phenomena in which flow acts directly on sandy ground, scouring can be reproduced to some extent by a relatively small-scale experiment using the same sand material as the site. Nonetheless, since flow velocity is low under permeable structure, it has not been confirmed with small-scale movable bed experiments. Additionally, in several movable bed experiments conducted in the past, the law of similarity of suspended sediment was not elucidated and the sand of the same material as the site was often used. This is another factor that has been hindering the clarification of the scouring phenomenon under a permeable structure. For example, in the past experiment on breakwater covered with wave-dissipating blocks,[4.2.3] the sedimentation phenomenon, which is the opposite of scour phenomenon, occurred because the scale of the model was small and the sand of the same scale as to the site was used.

4.2.2 Method of Reproducing Scouring Phenomenon with Small-to-Medium-Scale Model Experiments

If scouring can be reproduced with relatively small-scale hydraulics experiments, modeling can be relatively easy and experiments can be conducted efficiently. Here, we will introduce movable bed hydraulics model experiments of several scales one by one in subsequent sections and compare them with existing large-scale movable bed experiments with aim to establish the method of reproducing scouring phenomenon with small- and medium-scale experiments. Experiments are conducted with several scales and geomaterials to examine the validity of the law of similarity in existing movable bed experiments, the applicability of centrifuge model experiments to movable bed experiments, reproducibility of scouring in small-scale experiments, etc. Especially, the reproducibility of the scouring phenomenon in centrifuge model experiments has been unexplained; therefore, this is one of the focus points of this report.

The experiment was conducted using a super-large-scale experiment on breakwater covered with wave-dissipating blocks[4.2.1] and a large-scale experiment conducted in a flume of 105 m[4.2.4] for comparison. Table 4.2.1 provides scales of each experiment. The

TABLE 4.2.1

Each Experimental Scale and Materials Used in the Experiments

	Gravity (g)	For Actual Structure (Scale)	For Large-scale Model (Scale)	Number of Waves	Tetrapod (kg)	Rubble Mound D_{50} (mm)	Sand Particle d_{50} (mm)	Scour Depth (Model Scale) (mm)	Remarks
Actual Structure	1	1			64,000	425	0.14–0.2		Miyazaki Port
Super Large Model	1	1/4	1	2,000	500	73	0.2	750	PARI (Port and Airport Research Institute) 4.2.4
Large Model	1	1/16	1/4	3000	4.3	23	0.08–0.3	0-200	PARI 4.2.3
Middle Size -Lab A-	1	1/55.8	1/14	2,000	0.256	10.9	0.15	0	Nagoya University 4.2.5
				100,000				45	
Middle Size -Lab B-	1	1/55.8	1/14	3,200	0.184	10	0.11	44	Fudo Tetra Corporation 4.2.6
Middle-Small Size –Lab C-	1	1/100	1/25	10,000	0.032	4.6	0.11	0	Chuo University 4.2.7
Centrifuge Model -Lab D-	38	1/152	1/38	500	0.032	4.75-9.5	0.02	36	Toyo Construction Co. Ltd. 4.2.8
				500		4.75-9.5	0.15	0	
				2,000		9.5-19.5	0.15	0	

scales in comparison with the super-large-scale experiment are 1/4, 1/14, 1/25, and 1/38, respectively. Also, the super-large-scale experiment was conducted with a 1/4 scale of the site (the south breakwater at Miyazaki Port) and the scales were 1/55.8, 1/55.8, 1/100, and 1/152 of the site, respectively.

4.2.3 Mechanism of Subsidence of Blocks Based on a Large-scale Experiment (1/16 Scale of the Site)

This section describes the mechanism of the subsidence phenomenon of blocks on the basis of a large-scale experiment that is approximately 1/16 scale of the site. Movable bed experiment and fixed bed experiment were executed[4.2.4] with regular wave whose depth is 1.0 m and height is 0.20–0.55 m. The experiments were conducted in a large wave flume that is 105 m long, 0.8 m wide (overall width is 3 m), and 2.5 m deep, followed by a super-large-scale experiment on a scale of 1/4 (Figure 4.2.4). Figure 4.2.4 is a cross section of the experiment. In the movable bed experiment, sand with mean particle size (d_{50}) of 0.08, 0.15, and 0.3 mm was used as sandy ground and the mean diameter of crushed stones in the mound and stones in a gabion (D_{50}) was 23 mm (including rubble of 7.2 mm). Additionally, the mass of the wave-dissipating blocks used was 4.3 kg.

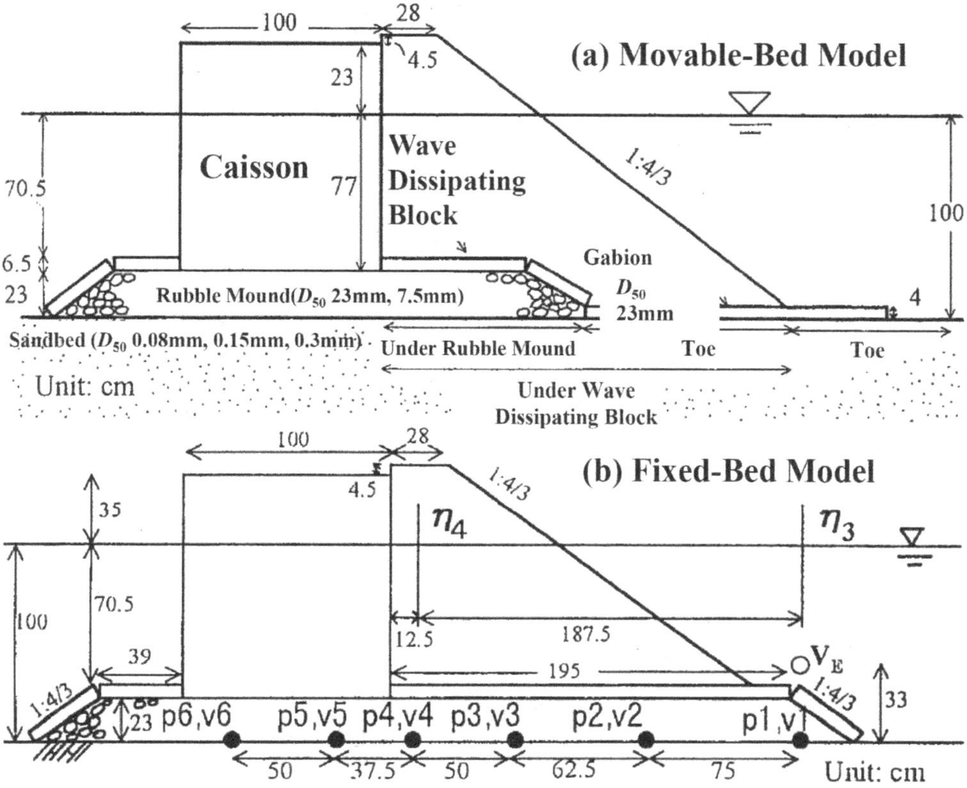

FIGURE 4.2.4
Cross section of the large-scale experiment.[4.2.4] (a) Movable-Bed Model and (b) Fixed-Bed Mode.

FIGURE 4.2.5
Sand suspension situation. *Left:* Wave trough. *Right:* Wave crest.[4.2.4]

4.2.3.1 Generation Status of Scouring

Figure 4.2.5 shows the suspension of sand when a regular wave whose wave period is 3.0 s (12 s on-site) and wave height is 55 cm (8.8 m on-site) is applied. When waves act on the structure, sand stirs up under the wave-dissipating blocks, and the backwash creates an upward flow at the foot of the slope of the wave-dissipating blocks, causing dense lumps of sand to drift up to the surface of the water. After the wave breaks at the slope of the wave-dissipating blocks when the next wave comes, overtopping water mass is carried behind the caisson. At this point, sand that has been floating is carried behind the caisson along with the overtopping water mass. When waves break at the slope of wave-dissipating blocks, the sand that floated up in the mound will gradually move out to the sea since the duration of flow velocity within the mound during backwash is longer than the duration during incoming waves.

Figure 4.2.6 shows the final scouring situation; the sand under the mound under the block is scoured, and the blocks have sunk. The amount of scouring outside of the wave-dissipating blocks that can be confirmed by the submersible survey is smaller than the amount of scouring inside of (under) the wave-dissipating blocks, and you can see that the wave-dissipating blocks sank due to this scouring. The reason why the cause of subsidence of wave-dissipating blocks could not be clarified in the past was that it was difficult to directly measure the phenomenon occurring within the blocks. Note that although scouring has been observed under caisson, it has not sunk since the amount of

FIGURE 4.2.6
Scoured sand bed and the subsidence of wave-dissipating blocks at the end of the experiment.[4.2.4]

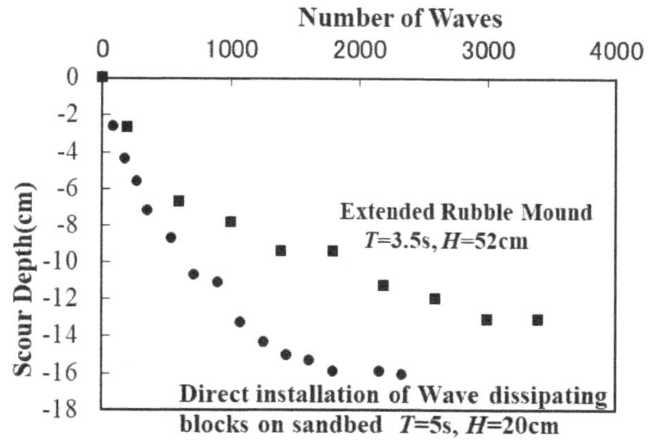

FIGURE 4.2.7
Relationship between the number of waves and scour depth.[4.2.4]

scouring at the front end of the caisson is small. Nevertheless, the stones of the mound under the caisson have sunk and the sliding resistance of caisson may become lower.

Figure 4.2.7 represents the time series of scour at the deepest scour location. The speed of scouring decreases as the number of wave actions increases and the scouring calms down at approximately 3,000 waves. The number of waves until scouring calms down depends on the differences of the waves and cross section, but it stabilizes at approximately 2,000–3,000 waves. At sites where wave-dissipating blocks have sunk, the subsidence observed is relatively minor after new blocks have been added for restoration. This is considered to be because such site has reached a stabilized state after getting much scouring. However, scouring may continue and result in subsidence of blocks if the site has not reached the stabilized state or is hit by larger waves.

4.2.3.2 Characteristics of Flow Within the Mound

Figure 4.2.8 shows the time-series data of the water level η, pressure p, and horizontal velocity V measured using a wave gauge installed in the front part of the caisson and a propeller current meter, and a pore pressure gauge installed in the mound. The hydraulic gradient i obtained from the difference of pressure is also shown along with the velocity. Figure 4.2.4 shows the installation position of each sensor. As you can see from velocity data V_2 and V_3, the duration time of flow toward the sea (under wave trough) is longer than that of the flow toward the shore (under wave crest). This time-series data show that sand is stirred up by incoming waves with high flow velocity and then gradually carried out to sea by the flow toward the sea (under wave trough) with long duration, causing scouring of sand within (under) the wave-dissipating blocks. Additionally, it shows that the flow within the mound is proportional to the hydraulic gradient.

4.2.3.3 Relationship between Wave Conditions and Scouring

We would like to focus on the relationship between wave conditions and scouring. Figure 4.2.9 shows the deformation of sandy ground and subsidence condition of wave-dissipating blocks in a cross section in which wave-dissipating blocks are installed

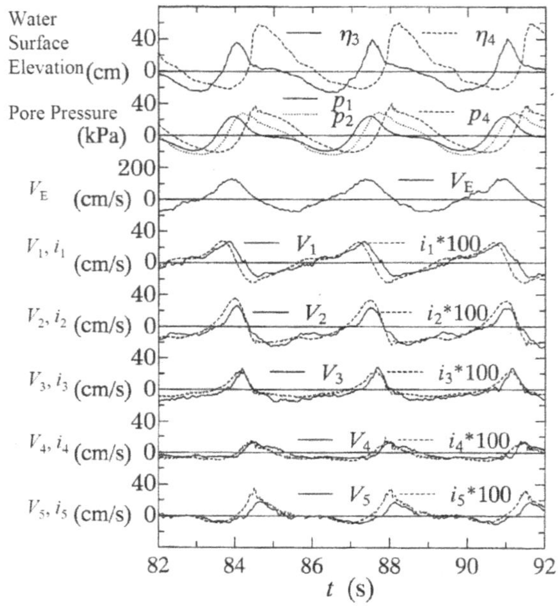

FIGURE 4.2.8
Time series of flow velocity, pressure, and hydraulic gradient inside the rubble mound.[4.2.4]

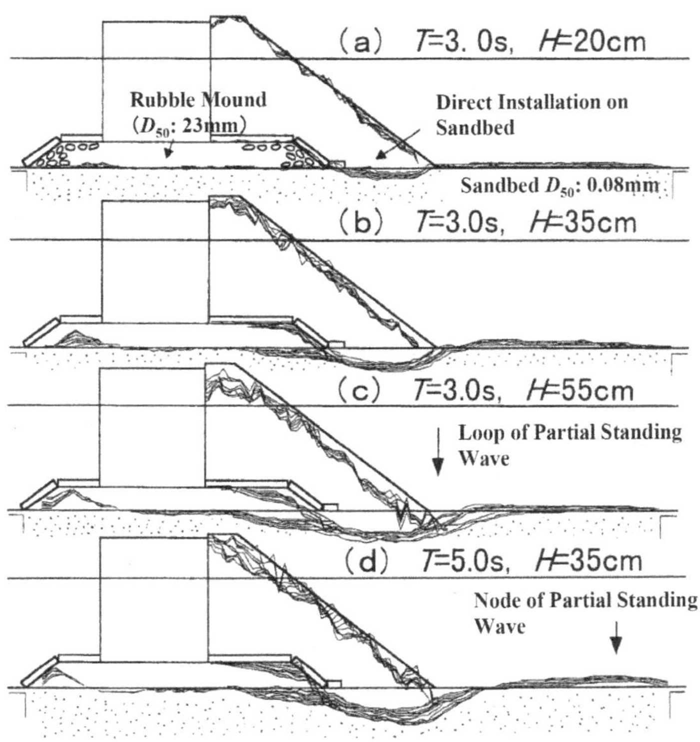

FIGURE 4.2.9
Scour under multiple wave conditions.[4.2.4] (a) T=3.0s, H=20cm, (b) T=3.0s, H=35cm, (c) T=3.0s, H=55cm, and (d) T=5.0s, H=35cm.

directly on the sandy ground (direct installation cross section). The sandy ground was composed of sand with a mean particle size (d_{50}) of 0.08 mm and regular waves with a wave period of 3.0 s were applied, starting with small waves with a height of 10 cm. Once the movement of sand was stabilized, larger waves (20, 35, and 55 cm) were applied. Naturally, larger waves caused a greater amount of scouring. The amount of scouring at the foot of the slope of the wave-dissipating blocks was especially large. With a wave height of 55 cm, sand was drawn out even from under the mound, increasing the amount of subsidence of wave-dissipating blocks.

Figure 4.2.9d shows the result with a wave period of 5.0 s and wave height of 35 cm. The amount of scouring is the same as that caused by waves with a wave period of 3.0 s and wave height of 55 cm shown in Figure 4.2.9c, suggesting that the longer the wave period, the larger the amount of scouring. The position of the sediment behind the caisson is closer to the shore than in the case of a wave period of 3.0 s since the water movement in the mound becomes larger with a longer wave period. Conversely, scouring was not observed with a wave period of 1.5 s even with waves that exceed the breaking wave height.

The moving distance of water particles becomes longer when the wave height is larger, or the wave period is longer. A phenomenon in which the amount of scouring becomes larger in step with this is similar to the phenomenon where the amount of scouring around circular cylinders increases in proportion to *KC* number, and this is distinctive to scouring.

4.2.3.4 Relationship of Particle Size of Sandbed and Diameter of Stones of the Mound with Scouring

Series of experiments were conducted with three different sandbed (d_{50} = 0.08, 0.15, and 0.3 mm). Consequently, the scouring amount decreased with larger particle size, and no scouring was observed with a diameter of 0.3 mm. As shown, conducting experiments with sand of large particle size creates a problem in which scouring cannot be reproduced. Nevertheless, a material that has a particle size smaller than 0.074 mm is silt and its property differs from that of sand. Nevertheless, since the particle size of sand at the site is 0.1–0.3 mm in general, the sand that is approximately the same particle size as the site has to be used in the experiment. Thus, it is necessary to decide the particle size of the sand used in the experiment according to the law of similarity given in Chapter 3. Additionally, regarding experiments in which the diameter of mound stones was changed to 7.2 and 23 mm, scouring did not occur in the experiment using smaller mound stones. Rubbles of small particle size function as a filter and prevent sand from being drawn out.

Scour prevention geotextile sheets or scour prevention mats may be laid under the mound in the general breakwater. In the series of experiments, it was shown that scouring does not occur in parts where scour prevention sheets or mats are laid, but the parts without the sheets or mats get scoured. There is no problem when scour prevention sheets or mats are laid under the entire breakwater, but in most cases, they are only laid under the toe of the slope or foot of the slope of the block. Additionally, since the installation of sand invasion prevention sheets or mats is difficult in the open sea, a gap can be formed in the joints between sheets or mats and scouring can expand from the gap. Therefore, it is important to use scour prevention sheet or mat with sufficient strength and to install them properly without making any gaps.

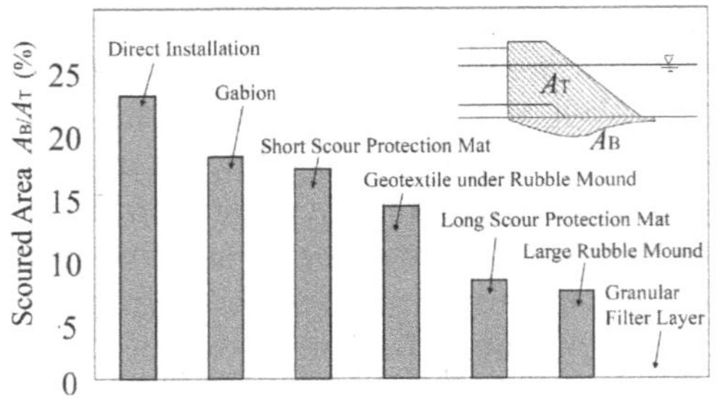

FIGURE 4.2.10
Effects of multiple scour protection works.[4.2.4]

4.2.3.5 Comparison between Scouring of Each Cross Section

Figure 4.2.10 shows a comparison of the amount of scouring between each cross section. The value shown is the cross-sectional area of scoured sandbed (represented by A_B) divided by the cross-sectional area of the mound in front of the caisson and the wave-dissipating block (represented by A_T). When compared with wave-dissipating blocks that are installed directly, the amount of scouring decreases when scouring prevention work such as installation of gabion is carried out, and the amount of scouring also decreases to approximately 8% with thicker rubble mounds or when mats are installed. Nevertheless, in consideration of blocks becoming disengaged by subsidence of blocks due to scouring and can cause scattering or breakage, 8% is no small value.

4.2.4 Reproduction of Subsidence of Blocks by Super-large-scale Experiment (1/4 Scale of the Site)

4.2.4.1 Experimental Flume and Experiment Condition

In 2001, a super-large-scale experiment was conducted at PARI's Large Hydro-Geo Flume (length: 183 m, width: 3.5 m, depth: 12 m; sandy ground with a depth of 4.0 m in the center) with a water depth of 4.0 m and a wave height of 2.5 m.[4.2.4] The experiment was conducted with a water depth of 4.0 m and a maximum wave height of 2.5 m. Regarding the scale of the experiment, if the revetment or breakwater is installed in a position where the water depth is 4.0 m, the scale is equivalent to 1/1 of the site, and if such position has a water depth of 16 m, the scale is equivalent to 1/4 of the site.

The main purposes of this experiment are as follows: (1) to solve the problem regarding scale (confirm if the same phenomenon can be observed by using sand only since the mixture of sand and silt of d_{50} = 0.08 mm was used in large-scale experiments[4.2.2]) and (2) to solve the problem regarding liquefaction (confirm if subsidence of blocks due to liquefaction can be observed using a scale of 1/1 to 1/4, since liquefaction of sandbed did not occur in the large-scale experiments).

Three types of experiments were conducted: (1) fixed bed experiment (an experiment to check the velocity within the mound, pore pressure inside seabed, and changes in water level in front of caisson), (2) reproductive experiment using a movable bed (experiment to reproduce the subsidence of blocks), and (3) experiment of countermeasure using movable bed experiment (experiment regarding scour prevention method that uses finely crushed stones).[4.2.2] Here, we will explain the reproductive experiment of (2).

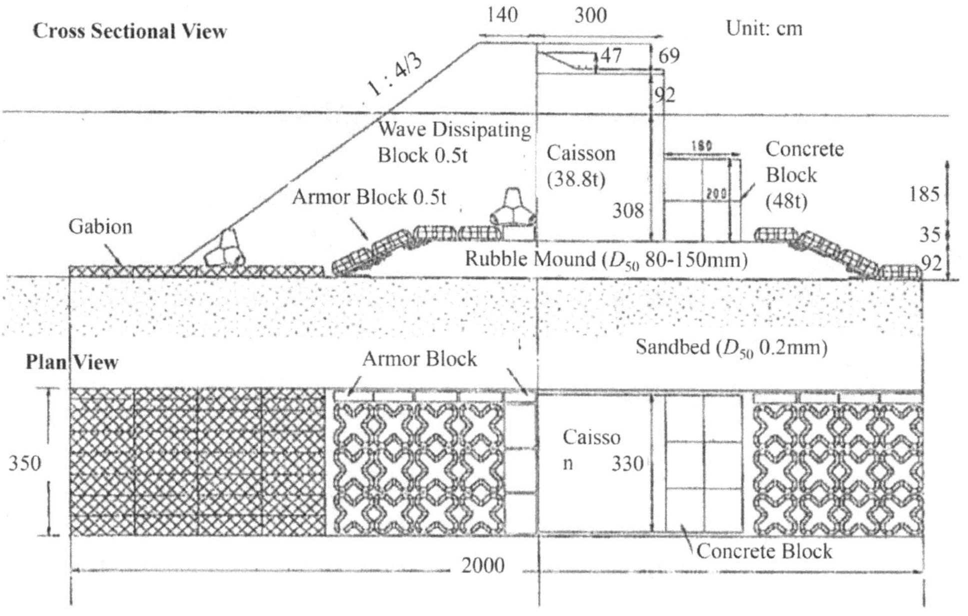

FIGURE 4.2.11
Cross section of the super-large-scale experiment.[4.2.1]

4.2.4.2 Reproductive Experiment Using a Movable Bed

Figure 4.2.11 shows a cross-sectional view of the experiment where gabion is used as a way for preventing scouring so that the subsidence of blocks can occur easily. Rubbles in the gabion are $D_{50} = 73$ mm, which is the same as the stones in the mound. The wave height was gradually increased and experiments with waves up to a maximum wave height of 2.5 m were conducted as shown in Table 4.2.2. Wave-making time was extended until the deformation of the ground stabilizes for the condition of each wave.

Figure 4.2.12 shows the action of a wave with a period of 5.0 s and a height of 2.5 m. Wave periods were 3.0 s, 5.0 s, and 7.0 s, and the wave breaking from offshore was observed in all cases when the wave height was 2.0 m. Additionally, with a wave period of 7.0 s, the wave could break at the loop of a partially standing wave when the wave

TABLE 4.2.2
Experimental Wave Conditions[4.2.1]

Wave Period (s)	Wave Height (m)	Number of Waves	Wave Period (s)	Wave Height (m)	Number of Waves	Wave Period (s)	Wave Height (m)	Number of Waves
3.0	0.5	200	5.0	0.5	200	7.0	0.5	200
	1.0	200		1.0	200		1.0	200
	1.5	200		1.5	200		1.5	200
	2.0	200		2.0	200		2.0	200
				2.5	200		2.5	2,200

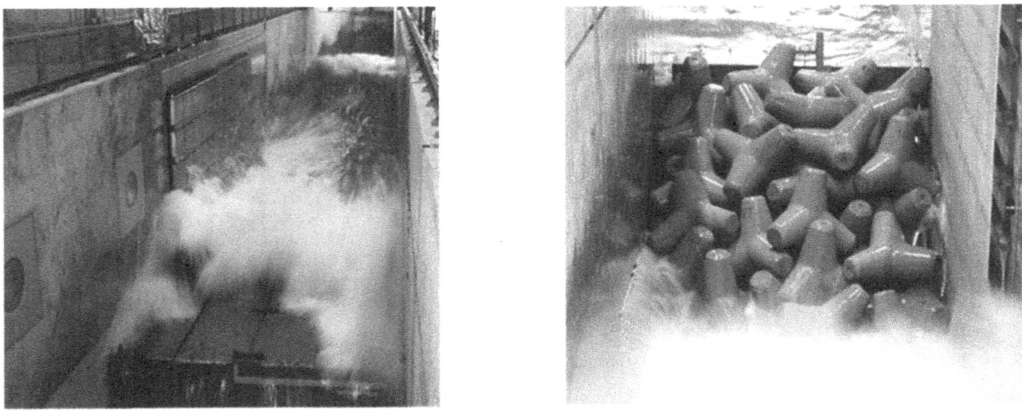

FIGURE 4.2.12
Wave action on a breakwater model in the super-large-scale experiment.[4.2.1]

height is 1.5 m. The suspension of sand got intense as the wave size increased and the water in front of wave-dissipating blocks became cloudy due to suspended sediment. Figure 4.2.13 shows the status of disturbance of sand under the mound under wave-dissipating blocks with a wave period of 7.0 s and a wave height of 1.5 m. As shown, a rapid current is generated within the rubble stones in the mound and sand is stirring up intensely because of the vortex generated behind the stones. Figure 4.2.14 shows the travel distance of water particles O_D measured from the acrylic observation window located on the sidewall of the flume near the foot of the slope of the wave-dissipating blocks. The amount of water movement is shown to increase as the wave period extends, and the movement gets larger than the width of the observation window (2.5 m) when the wave period is 7.0 s. Sand stirred up by such water movement was drawn from under the mound and from the gabion and floated up near the still water surface. Some of them were carried to the back of the caisson by wave overtopping.

Thus, the sand under the mound was gradually drawn out, and when a wave with a wave period of 7 s and a wave height of 2.5 m was applied, the mound sank by 0.75 m and the wave-dissipating block sank by 1.4 m, as shown in Figure 4.2.15. This means that the blocks sank by 5.6 m in a breakwater located at a position with a water depth of 16 m. Additionally, sand is drawn from under the caisson as well and the caisson sunk at least 0.1 m.

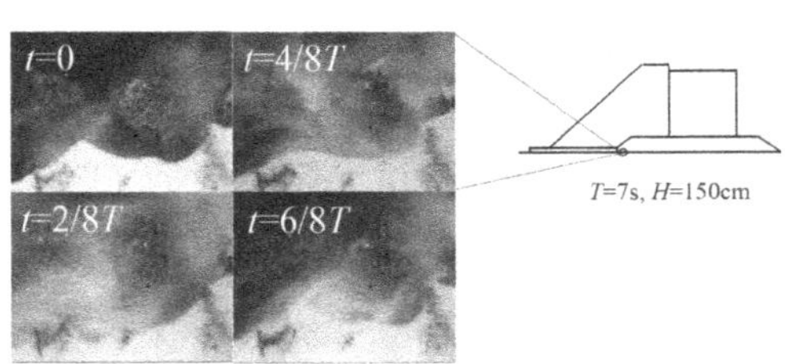

FIGURE 4.2.13
Suspension of sand inside rubble mound in the super-large scale experiment.[4.2.1]

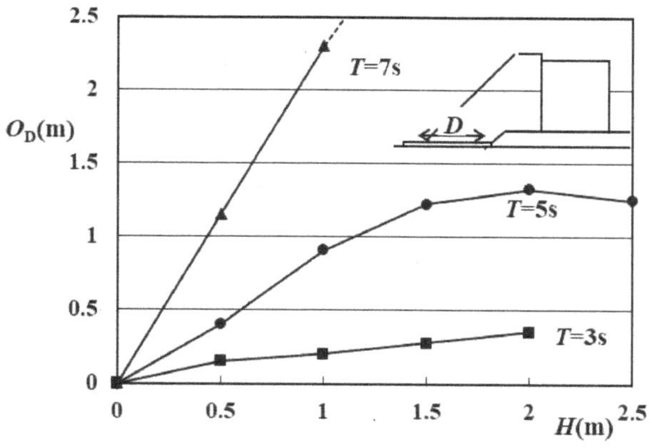

FIGURE 4.2.14
Travel distance of water particle.[4.2.1]

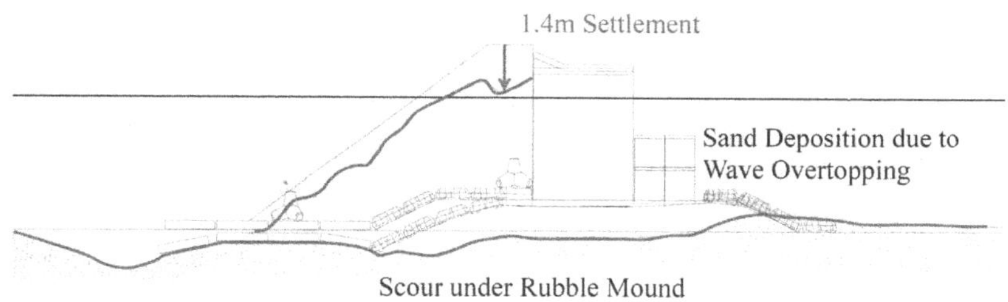

FIGURE 4.2.15
Scour of sandbed and subsidence of wave-dissipating blocks at the end of the experiment.[4.2.1]

Figure 4.2.16 shows the initial cross section and the condition after completion of the experiment. You can see that the blocks have sunk without changing their position.

4.2.5 Small-to-medium-scale Model Experiment (1/14 of Super-large-scale Experiment and 1/55.8 of the Site): Institution A

4.2.5.1 Overview of the Experiment

The experiment to reproduce the scouring phenomenon under the mound using a small-to-medium-size flume[4.2.5, 4.2.6] is as follows. This experiment is a 1/14 scale model of the super-large-scale experiment (see Section 4.2.4). Details regarding the experimental flume and the model parts are shown in Figure 4.2.17. The size of the flume was 30 m long, 0.7 m wide, and 0.9 m high. The sand material used for the sandbed was silica sand No. 6 (Toyo Silica Sand No. 6-2) with $d_{50} = 0.15$ mm. Mound material was gravel material (Mino-Shirakawa, size 3) with $D_{50} = 11$ mm. The mass of the wave-dissipating block was 184.3 g and the mass of the covering block was 188.7 g.

Sand of $d_{50} = 0.2$ mm was used in the super-large-scale experiment, but when Froude's law (Section 3.4.1) is applied to fall speed, it is calculated to be $d_{50} = 0.088$ mm for the scale of this small-to-medium-scale model experiment. Thus, from the aspect of scale, sand that is larger in size (approximately twice the size) is used in this experiment.

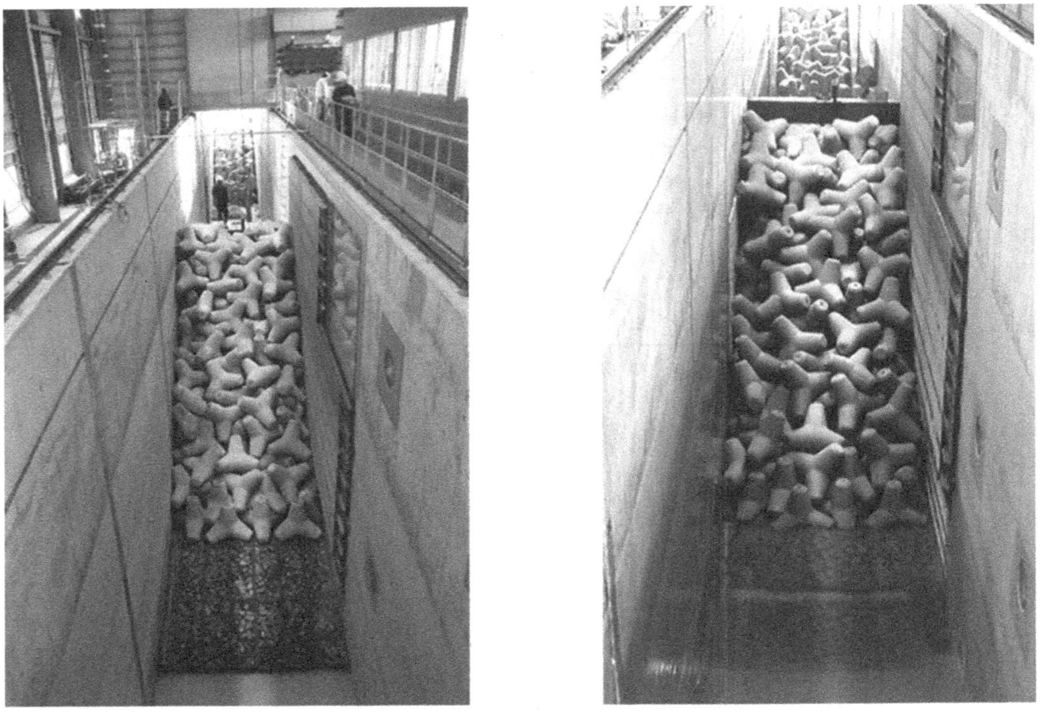

FIGURE 4.2.16
Initial cross section (*left*) and the cross section at the end (*right*) of the experiment.[4.2.1]

Additionally, Rubey's formula[4.2.7] is used to calculate the fall speed. As for the gravel material, the gravel used is also approximately twice the size from the aspect of the scale compared with the particle size of 5.26 mm in the super-large-scale experiment. The particle size ratio (mound material D_{50}/geomaterial d_{50}) is approximately 1/5 of that of the super-large-scale experiment. As you can see from the wave condition shown in Table 4.2.3, in the final wave, 100,000 incident waves that are the same as the super-large-scale experiment were applied.

4.2.5.2 Fabrication of the Ground

The fabrication method detailed by Nakamura et al.[4.2.5] and Muraoka et al.[4.2.6] is explained below. The procedure is as follows:

1. Install the measuring instrument within the ground in advance and fill the flume with water.
2. Put sand by water pluviation method. At this time, sand should be put in carefully and gradually to prevent the incorporation of air into the ground.
3. Apply waves, starting from small waves to large waves, to compact the ground.
4. Fabricate the mound using graving material on top of the sandy ground. Install caisson and blocks on the mound and then place wave-dissipating blocks.

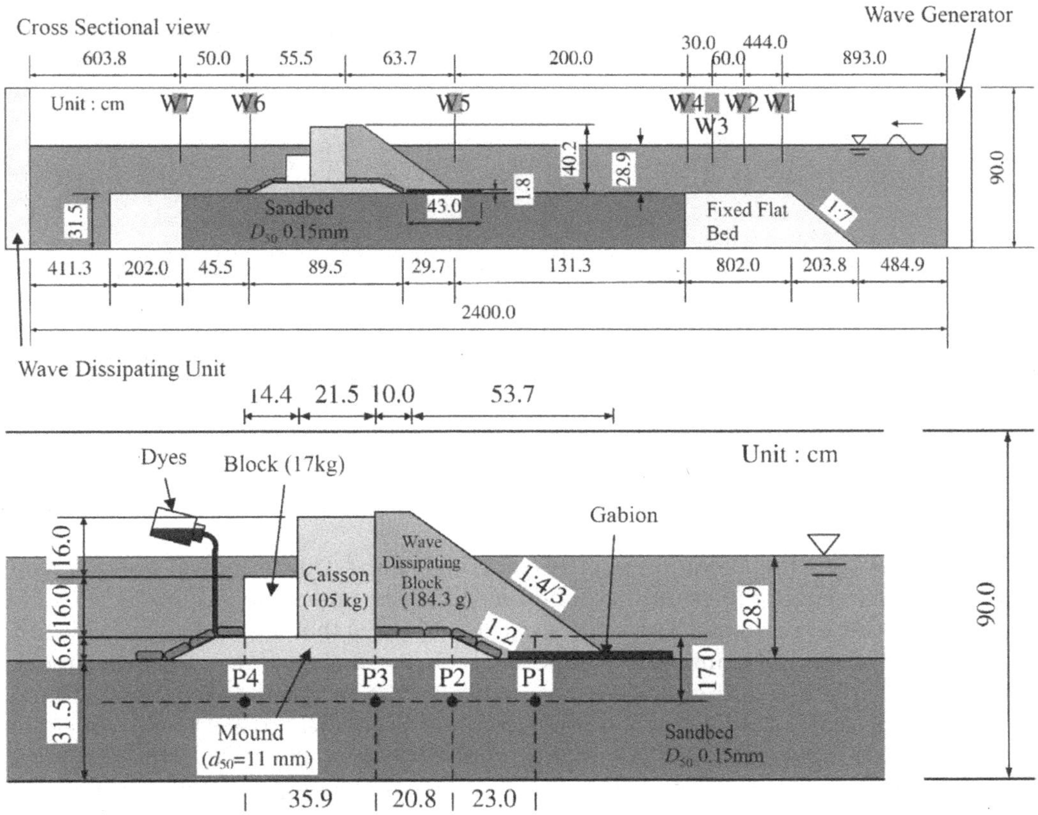

FIGURE 4.2.17
Cross section of the experimental wave flume (*top*) and the detail of model section (*bottom*).[4.2.6]

The ground fabricated in this method is shown in Figure 4.2.18. If you look at it in detail, you can see that thin ground layers are sequentially deposited, creating a uniform ground. Such thin sequential layers cannot be created if a large amount of sand is put at once during the fabrication of the ground. Thus, in this experiment, the sandy ground is fabricated by water pluviation in which sand is poured carefully and gradually in small amounts.

TABLE 4.2.3
Experimental Wave Conditions

Wave Period (s)	Wave Height (cm)	Number of Waves	Wave Period (s)	Wave Height (cm)	Number of Waves	Wave Period (s)	Wave Height (m)	Number of Waves
0.80	3.6	200	1.34	3.6	200	1.87	3.6	200
	7.2	200		7.2	200		7.2	200
	10.8	200		10.8	200		10.8	200
	14.3	200		14.3	200		14.3	200
				17.9	200		17.9	100,000

FIGURE 4.2.18
Initial condition of sandbed: A series of thin layers of sand deposited by careful underwater drops.[4.2.6]

4.2.5.3 Overview of Experimental Result

Figure 4.2.19 shows the alteration of cross-sectional shape caused by wave action. You can see that scouring of the sandy ground has occurred under the mound as with the super-large-scale experiment. The interesting point here is that it took longer for scouring to reach an equilibrium state compared with the super-large-scale experiment. Specifically, while it required 2,000 waves to scour 3 m of ground under the mound (four times the scale of the site when converted) in the super-large-scale experiment, it required 100,000 waves to scour 2.5 m of ground (55.8 times the scale of the site when converted) in this medium-scale experiment. Although the experimental model and specifications of the wave in this experiment were determined based on Froude's law, this result implies that the timescale for the scouring phenomenon at the sandy ground under the mound may be larger than the timescale based on Froude's law. Further research on this point is expected.

4.2.6 Small-to-medium-scale Model Experiment (1/14 of Super-large-scale Experiment and 1/55.8 of the Site): Institution B

4.2.6.1 Overview of the Experiment

This experiment is a 1/14 scale model of the super-large-scale experiment. Details regarding the experimental flume and the model parts are shown in Figure 4.2.20. The size of the flume was 29 m long, 0.5 m wide, and 1 m high. Silica sand No. 8 (Tohoku Keisya No. 8) with $d_{50} = 0.11$ mm was used for the sandy ground. Mound material was

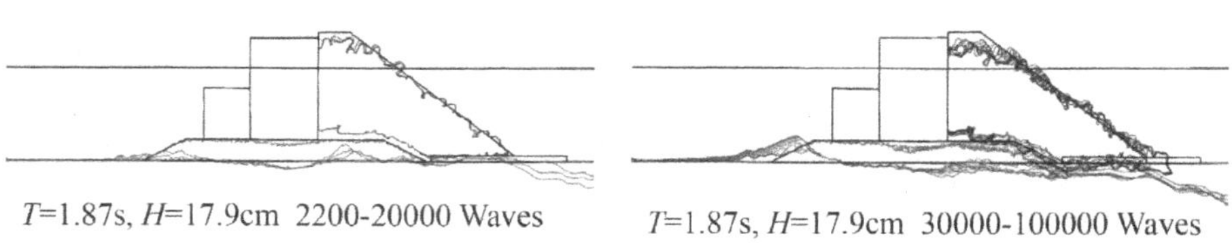

FIGURE 4.2.19
Progress of scouring of sand under the rubble mound.[4.2.6]

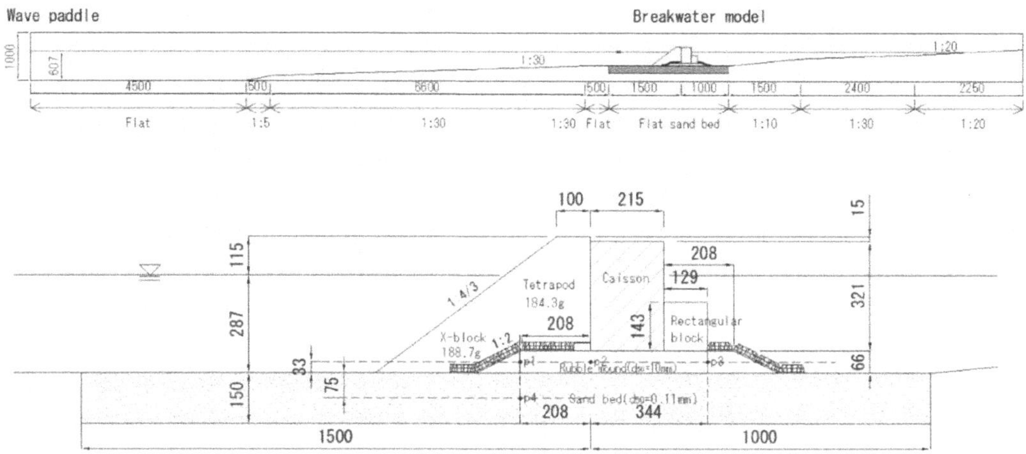

FIGURE 4.2.20
Cross section of the experimental wave flume (*top*) and the detail of model section (*bottom*).

gravel material with $D_{50} = 10$ mm. The mass of the wave dissipating was 184.3 g and the mass of the covering block was 188.7 g. Regarding the wave condition, as shown in Table 4.2.4, the same number of waves as the super-large-scale experiment was applied and 3,200 waves were applied in the final wave.

4.2.6.2 Liquefaction of the Initial Ground

The model bed for the experiment is fabricated by loosely depositing sand in a flume filled with water (water pluviation method). It is fabricated so as not to enclose air as it is assumed that air in the sand may change the pore pressure. Liquefaction was observed during the passing wave test (Section 2.2.4) conducted before the experiment. No changes in the sandy ground were observed with waves of short intervals (0.8 s) and sand ripples were observed on the surface. However, when the wave period was 1.34 s and the wave height was gradually increased, liquefaction of the sandy ground occurred at a wave height of 10.8 cm and the sandy ground subsided in the end because the sandy ground was compacted (Figure 4.2.21). Liquefaction was not observed when a wave was applied after sand was added, but the sand ground sunk by a few centimeters.

TABLE 4.2.4
Experimental Wave Conditions

Wave Period (s)	Wave Height (cm)	Number of Waves	Wave Period (s)	Wave Height (cm)	Number of Waves	Wave Period (s)	Wave Height (m)	Number of Waves
0.80	3.6	200	1.34	3.6	200	1.87	3.6	200
	7.2	200		7.2	200		7.2	200
	10.8	200		10.8	200		10.8	200
	14.3	200		14.3	200		14.3	200
				17.9	200		17.9	3,200

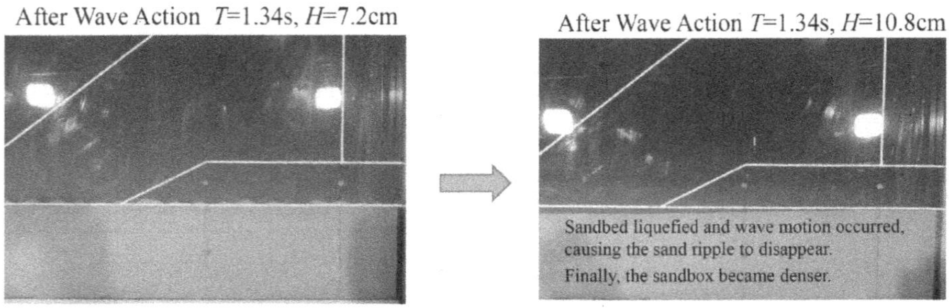

FIGURE 4.2.21
Before and after the liquefaction of sandbed due to wave action.

As you can see, liquefaction can occur in a movable bed experiment due to the deposition state of the sand, and it is believed that it is appropriate to generate waves beforehand to compact the ground. Liquefaction tends to last especially with sand of small particle size.[4.2.8, 4.2.9]

4.2.6.3 Experimental Result

Figure 4.2.22 shows the final deformation state after waves were applied in the same manner as to the super-large-scale experiment. Two cases of experiments were conducted. The foot of the slope of the blocks was placed on the sandy ground in the first case, and gabion was installed in the second case. In case 1 in which the foot of the slope

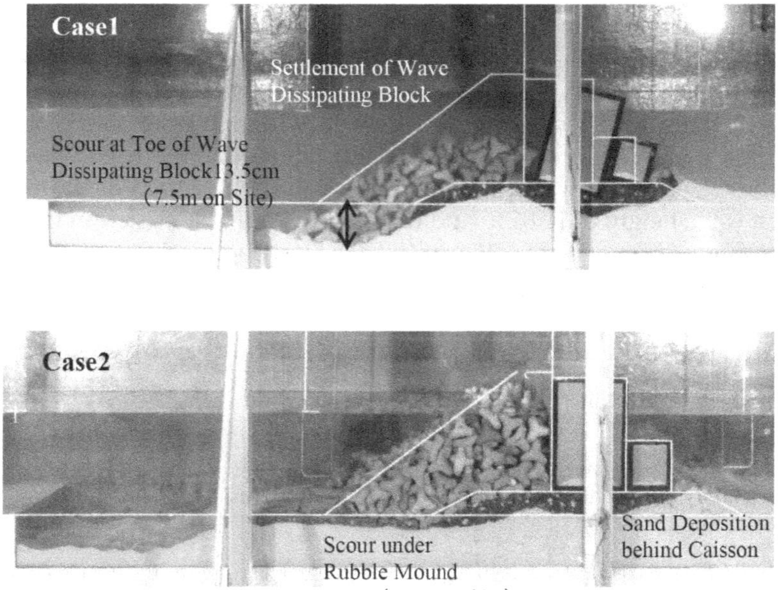

FIGURE 4.2.22
Scouring in Case 1 and Case 2.

is located on the sandy ground, scouring at the foot of the slope was intense; you can see that the scouring amount was 7.5 m in actual scale and the amount of subsidence of the wave-dissipation blocks was also large. In case 2 in which the foot of the slope is made using gabion, the cross section is the same as that of the super-large-scale experiment. Since the gabion works as a filter, less scouring was observed compared with case 1. The scouring amount was 2.6 m on an actual scale, and this was roughly the same as the super-large-scale experiment in which the subsidence was 3 m. However, the number of wave action was larger; while 2,200 waves were applied in the super-large-scale experiment, 3,200 waves were applied in case 2. When 2,200 waves were applied as with the super-large-scale experiment, the scouring amount was only 2.1 m in actual scale (Figure 4.2.23).

Given the above, it can be said that the depth of scouring can be reproduced to a certain extent in small- to-medium-scale model experiments by using sand of small particle size and that a larger number of wave actions must be set in a small-scale experiment because of the scale effect.

4.2.6.4 Raising of Water Level behind the Caisson Due to Wave Overtopping

In case 1, the water level increased behind the caisson because of wave overtopping, and a flow from the back of the caisson toward the sea through the mound under the caisson was generated. Therefore, scouring under the caisson was intense and the caisson itself shifted massively. Such a rise in water level will not occur at the actual site since there is a breakwater opening. Since there is a secondary flume (circuit) in the super-large-scale experiment, overtopping water mass flowed offshore through the secondary flume, reproducing the situation at the actual site. Therefore, if there is only a main flume, it is necessary to return the overtopping water mass offshore using pumps and other means. Nearly identical result as the super-large-scale experiment was obtained in case 2 by returning overtopping water mass offshore using a pump.

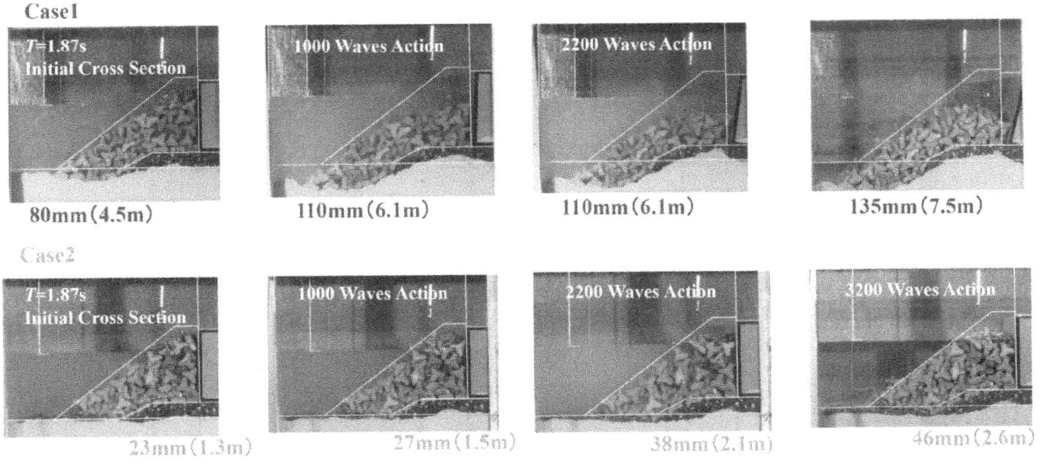

FIGURE 4.2.23
Progress of scouring.

4.2.7 Small-to-medium-scale Model Experiment (1/25 of Super-large-scale Experiment and 1/100 of the Site): Institution C

4.2.7.1 Overview of the Experiment

This experiment[4.2.10] is a 1/25 scale model of the super-large-scale experiment. Details regarding the experimental flume and the model parts are shown in Figure 4.2.24. The sand material used for the sandy ground was silica sand No. 8 (Tohoku Keisya No.8) with d_{50} = 0.11 mm. Mound material was gravel material (silica sand No.1) with D_{50} = 4.6 mm. Although silica sand No. 8 was used as sand to reduce particle size as much as possible, considering the scale, it is larger by approximately one order than the scale of the super-large-scale experiment. The mound material used corresponds to the 1/25 scale of the super-large-scale experiment. Therefore, the particle size ratio (mound material D_{50}/geomaterial d_{50}) was smaller than that of the super-large-scale experiment by approximately one order. Chen et al.[4.2.10] mentioned later consider that scouring occurs when this particle size ratio is larger than 100. Two cases of the experiment were conducted: a case installing a gabion at the lower part of the foot of the slope of wave-dissipating blocks, and a case without installing a gabion. Since gabion was installed in the super-large-scale experiment, first, we would like to show the experimental result with gabion installed. Then, we would like to present an interesting result obtained in an experiment without a gabion.

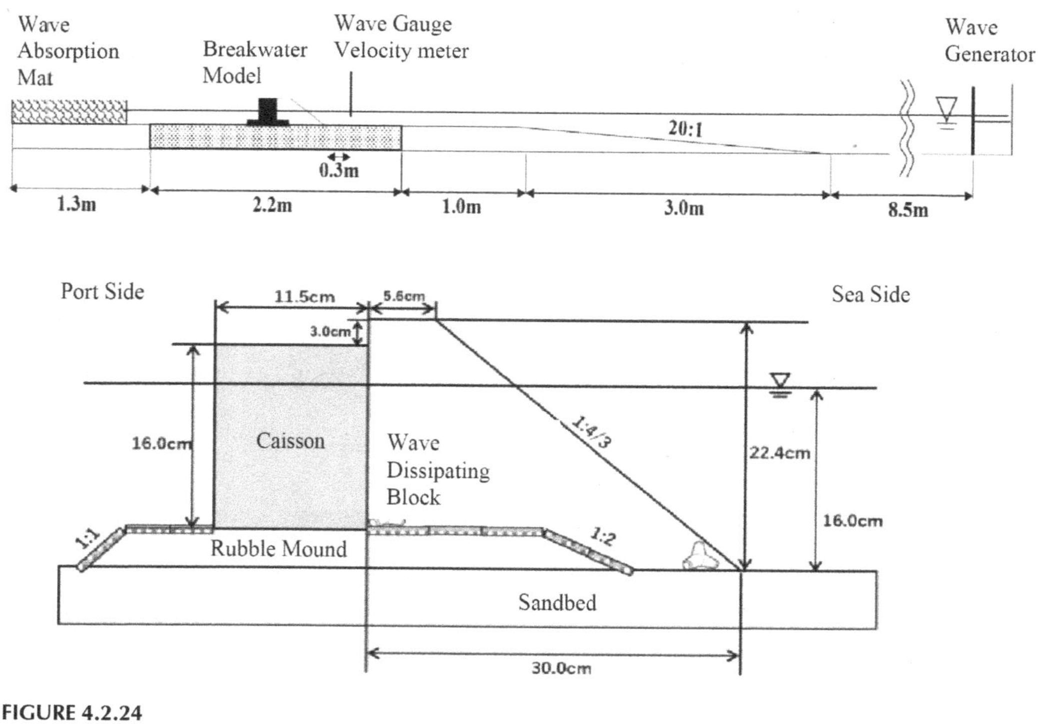

FIGURE 4.2.24
Cross section of the experimental wave flume (*top*) and the detail of model section (*bottom*).[4.2.10]

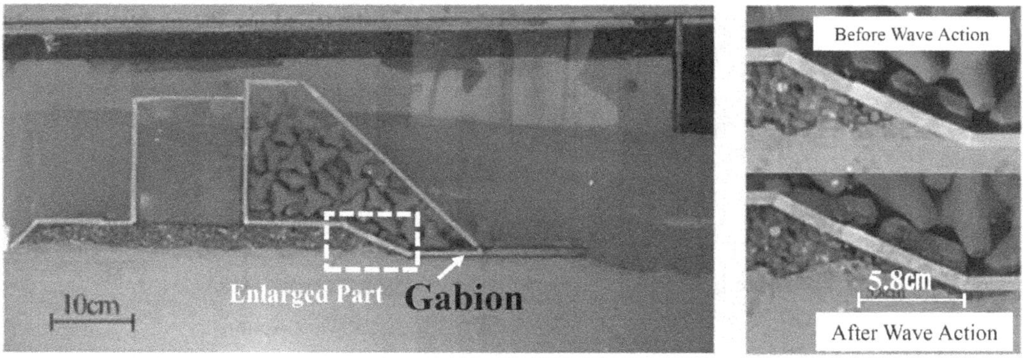

FIGURE 4.2.25
Sandbed at the end of the experiment (after 10,000 waves of action).[4.2.10]

4.2.7.2 Overview of Experimental Result

The result of the experiment with gabion installed is shown below. Unlike the super-large-scale experiment, scouring of sand from under the mound did not occur in the experiment of this scale (Figure 4.2.25). Instead of scouring, a sedimentation phenomenon was observed at the foot of the slope of the mound. When the flow velocity within the mound was estimated based on the measurement result of pore pressure within the mound, although the characteristics of velocity waveform were the same as those of the super-large-scale experiment, low velocity is pointed to as one of the reasons why scouring did not occur. Additionally, the fact that the particle ratio (mound material D_{50}/geomaterial d_{50}) is around 40, which is approximately 1/10 of that of the super-large-scale experiment (whose particle ratio is 360), is also considered to be the most important reason why scouring did not occur.

In the result of the experiment without the gabion, the sand of the ground directly under the caisson got scoured and a massive tilt of the caisson toward the shore was observed (Figure 4.2.26). In this case, gravel material of larger size than those used in the previously mentioned experiment was used, and the particle ratio (mound material D_{50}/geomaterial d_{50}) exceeded 100, making a condition in which scouring can happen easily from the viewpoint of particle ratio. Additionally, wave-dissipating blocks moving over the caisson by the wave were observed in the experiment of this scale. Deformation of wave-dissipating blocks tends to be greater with an experiment of smaller scales. One reason for this is the fact that, in a small-scale experiment, Reynolds number lessens and resistance increases proportionally[4.2.11, 4.2.12] (Section 3.2.5).

4.2.8 Centrifuge Model Tests (1/38 of Super-large-scale Experiment and 1/152 of the Site): Institution D

4.2.8.1 Overview of the Centrifuge Model Test

Here, an example in which the scouring phenomenon occurred under the mound is reproduced with the centrifuge wave testing.[4.2.13] Wave–soil–structure model tests were performed in a drum-type geotechnical centrifuge. The appearance of the apparatus is

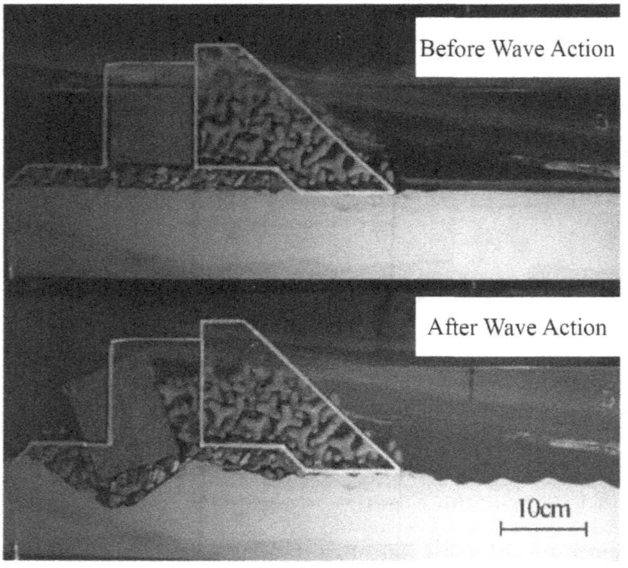

FIGURE 4.2.26
Large tilt of caisson due to scouring under rubble mound.[4.2.10]

shown in Figure 4.2.27. Centrifugal acceleration acts on the entire flume by rotating the cylindrical flume at high speed. Gravitational acceleration applied to the flume can be increased virtually by this centrifugal acceleration, and wave tests are conducted under this condition. The wave conditions in the centrifugal test model also follow Froude's law (see Table 3.2.2). The cross section of the cylindrical channel and the cross section of the soil–structure model used in the wave test are shown in Figure 4.2.28. Although the model is small (1/38 scale of the super-large-scale model experiment[4.2.1]), since the tests were performed in a centrifugal acceleration of 38*g*, the centrifuge model has the same stress field in the ground and wave pressure as those of the model of the super-large-scale experiment (see Table 3.3.2).

Araki et al.[4.2.13] performed three cases of wave tests by changing the particle size of the soil material and changing the viscosity of pore fluid and external fluid (see

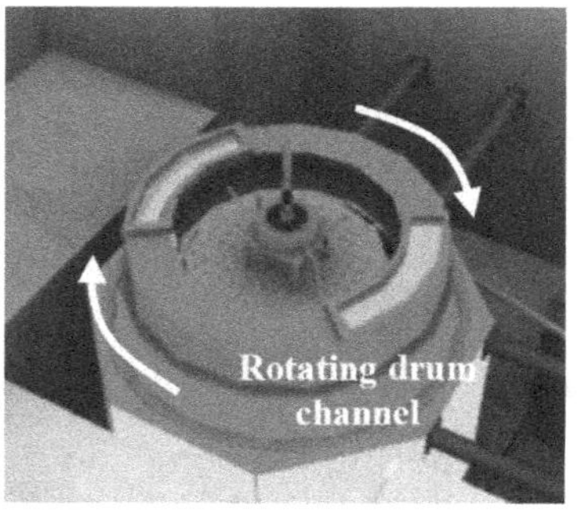

FIGURE 4.2.27
Drum-type geotechnical centrifuge.

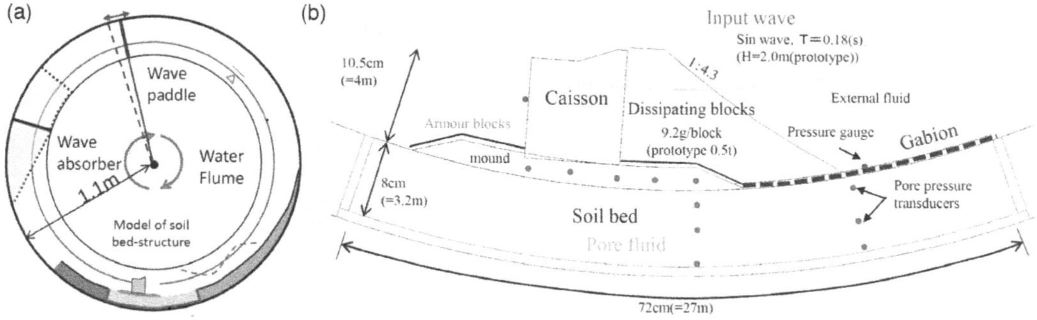

FIGURE 4.2.28
(a) Cross section of (a) a water channel installed inside a drum centrifuge and (b) a model of breakwater covered with wave-dissipating blocks[4.2.13]

Table 4.2.5). The characteristic of Case 1 is that, although soil material of the same particle size as that of the super-large-scale model experiment (silica sand No.7, $d_{50} = 0.15$ mm) is used, it uses viscous fluid (fluid with a viscosity of 38 times that of water) to achieve the same fall speed of a sand particle as that of the super-large-scale model experiment. Here, for example, fall speed can be represented with Equation (4.2.1) (Hallemeier equation).[4.2.14] The same fall speed w_s as that of the prototype is maintained by multiplying the viscosity by N, where N represents the Ng centrifugal field.

$$w_s = \frac{\rho' g (d_{50})^2}{18\nu} (\mathrm{A} < 39) \quad A = \frac{\rho' g (d_{50})^3}{\nu^2} \tag{4.2.1}$$

Here, ρ' is the submerged density of soil particle, ν is the coefficient of kinematic viscosity of the external fluid, d_{50} is the median particle size of the sand, and g is the gravitational acceleration (38 g in this experiment). The important point in this case is that the Reynolds number within the pore fluid of mound or blocks will decrease because of the use of viscous fluid (refer sub-section 3.2.5). The effect of the use of viscous fluid on the result will be described later.

The characteristics of Case 3 are that although water is used as fluid, soil material whose diameter is approximately 1/10 of that of the super-large-scale experiment (nonplastic silt of $d_{50} = 0.02$ mm) is used to match the fall speed with that of the super-large-scale model experiment. That is, fall speed is maintained to be the same as that of the super-large-scale model experiment by multiplying the second power of particle size (d_{50}^2) by $1/N$, where N times represents the amount of gravity increase due to centrifugal force field according to Equation (4.2.1). When reducing the particle size of the prototype sand material to use in the experiment, the sand used may be fine-grained soil with a small particle size such as silt or clay. Nevertheless, using fine-grained soil in the experiment can cause difficulty in the reproduction of the result of the experiment with an actual object because fine-grained soil shows plasticity in general. Therefore, nonplastic silt was used in this study. The characteristics of Case 2 are that it uses the same geomaterial as the super-large-scale experiment and uses water as fluid and thus has the same condition as general hydraulic model experiments. It is important that, since the gravity increases N times due to the centrifugal force field, fall speed also increases N

TABLE 4.2.5

Centrifuge Wave Tests: Test Cases and Conditions of Soil and Fluid

		Experimental Conditions				Sedimentation Velocity of Soil Particle (Equation 4.2.1)	Reynolds Number within Mound	Results
		Soil Material	Mound Material	Pore Fluid	External Fluid			Scouring
Centrifugal acceleration: 38g	Case 1	Silica sand 7 ($D_{50} = 0.15$ mm)	Rubble 4.75–9.5 mm	Viscous fluid (38 times)	Viscous fluid (38 times)	2.0 cm/s ○	47 (×)	No
	Case 2	Silica sand 7 ($D_{50} = 0.15$ mm)	rubble 9.5–19 mm	Water	Water	23 cm/s ×	2,280 (Δ–○)	No
	Case 3	Nonplastic silt ($D_{50} = 0.02$mm)	rubble 4.75–9.5mm	Water	Water	1.3 cm/s ○	1,020 (Δ)	Yes (○)
Super-large-scale model 4.2.1)		Sand ($D_{50} = 0.2$ mm)	Stone 80–150mm	Water	Water	2.5 cm/s	9,750	Yes

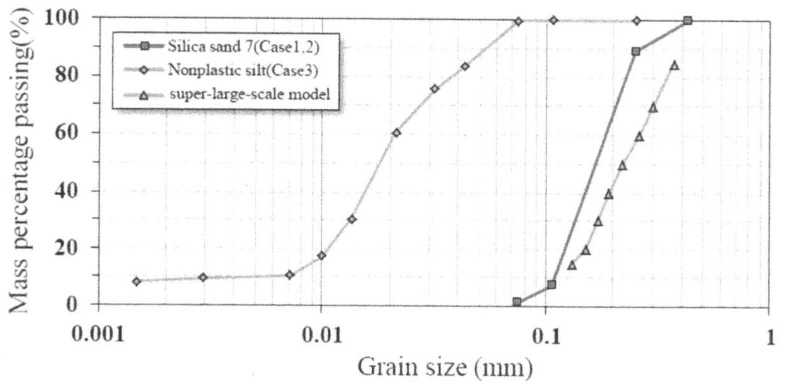

FIGURE 4.2.29
Grain size distribution curves of soil material used.

times under this condition. The particle size distribution of the soil material used in the model test is shown in Figure 4.2.29. Here, nonplastic silt is an industrial powder product whose raw stones are silica rock, pagodite, and kaolin. Although they are classified as fine-grained soil according to particle size distribution shown in Figure 4.2.29, they can also be called sand with small particle size since they do not present plasticity such as cohesion.[4.2.15, 4.2.16] Therefore, this nonplastic silt is convenient for use in model tests for which the particle size of the prototype sand material is scaled.

Theoretically, matching the fall speed of a sand particle in a centrifuge to that of the prototype one in the field is equivalent to matching the Dean number of the model to that of the prototype (Equation (3.4.28)). Thus, in Case 1 and Case 3, the Dean number is almost the same between the model and the prototype. Additionally, the similitude ratio of the experiment conducted under the condition of Case 3 is shown in Table 3.4.3. In this case, soil materials of the size around $1/N^{0.5}$ are used so that the similitude ratio of fall speed will be 1.

4.2.8.2 Waves in a Drum Centrifuge: Coriolis Effects

In wave tests in a drum centrifuge, because of rotating the cylindrical flume shown in Figure 4.2.27, the Coriolis effect (refer sub-section 2.3.2) will have an impact on wave generation and propagation. Propagation of gravity waves and the Coriolis effect generated in the cylindrical flume in a centrifugal force field have been studied theoretically by Sekiguchi and Phillips.[4.2.17] Araki et al.[4.2.13] performed wave tests in a drum centrifuge and confirmed the Coriolis effect (Figure 4.2.30). In the wave tests in this figure, the wave height as well as the water pressure change at the rigid bottom of the same location was measured. Measurement values of wave height (prototype) and estimated wave height from pressure measurements at the bottom using the Airy wave theory are plotted in Figure 4.2.30. From this figure, you can see that there is a difference between the value measured with the wave gauge and the wave height estimated based on the pressure at the seabed. This is due to the Coriolis effect. It is presented in the theory[4.2.17] that when the direction of rotation of the cylindrical flume and the travel direction of the wave is the same, the bottom pressure fluctuation brought by waves increases than

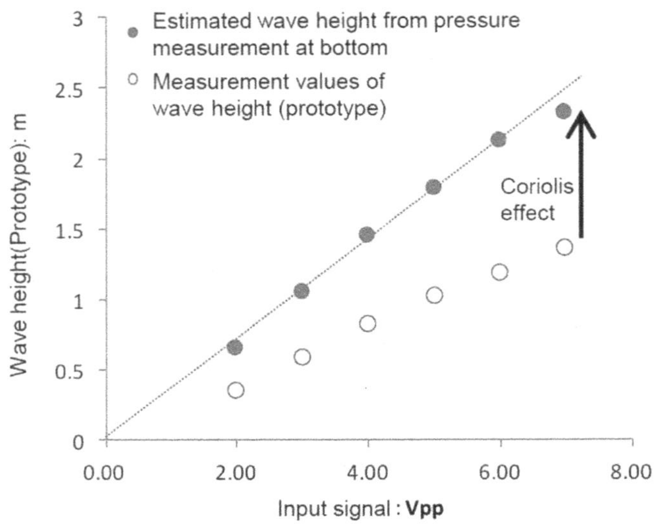

FIGURE 4.2.30
Coriolis effect: difference between the measured wave heights and those calculated.

when the Coriolis effect does not occur, and this is shown in the experimental result provided in Figure 4.2.30.

The Coriolis effect decreases when the radius of the cylindrical flume is large. The difference between the relationship of wave height and wave pressure of the actual wave, and the relationship of wave height and wave pressure of the centrifuge model will increase if the Coriolis effect is significant. Thus, it can be said that when handling problems regarding bottom pressure fluctuation in a model test in a centrifuge (problems focusing on pore pressure fluctuations within the ground or mound or ground surface wave pressure), the input wave of the experiment must be determined based on bottom pressure rather than the geometric scale of the actual wave height. In reproducing the super-large-scale model test with a centrifuge model, Araki et al.[4.2.13] determined the height of the input wave of the test so that the wave height converted from the bottom pressure and the wave height of the super-large-scale model experiment (actual measurement was slightly less than 2.0 m) match.

4.2.8.3 Overview of Experimental Results

The experimental results, that is occurrence of scouring under the mound in Cases 1, 2, and 3, are shown in the right column of Table 4.2.5. Occurrence of scouring phenomenon under the mound was only observed in Case 3, in which nonplastic silt (whose particle size is 1/10 of the prototype one) is used as soil material. Deformation of breakwater model after wave loading of 2,000 waves in Case 3 is shown in Figure 4.2.31. The sandy ground under the mound is observed to be scoured, causing the subsidence of the mound. This is the same as the observation in the super-large-scale model experiment, and it is shown that the scouring phenomenon observed in the super-large-scale model experiment was reproduced in the centrifugal wave test. Scouring under the mound did not occur in Case 1 and Case 2. Regarding Case 2, fall speed being smaller than the

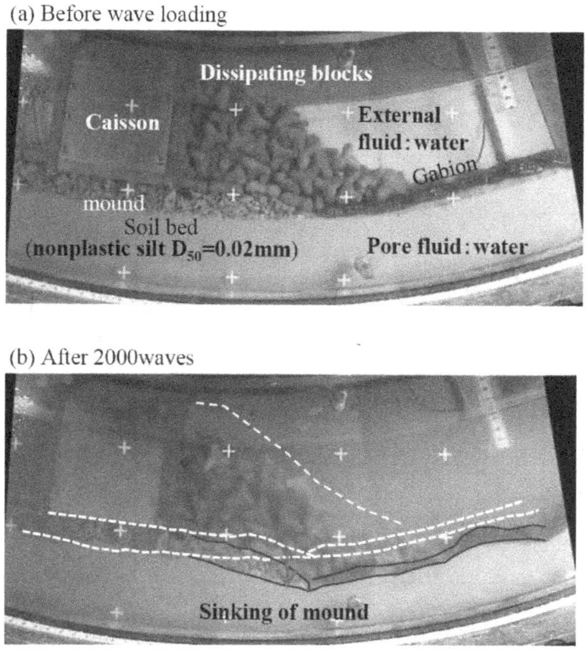

FIGURE 4.2.31
Observation of scouring under the mound and sinking of the mound (Case 3).[4.2.13] (a) Before wave loading and (b) After 2000 waves.

large-scale model experiment by one order is considered to be the reason why scouring did not occur (see Table 4.2.5). In Case 1, it is considered that although fall speed matched to that of the super-large-scale model experiment, scouring did not occur because the Reynolds number within the mound has decreased significantly because of the use of viscous fluid and the turbulence within the mound was reduced (see Section 3.4.3). Thus, the fall speed and the Reynolds number (turbulence within the mound) prove to be important for the scouring phenomenon that occurs under a permeable structure.

Additionally, Araki et al.[4.2.13] showed the flow velocity forms within the mound of the centrifuge model tests which were calculated from the pressure fluctuation within the mound as with the super-large-scale model experiment. Its magnitude and form of fluctuation are consistent with those of the super-large-scale model experiment. Therefore, it is shown that the flow velocity within the mound could be reproduced in a drum centrifuge.[4.2.13]

4.2.9 Dean's Law (Dean Number) and Results of Experiments of Each Scale

The scouring phenomenon that occurs under wave-dissipating blocks covering the breakwater is a phenomenon in which scouring progresses due to sand under the mound or gabion floating violently. Therefore, in this section, we would like to evaluate the experimental result using the Dean number which represents the law of similarity related to fall speed of suspended sediment.

In the experiment in a gravitational force field of 1*g*, Dean number and Froude's law are combined, as shown in Figure 4.2.32. For example, when conducting an experiment

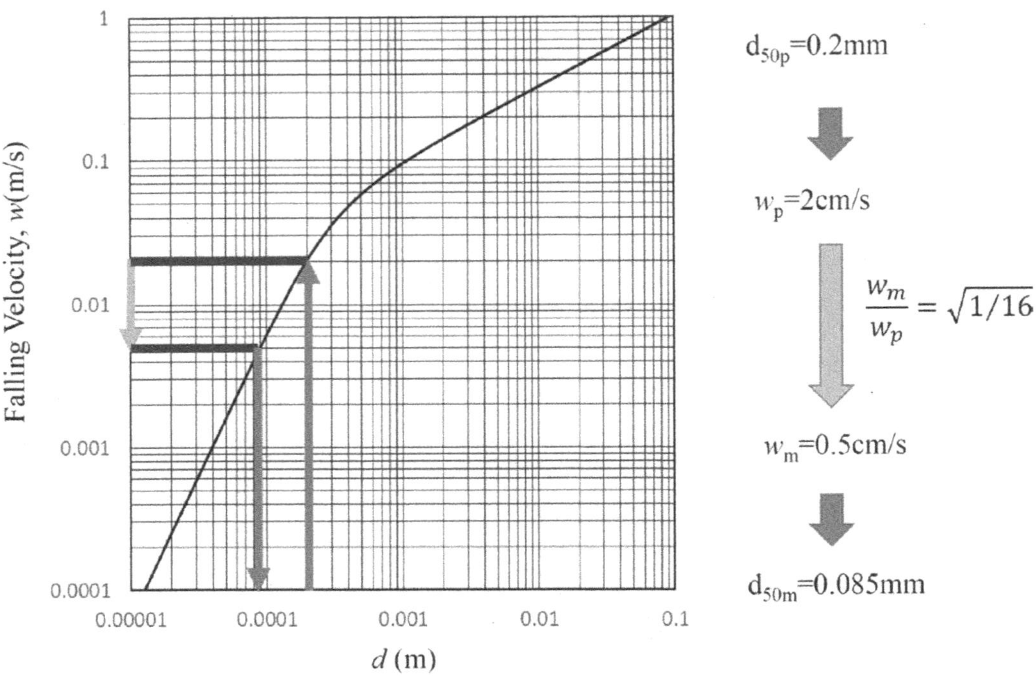

FIGURE 4.2.32
A method for determining the grain size of sand by combining the Dean number and Froude's law.

of 1/16 scale on the sand of 0.2 mm (with fall speed of 2.6 cm/s), since the fall speed will be 1/4 based on Froude's law, the sand used in the experiment will be 0.09 mm (with fall speed of 0.65 cm/s). Conversely, in the centrifuge model experiment of Ng, since the ratio of wave height H and wave period T is 1 as described in Section 3.4.3, the similitude ratio of fall speed translates directly into the similitude ratio of the Dean number. Araki et al.[4.2.13] obtained the fall speed of sand at the field of Ng by assuming g of the Hallemeier equation shown in Table 4.2.6 as Ng.

Figure 4.2.33 represents the relationship between the particle size and fall speed of sand and shows the fall speed in fields of 1g to 38g. Since the flow velocity does not change in a centrifugal force field (see Table 3.2.2), the corresponding particle size for 1–38g fields must be determined by assuming that the fall speed of sand in the super-large-scale experiment (2.6 cm/s) is constant. Additionally, when gravity is assumed to

TABLE 4.2.6
Hallemeier Formula[4.2.14]

	Equation (4.2.2)	Equation (4.2.3)	Equation (4.2.4)
$A = \dfrac{\rho' g (d_{50})^3}{\nu^2}$	$A < 39$	$39 < A < 10^4$	$10^4 < A < 3\times10^6$
W_s	$\dfrac{\rho' g (d_{50})^{20}}{18\nu}$	$\dfrac{(\rho' g)^{0.7}(d_{50})^{1.1}}{6(\nu)^{0.4}}$	$\dfrac{(\rho' g)^{0.7}(d_{50})^{0.5}}{0.91}$

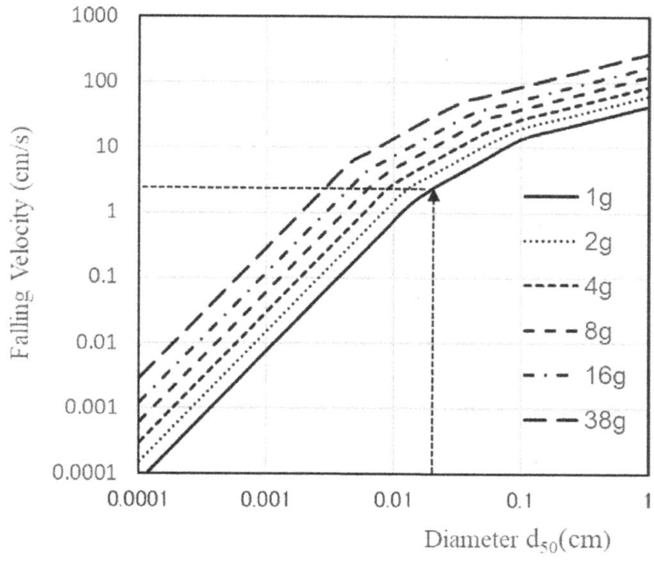

FIGURE 4.2.33
Falling velocity for each gravity field when combining the Hallemeier equation and the ground similarity law.

be *Ng*, some ranges within coefficient *A* show slight nonconformity, and that part has been excluded.

Figure 4.2.34 shows the corresponding particle size calculated using the Dean number and experimental results of each institution. Black squares indicate the particle sizes of which reproduction of scouring was difficult (or required a long period of time for scouring to occur), and white circles indicate the particle sizes that reproduced scouring for the most part. The super-large-scale experiment was conducted using

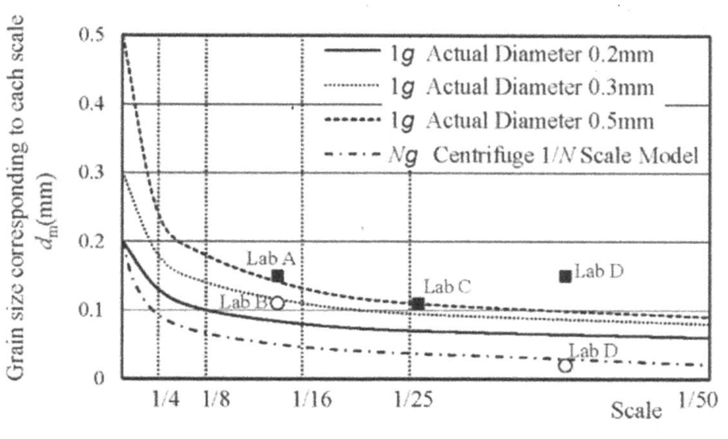

FIGURE 4.2.34
Grain size corresponding to each scale based on Dean number (black squares are those where the scour was difficult to reproduce; white circles are those where the scour was almost reproduced).

sand of $d_{50} = 0.2$ mm, and this corresponds to the phenomenon with sand of 0.3 mm at the actual site. The experiment conducted by Institution B is on the same line as the super-large-scale experiment, corresponding to the fact that the experimental result was almost the same as the result of the super-large-scale experiment. Additionally, since the centrifuge model experiment conducted by Institution D also corresponds to the phenomenon with sand of 0.2 mm at the actual site, it can be assumed that it corresponds to the experimental result.

Conversely, according to the experimental result of Institution B, the progress of scouring was slightly slow compared with the large-scale and super-large-scale experiment, and it required a longer period of time for scouring to become stable. With the method of using the Dean number, although the fall speed is similar and the form of the scouring can be reproduced, the concentration of suspended sediment may not be similar, and this could have affected the rate of progress of the scouring.

4.2.10 Conclusion

In this section, we compared various experiments of different scales and geomaterials with the super-large-scale experiment[4.2.1] aiming to establish a method of reproducing scouring phenomenon in small- to-medium-scale experiments. Consequently, the following results were obtained:

- It was found that the form of the scouring can be reproduced almost accurately with the law of similarity (Dean number) of fall speed. Additionally, regarding the centrifuge model test, the validity of the law of similarity of scouring phenomenon that is based on the Dean number was presented by Araki et al.[4.2.13]
- Regarding the reproducibility of the super-large-scale experiment, the 1/14 scale experiment using sand of 0.11 mm presented the highest reproducibility, followed by the centrifuge model test of 38*g* (1/38 scale) using nonplastic silt of 0.02 mm. This is a result that corresponds to the result of the examination based on the law of similarity presented by Araki et al.[4.2.13] and supports the reproducibility of the scouring phenomenon using the Dean number. However, with the centrifuge model experiment of 38*g*, although the reproducibility of the suspension phenomenon was high, the reproducibility of wave breaking phenomenon at the wave-dissipating blocks was low.
- Regarding small-to-medium-scale experiments, in the 1/14 scale experiment conducted by Institution A using sand of 0.15 mm, the scouring progressed when the number of waves increased, and Murakami et al.[4.2.6] have presented that the law of similarity is also required regarding the time it takes until scouring will be stabilized. For example, as the required time was reduced to 1/4 of the actual time in the 1/16 scale experiment, the time required for stabilization appears as if it has been extremely shortened. However, from the viewpoint of the velocity of scouring, if the scouring speed is 1/4 of the actual speed, the rate of progress of the scouring would be quadrupled.
- In the result of a small-to-medium-scale experiment conducted by Institution B, although the form of scouring was similar to that of the large-scale and super-large-scale experiment, the rate of progress of scouring was slightly slower than

the large- and super-large-scale experiment and the time taken for stabilization was also longer. Although the fall speed was similar based on Dean number and the form of scouring can be reproduced, since the concentration of suspended sediment may not be similar, the effect of this could have possibly affected the rate of progress of scouring.

- The result of the small-to-medium-scale experiment conducted by Institution C shows that turbulence within the mound is also an important point, and it became evident that it is advisable to conduct experiments of larger scale in terms of Reynolds number and particle ratio between crushed stones in the mound and sand. Additionally, under the condition in which the Reynolds number is small, resistance increases proportionally,[4.2.11, 4.2.12] and it became clear that one must note that the wave-dissipating block model can become unstable.
- It also revealed that it is important to prevent the water level behind the caisson from rising by using pumps or secondary flumes in case the water level behind the caisson rises because of wave overtopping.

4.3 Case (2) Erosion of Beach Nourishment Filling

4.3.1 Introduction

Beach nourishment that is carried out as a measure against coastal erosion is classified roughly into two types: static beach nourishment in which sand beach that is necessary from the viewpoint of protection is constructed and dynamic beach nourishment in which earth and sand that replenishes the amount that is sufficient to maintain the necessary sand beach are added. Additionally, it can be classified into onshore beach nourishment in which earth and sand are brought to the shore and offshore beach nourishment in which earth and sand are charged into the sea. When carrying out dynamic beach nourishment on shore, it is often done in the form of embankment because of reasons such as construction management. This embankment is eroded during high waves and its earth and sand are fed to the coast downstream of the longshore drift. In many cases, the location of beach nourishment and the amount of sediment are determined on the basis of the calculated estimate of long-term beach deformation considering the particle size of earth and sand used for beach nourishment. This calculation estimates the deformation of the beach across a wide area that corresponds to the mean wave energy of a set period of time (mean energy wave) but does not accurately estimate the deformation of the beach nourishment filling itself. Nevertheless, especially in the dynamic beach nourishment, as it may require repetitious filling depending on erosion after beach nourishment, the erosion process of beach nourishment filling must also be examined.

The erosion process of beach nourishment filling is often regarded as the process of vertical 2D beach deformation, and therefore, it depends on not only waves but also the material used for beach nourishment. Although the effect of the material used for beach nourishment on the advancement and recession of shoreline and classification of the material is studied, examination with consideration given to characteristics of the land that has the possibility of unsaturation has not been made, unlike offshore beach nourishment. For example, although Noguchi et al.[4.3.1] have presented that erosion amount of beach nourishment filling can change depending on the material used for beach nourishment or the layer structure through hydraulic model experiments, since they were focusing on the erosion of beach nourishment filling due to tractive force of run-up waves, the effects related to suction and degree of ground compaction were not considered. Additionally, although Serizawa et al.[4.3.2] have examined the relationship between the gradient of the slope of beach nourishment filling and the amount of sand supplied at the sea area, they have not discussed the condition to determine the gradation of the slope.

Although various studies regarding phenomena that are similar to erosion of beach nourishment filling, such as erosion of sand dune or formation of beach scarp, have also been conducted, most of them are focused on tractive force of run-up waves. For example, Hattori and Kakegawa[4.3.3] found that a beach scarp is formed when a steep beach is eroded rapidly by bore-type run-up waves. Also, Uda and Serizawa[4.3.4] indicated that the interruption of longshore drift sand affects the formation of beach scarp. You can find other studies focused on the soil conditions of sand dunes and so on as well. For example, Nishi and Kraus[4.3.5] established a vertical 2D beach deformation model that incorporates the drift sand occurrence mechanism due to destruction of sand dune caused by the impact of run-up waves and indicated that the coefficient related to the impact depends on whether the compacting of the sand dune model was done or not. Honma et al.[4.3.6] analyzed the erosion mechanism of the slope of backshore of shore roads due to high waves, focusing on relative density and degree of saturation. Ericson et al.[4.3.7] simulated the mechanism of sand dune moving backward by repeated formation and collapse of notches, considering suction. Also, van Bemmelen[4.3.8, 4.3.9] simulated the process from the formation of the beach scarp to its collapse. Regarding erosion of beach nourishment filling composed of sand of different particle sizes, not only the tractive force of run-up waves but also the impact of suction must be considered.

In this section, we will focus on the impact of filling material on the erosion process of beach nourishment filling and based on existing knowledge regarding the mechanism of sand dune and beach scarps moving backward, we will examine the change in cross section of filling immediately after starting wave generation in the tests conducted by Noguchi et al.[4.3.1] and present the law of similarity that should be applied.

4.3.2 Existing Knowledge Regarding Recession Mechanism of Sand Dune and Beach Scarp

We would like to introduce knowledge from existing research that analyzed the recession mechanism of sand dune, etc., focusing on ground conditions. Honma et al.[4.3.6] applied solitary waves that run-up to the crown of the slope (with a representative grain diameter of 0.28 mm) that is in the shape of road embankment and presented that the road width fracture pattern when the slope of the embankment is loose (pattern 1 in Figure 4.3.1) is different from when the slope of the embankment is dense (pattern 2 in Figure 4.3.1) (Figure 4.3.1) and that regarding unsaturated slope, when relative density is high, not only the shear strength but also the suction between soil particle will increase and the amount of sediments that flow from the collapse will decrease (Figure 4.3.2).

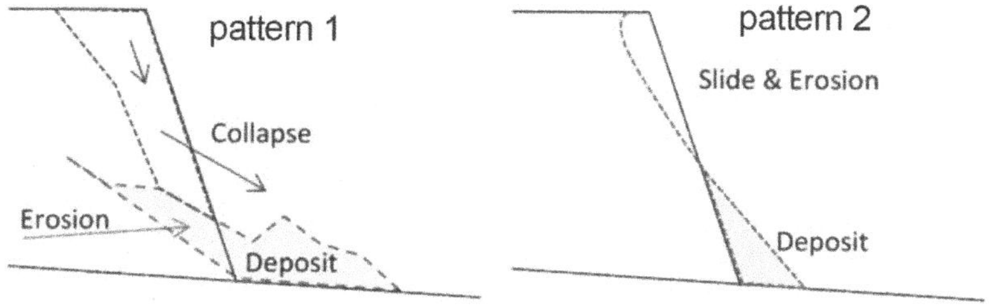

FIGURE 4.3.1
Damage patterns on the slope that model the road embankment.[4.3.6]

Ericson et al.[4.3.7] classified the collapse mechanisms of the sand dune into two categories: "Shear-type failure" in which fracture occurs when the weight of the overhang exceeds the shear strength of soil particles and "Beam-type failure" in which slip or collapse occurs when crack due to tension forms in the front of the sand dune (Figure 4.3.3). In shear-type failure, the base of the sand dune is eroded as shown in Figure 4.3.3a and the overhang collapses as shown in Figure 4.3.3b. Conversely, in beam-type failure, when the crack on the sand dune base caused by erosion reaches the slide plane in the front of the sand dune as shown in Figure 4.3.3c), the clod in the front of the sand dune will slide down while spinning as Figure 4.3.3d) or slip as it is as Figure 4.3.3e. Additionally, the elements of the governing equation of each collapse mechanism are presented (Figure 4.3.4). Regarding Figure 4.3.4a, it is assumed that collapse occurs on the cross section extended vertically from the erosion point at the base of the sand dune (z-axis in the figure) at the moment when the weight of the clod in the front of it comes into balance with shear strength. Also, regarding Figure 4.3.4b, the crack caused by erosion of the sand dune base as well as sliding on the slide plane is inspected. Suction is considered in calculating the shearing behaviors of both collapse mechanisms.

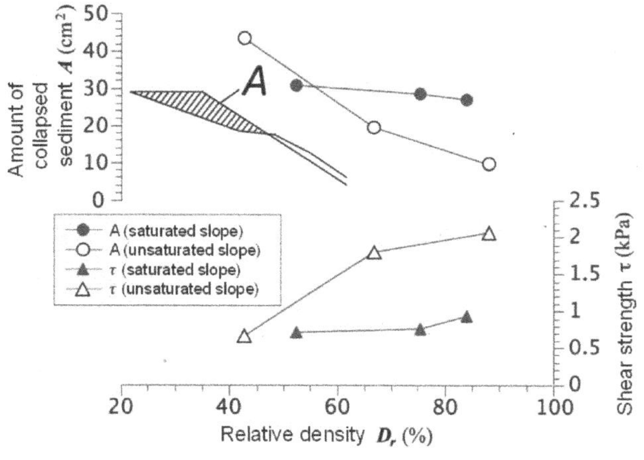

FIGURE 4.3.2
Relationship between relative density, collapsed sediment volume, and shear strength. (Modified from Honma et al.[4.3.6])

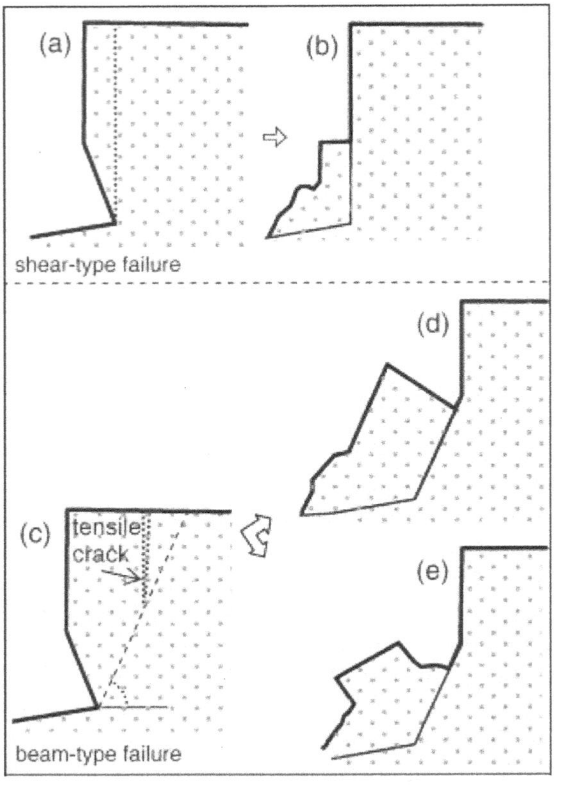

FIGURE 4.3.3
Type of sand dune failure mechanisms.[4.3.7] (a) and (b) show shear-type failure and (c) to (e) show beam-type failure.

Van Bemmelen[4.3.8] obtained the limit height $S_{h,cr}$ of the beach scarp from gradient i of the beach scarp and suction σ^s as shown in Figure 4.3.6 by Equation (4.3.1) that replaced the tensile stress in Karman-type slope collapse analysis (Figure 4.3.5).

$$S_{h,cr} = \frac{4\sigma^s}{\gamma}\left[\frac{\sin i \cos\varphi'}{1-\cos(i-\varphi')}\right] \tag{4.3.1}$$

Here, γ is the unit weight of the soil, and φ' is the internal friction angle. Additionally, van Bemmelen[4.3.8] sorted out the conditions of formation and collapse of the beach scarp as shown in Figures 4.3.7 and 4.3.8. The formation condition of the beach scarp is assumed to be when the initial gradient of the slope β_i is steeper than the equilibrium beach gradient β_{eq}, which is a function of wave height, cycle, and particle size, and when the gradient of the upper part of the slope β_u is steeper than the internal frictional angle φ'. It is assumed that β_u is calculated with Equation (4.3.2):

$$\beta_u = \frac{R'_{2\%} - R'_{15\%}}{\dfrac{R'_{2\%} - R'_{15\%}}{\beta_i} - \Delta x} \tag{4.3.2}$$

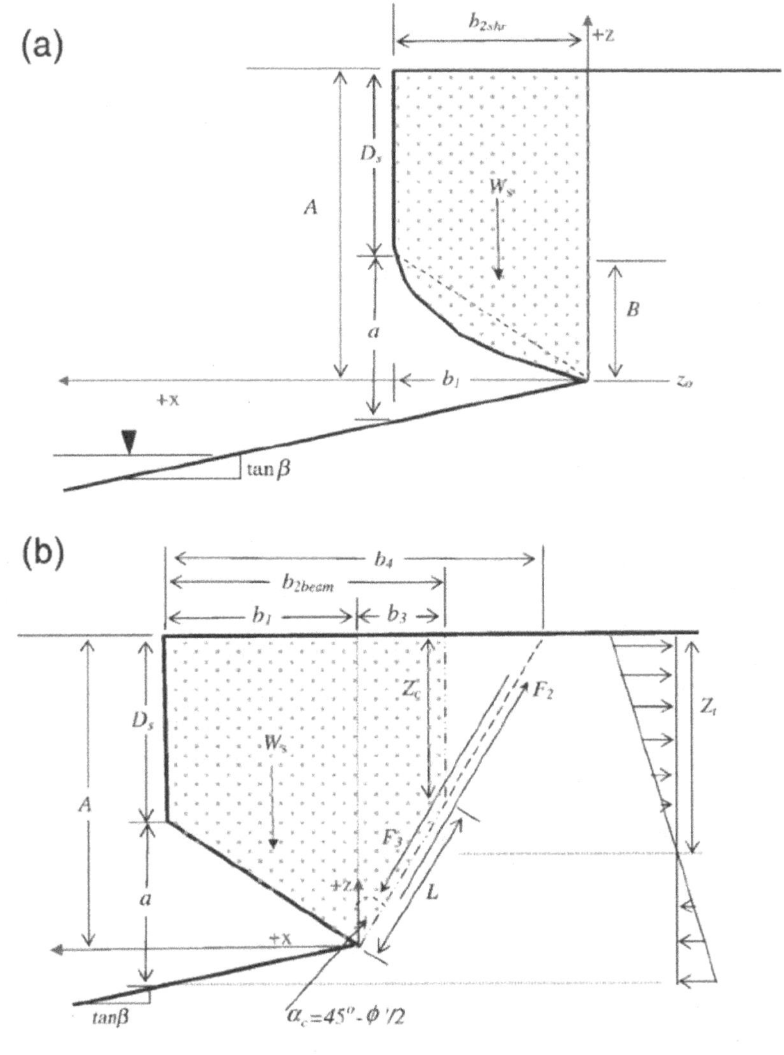

FIGURE 4.3.4
Definition sketch of (a) shear-type and (b) beam-type mass failure.[4.3.7]

Here, $R'_{2\%}$ and $R'_{15\%}$ are run-up height with the probability of exceedance of 2% and 15%, respectively, and Δx is the retreat amount of the shoreline. $R'_{2\%}$ is assumed to be the ultimate height S_t of the base of the beach scarp.

Conversely, the conditions in which the beach scarp disappear are assumed to be (1) when waves reach the top of the beach scarp (the upper surface of beach nourishment filling (S_c)), (3) when waves rise to or above the height of the beach scarp base (S_t) and accumulation occurs in front of the beach scarp, and (4) when waves do not reach the height of the beach scarp base (S_t) and dry out. Additionally, it is assumed that (2) when waves do rise to or above the height of the beach scarp base (S_t) but do not reach to the top of the beach scarp (the upper surface of beach nourishment filling (S_c), beach scarp will move to the land side.

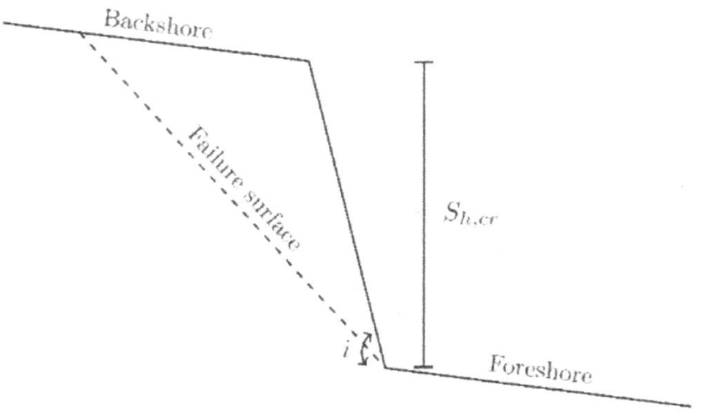

FIGURE 4.3.5
Culmann-type failure applied to stability of a beach scarp[4.3.8]

4.3.3 Test Method

Here, we would like to show the tests conducted by Noguchi et al.[4.3.1] to discover the effect of the use of different beach nourishment materials and banking structures on the conservation of foreshore topography. The tests were conducted using a small-scale wave flume that has a length of 33 m, a width of 0.3 m, and a depth of 0.8 m with a 1/10 gradient fixed bed made of mortar installed as shown in Figure 4.3.9. A beach nourishment filling that has a height of 0.2 m, a crown depth of 0.4 m, and a slope gradient of 1:1 was installed on the land portion of the fixed bed. The initial shoreline is at the foot of

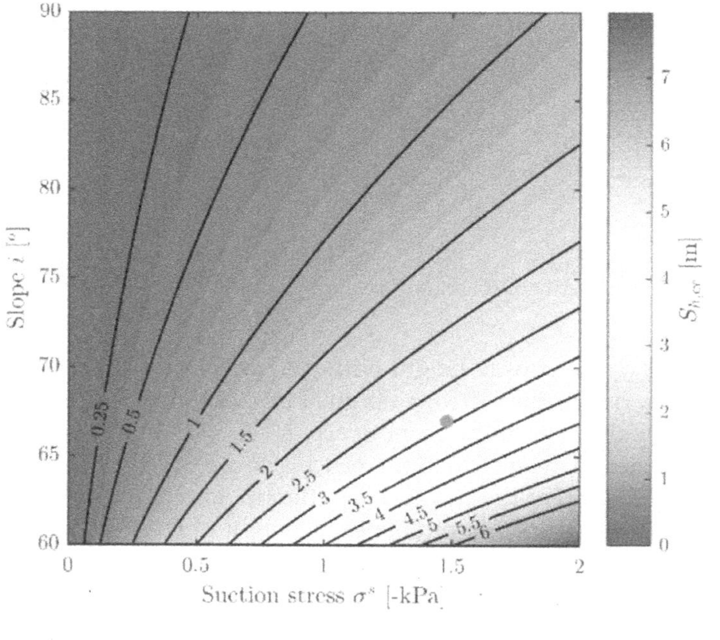

FIGURE 4.3.6
Maximum beach scarp height as a function of the suction and the scarp slope.[4.3.8]

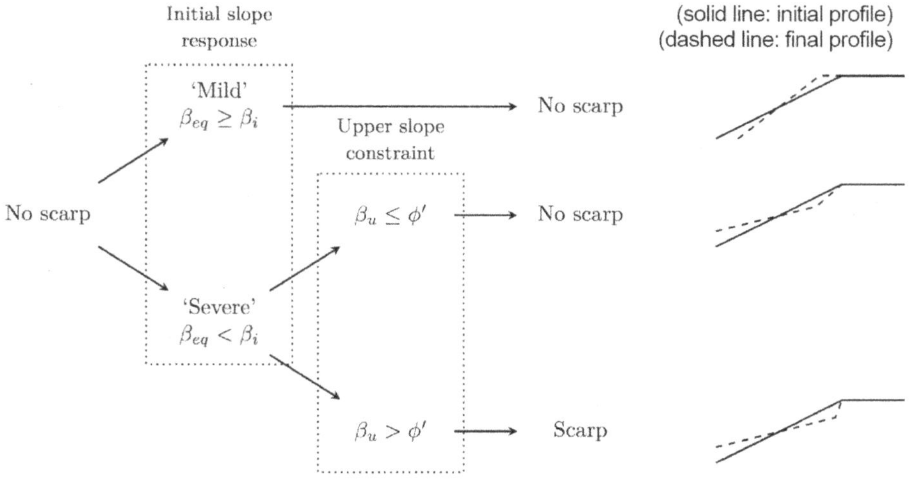

FIGURE 4.3.7
Conceptual diagram of the conditions required for beach scarp formation. (Modified from van Bemmelen.[4.3.8])

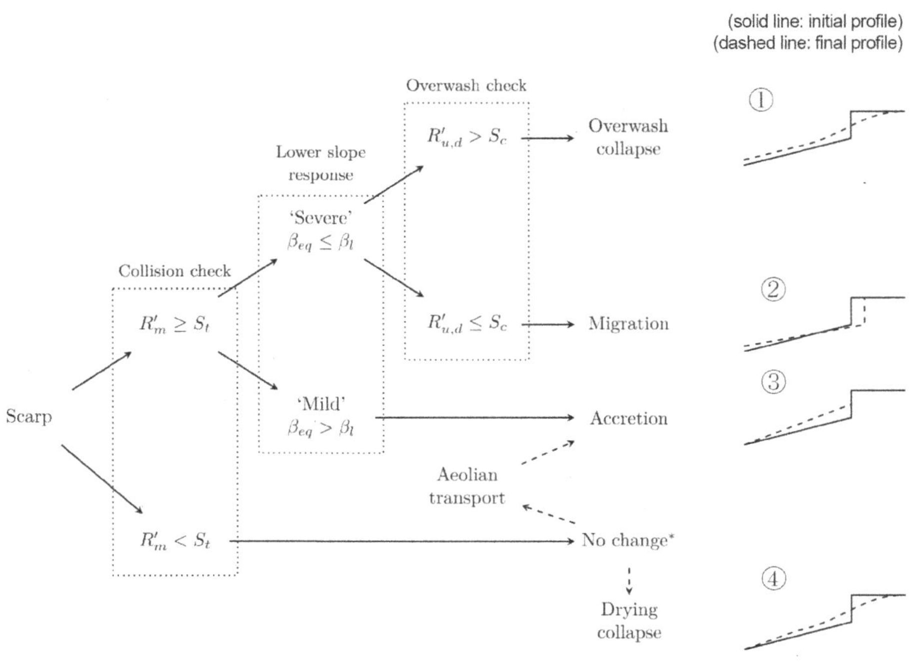

FIGURE 4.3.8
Conceptual diagram of the conditions causing beach scarp migration and eventual destruction. (Modified from van Bemmelen.[4.3.8])

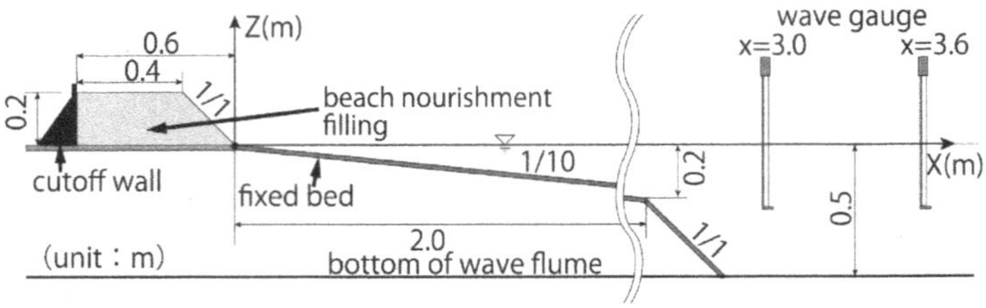

FIGURE 4.3.9
Setup of test model and gauses.

the slope of the beach nourishment filling, and the underwater portion is the fixed bed. The tests were conducted right after the beach nourishment filling was shaped using a trowel so that it is in the predetermined shape. Although the water content rate at the start of wave generation was not measured, it is assumed that it was in an unsaturated condition. Since the tests are not a model of a specific coastal area, the model scale was not established.

Since the nature of the collapse of filling and the gravel movement is important, materials used for beach nourishment filling were pea gravel and sand collected from the river channel in the vicinity of the coastal area. Figure 4.3.10 shows that sieved materials were used so that particle size distribution does not overlap. The basic materials are pebbles (material a) (D_{50}: 8.2 mm), granule (material b) (D_{50}: 2.3 mm), and sand (material c) (d_{50}: 0.82 mm). Material d was made by mixing materials a and b and material e was made by mixing materials a, b, and c. As shown in Figure 4.3.11, eight cases of beach nourishment filling were used in the tests: three cases using each of the basic materials, two cases using each of the mixtures of these base materials, and three cases of banking structures in which the gravel layer is created between the sand. The ratio (volume ratio) of the basic materials in each of the cases is as shown in Table 4.3.1. Among Cases 4–8, which use mixed materials, Case 8 contained the most amount of sand (material c) with a ratio of 73.3%, and Case 4 contained the most amount of pebble (material a) with a ratio of 50%.

A regular wave with a height of 0.1 m and a wave period of 1.3 s was applied during the test. Unlike the test conducted by Honma et al.,[4.3.6] the waves do not run-up to the

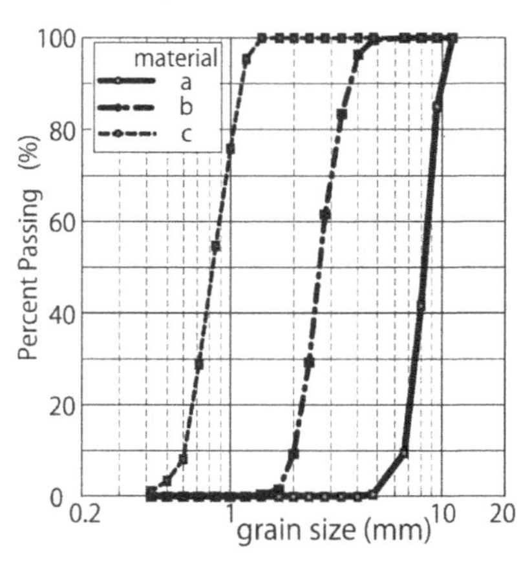

FIGURE 4.3.10
Grain size distribution of beach nourishment filling.

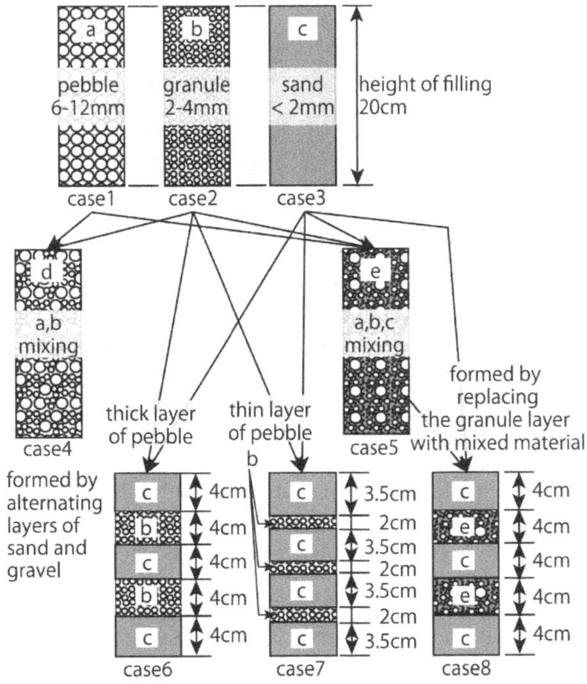

FIGURE 4.3.11
Structure of beach nourishment filling.

crown of the beach nourishment filling. The C values of Horikawa et al.,[4.3.10] which are obtained from wave height, wave period, sea-bottom slope, and median particle size, were 1.0 (accumulation type) for material a, 2.3 (accumulation type) for material b, and 4.6 (intermediate type) for material c.

4.3.4 Test Result

4.3.4.1 Case 1 (Material a; 100% Pebble)

The deformation process of beach nourishment filling in Case 1 is shown in Figure 4.3.12a–d. Right after starting wave generation, waves ran up to approximately 7 cm from the foot of the slope on the slope face of the filling, and pebbles near the foot of the

TABLE 4.3.1
Material Composition of Beach Nourishment Filling

Percentage Contained in the Beach Nourishment Filling (%)	Case							
	1	2	3	4	5	6	7	8
Material a (pebble)	100	0	0	50	33.3	0	0	13.3
Material b (granule)	0	100	0	50	33.3	40	30	13.3
Material c (sand)	0	0	100	0	33.3	60	70	73.3

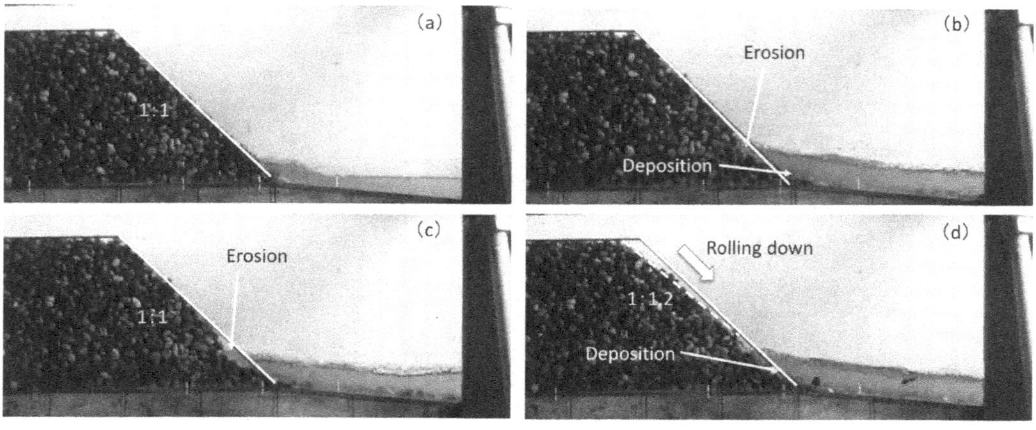

FIGURE 4.3.12
Deformation of beach nourishment filling in Case 1 (material a (pebble) 100%): (a) 0 min, (b) 5 min, (c) 10 min, and (d) 15 min after the start of wave generation.

slope were moved by the backwash and accumulated immediately offshore of the initial shoreline. Consequently, 5 min after the start of wave generation, the upper half of the portion between the wave run-up point and the initial position of the foot of the slope was eroded and the lower half of the slope became gentler than the initial slope gradient (a). Although the speed of topographical change was slow after that, the pebbles on the slope surface above the wave run-up point started to roll down after 11 min from the start of wave generation, burying the eroded portion around the foot of the slope. Conversely, the slope gradient of the portion above the wave run-up point became approximately 1:1.2 (d). Afterward, in contrast to immediately after the start of wave generation, erosion of the filling around the foot of the slope that was backfilled was not observed. It is considered that this is the effect of some pebbles accumulated immediately offshore of the initial shoreline. As shown above, in the case of beach nourishment filling composed of pebbles only, the foot of the slope of the beach nourishment filling was eroded right after the start of wave generation. After the washed-away pebbles have accumulated on the offshore side of the filling, the pebbles at the upper portion of the filling fell into the lower part of the filling that was eroded, and the upper slope surface of the filling became slightly gentler than before the wave generation. Thus, a beach scarp was not formed in this case.

4.3.4.2 Case 2 (Material b; 100% Granule)

The deformation process of beach nourishment filling in Case 2 is shown in Figure 4.3.13a–d. Right after starting wave generation, at the timing when the waves that ran up the slope surface of the filling receded, the granules around the foot of the slope were moved and accumulated on the offshore side of the initial shoreline. Since the amount moved to the offshore side was larger than in Case 1, granules on the slope surface above the wave run-up point started to roll immediately after the start of wave generation, and the amount accumulated on the offshore side of the initial shoreline was larger than in Case 1. Consequently, 5 min after the start of wave generation, the slope gradient

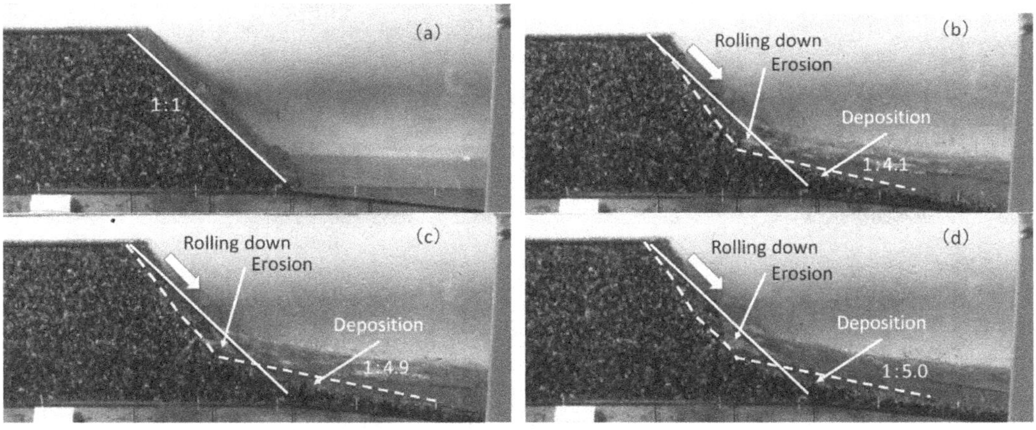

FIGURE 4.3.13
Deformation of beach nourishment filling in Case 2 (material b (granule) 100%): (a) 0 min, (b) 5 min, (c) 10 min, and (d) 15 min after the start of wave generation.

of the portion on the seaside of the wave run-up point became approximately 1:4, and the slope surface of the portion above this became slightly steeper than at the beginning (b). Although the speed of topographical change was slow compared with immediately after the increase, because of erosion of the portion below the wave run-up point, the rolling of granules on the slope surface above the wave run-up point was frequently observed. Consequently, from 10 min after the start of wave generation, the slope gradient of the offshore side of the wave run-up point was approximately 1:5 (d). As shown above, in the case of beach nourishment filling composed of granule only (Case 2), as with that composed of pebbles only (Case 1), the lower part of the filling was eroded and due to this, granules on the slope surface above the wave run-up point began to roll. Consequently, the slope surface below the wave run-up point became gentler than in the beginning while the slope surface above it became steeper. Thus, it could not be determined that a beach scarp was formed in this case.

4.3.4.3 Case 3 (Material c; 100% Sand)

The deformation process of beach nourishment filling in Case 3 is shown in Figure 4.3.14. As with Cases 1 and 2, right after the start of wave generation, the sand around the foot of the slope was moved at the time when the waves that ran up the slope surface of the filling receded, and sand accumulated on the offshore side of the initial shoreline. However, even if the slope surface below the wave run-up point got eroded, the slope surface above such point did not collapse and kept its shape, and a phenomenon in which the upper portion collapses after reaching an overhanging state was observed. The collapsed sand was moved to the offshore side by run-up waves, and the erosion of the point around the wave run-up point, formation of overhang, and collapse of the upper portion of the overhang were repeated. Consequently, 5 min after the start of wave generation, the slope gradient of the portion on the seaside of the wave run-up point became approximately 1:3.4, and the slope surface of the portion above such point became a shape that stands almost vertically (4.3.14b). The speed of topographical

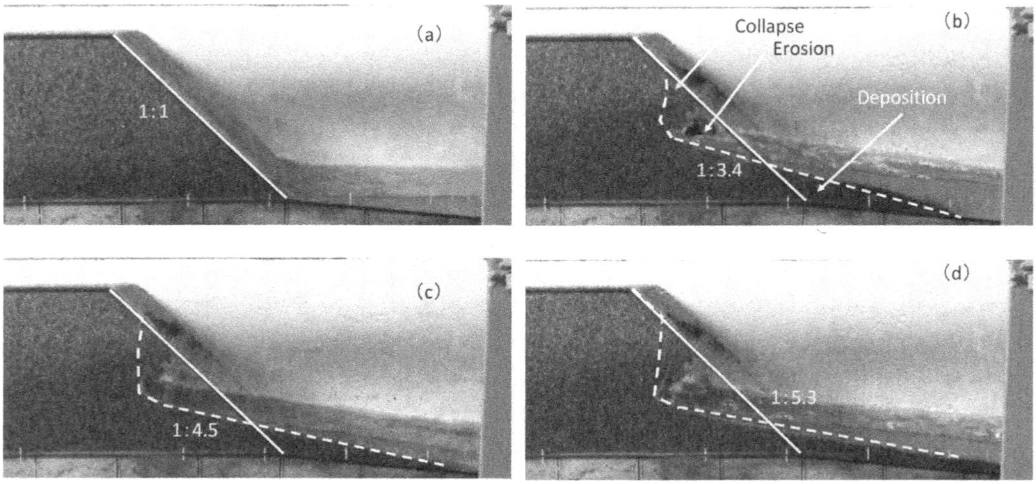

FIGURE 4.3.14
Deformation of beach nourishment filling in Case 3 (material c (sand) 100%): (a) 0 min, (b) 5 min, (c) 10 min, and (d) 15 min after the start of wave generation.

change was slow after that, and the slope gradient of the offshore side of the portion below the run-up point of waves became gentler gradually. Additionally, the topographical change of the portion above the wave run-up point did not reach the crown of the filling even after 15 min from the start of wave generation. As shown above, in the case of beach nourishment filling composed of sand only, the area around the wave run-up point was eroded right after the start of wave generation, and after that erosion of beach nourishment filling progressed while repeating the process of formation of overhang, the collapse of the upper portion of the overhang, and creation of a slope surface that rises steeply. Consequently, an almost upright beach scarp was formed.

4.3.4.4 Case 4 (Material a + Material b; 50% Pebble + 50% Granule)

The deformation process of beach nourishment filling in Case 4 is shown in Figure 4.3.15. As with Cases 1–3, right after the start of wave generation, the sand around the foot of the slope was moved at the time when the waves that ran up the slope surface of the filling receded, and sand accumulated on the offshore side of the initial shoreline. Additionally, as with Case 2 (material b; 100% granule), when the slope surface below the wave run-up point got eroded, gravel on the slope surface above the wave run-up point began to roll and accumulated on the offshore side of the initial shoreline. Consequently, 5 min after the start of wave generation, the slope gradient of the portion on the seaside of the wave run-up point became approximately 1:4.2, and the slope surface of the portion above this became slightly steeper than at the beginning (b). Other topographical changes were small: the gradient of the slope surface above and below the wave run-up point both remained almost the same (d). As shown above, the lower part of the filling was eroded and due to this, granules on the slope surface above the wave run-up point began to roll. Consequently, the slope surface below the wave run-up point became gentler than in the beginning while the slope surface above it became steeper.

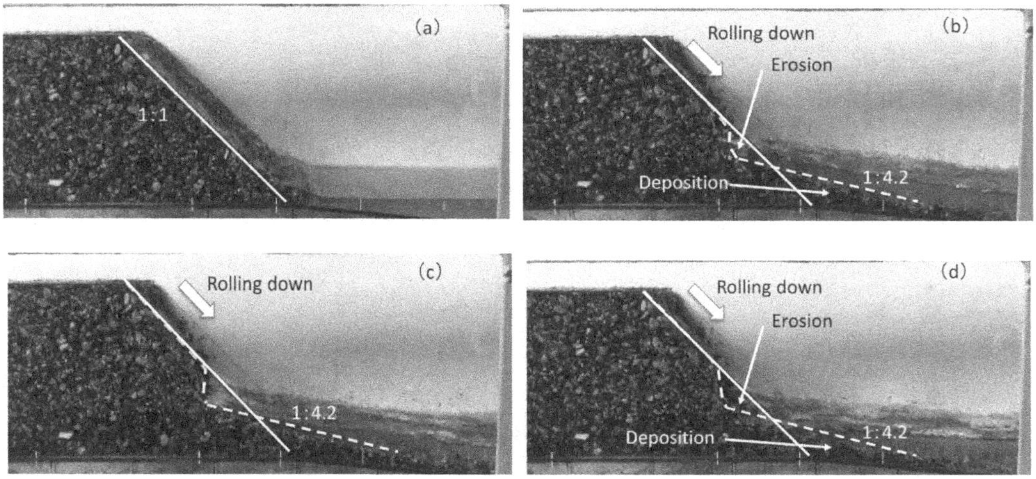

FIGURE 4.3.15
Deformation of beach nourishment filling in Case 4 (material a (pebble) 50% + material b (granule) 50%): (a) 0 min, (b) 5 min, (c) 10 min, and (d) 15 min after the start of wave generation.

Although the gradient of the slope surface near the wave run-up point was steeper than in Case 2 (material b; 100% granule), it could not be determined that beach scarp was formed in this case.

4.3.4.5 Case 5 (Material a + Material b + Material c; 33.3% Pebble + 33.3% Granule + 33.3% Sand)

The deformation process of beach nourishment filling in Case 5 is shown in Figure 4.3.16. As with Cases 1–4, right after the start of wave generation, the sand gravel around the foot of the slope was moved at the time when the waves that ran up the slope surface of the filling receded, and sand accumulated on the offshore side of the initial shoreline. However, as with Case 3 (material c; 100% sand), even if the slope surface below the wave run-up point got eroded, the slope surface above such point did not collapse and kept its shape, and a phenomenon in which the upper portion collapses after reaching an overhanging state was observed. The collapsed sand was moved to the offshore side by run-up waves, and the erosion of the point around the wave run-up point, formation of overhang, and collapse of the upper portion of the overhang were repeated. Consequently, 5 min after the start of wave generation, the slope gradient of the portion on the seaside of the wave run-up point became approximately 1:4, and the slope surface of the portion above such point became a shape that stands almost vertically (b). Although the speed of topographical change was slow compared with immediately after increase and the slope gradient of the offshore side of the portion below the wave run-up point became gentler gradually, after approximately 11 min from the start of wave generation, the upper part of the overhang collapsed and deformation reached the crown of the filling (d). As shown above, in the case of beach nourishment filling composed of a mixture of pebble, granule, and sand of the same ratio, as with Case 3 (material c; 100% sand), erosion of beach nourishment filling progressed while repeating the

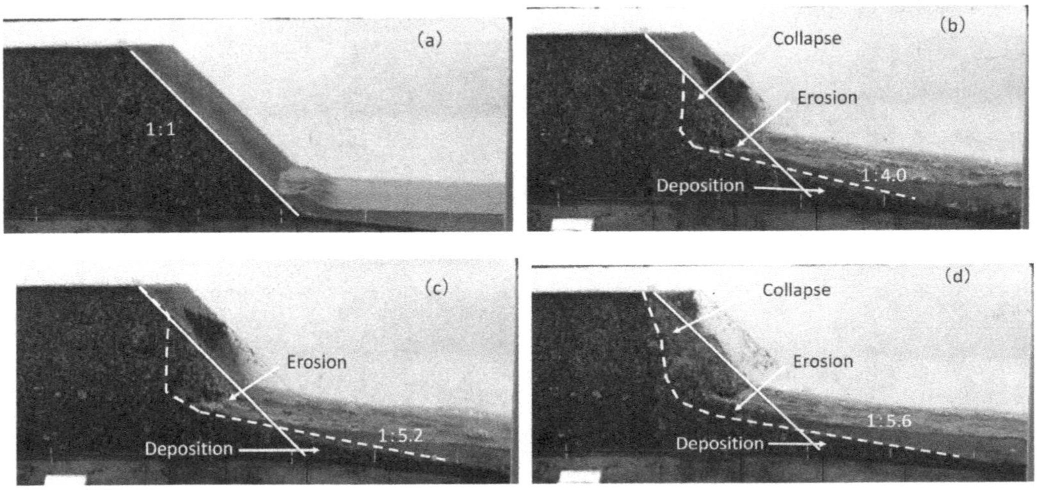

FIGURE 4.3.16
Deformation of beach nourishment filling in Case 5 (material a (pebble) 33.3% +material b (granule) + material c (sand) 33.3%): (a) 0 min, (b) 5 min, (c) 10 min, and (d) 15 min after the start of wave generation

process of erosion around the wave run-up point, the collapse of the upper portion of the overhang, and creation of a slope surface that rises steeply.

4.3.4.6 Case 6 (Material b + Material c; 40% Granule, 60% Sand)

The deformation process of beach nourishment filling in Case 6 is shown in Figure 4.3.17. Right after starting wave generation, at the timing when the waves that ran up the slope surface of the filling receded, the granules around the wave run-up point were moved

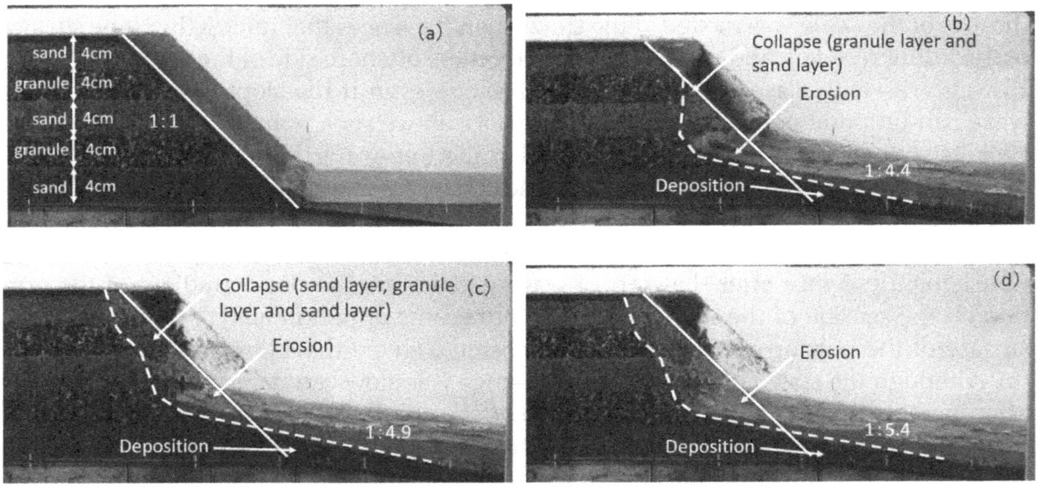

FIGURE 4.3.17
Deformation of beach nourishment filling in Case 6 (material b (granule) 40% + material c (sand) 60%): (a) 0 min, (b) 5 min, (c) 10 min, and (d) 15 min after the start of wave generation.

to the offshore side, and as the result of this erosion, the sand layer above it partially collapsed and the slope surface rose perpendicularly. Such erosion of granule layer and collapse of the sand layer above it repeated and eventually, not only the sand layer but also the gravel layer above it started to collapse. Consequently, after 5 min from the start of wave generation, while the sand layer in the middle level and the gravel layer above it formed a beach scarp with perpendicular slope surface, the gradient of the area on the seaside of the wave run-up point became approximately 1:4.4 (b). Erosion of granule layer around the wave run-up point continued thereafter. After 6 min from the start of wave generation, even a portion of the crown of the filling collapsed (c), and the gradient of the slope surface above the wave run-up point got slightly gentler (d). As shown above, in the case of beach nourishment filling of alternating sand and granule layers of even thickness, after the process of erosion of gravel layer in the lower portion of the filling that is on the wave run-up point and collapse of sand and gravel layers above it was repeated, a beach scarp with a gradient of 1:0.56 (approximately 61 degrees) was formed.

4.3.4.7 Case 7 (Material b + Material c; 30% Granule, 70% Sand)

The deformation process of beach nourishment filling in Case 7 is shown in Figure 4.3.18. Right after starting wave generation, at the timing when the waves that ran up the slope surface of the filling receded, the layer of granule in the lower part got eroded, and the sand layer above it got eroded by the collision of the run-up waves and became steep. Such erosion of granule layer and collapse of the sand layer above it were repeated, and eventually, not only the sand layer but also the granule layer above it as well as the sand layer and granule layers above it started to collapse. Consequently, after 5 min from the start of wave generation, while beach scarp with perpendicular slope surface was formed across the granule layer in the middle level and the granule layer in the upper level, the gradient of the portion on the seaside of the wave run-up point became approximately 1:4.8 (b). Erosion of granule layer around the wave run-up

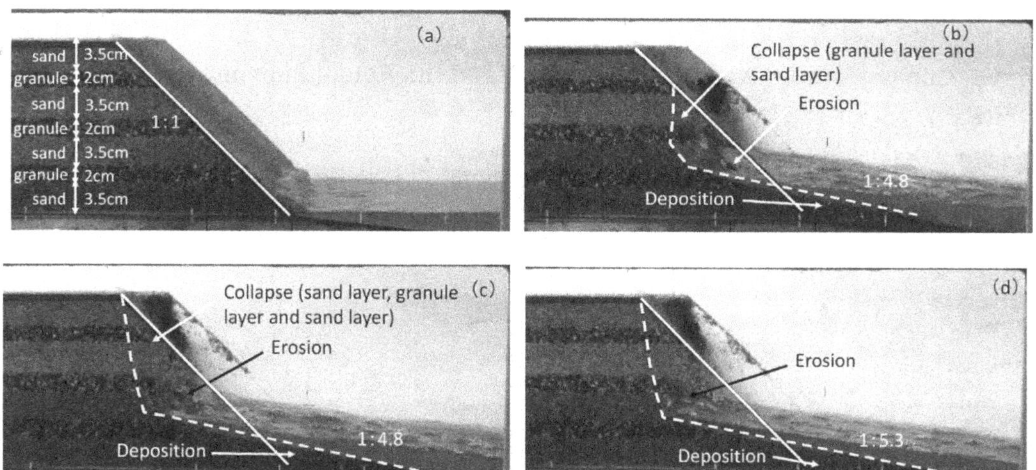

FIGURE 4.3.18
Deformation of beach nourishment filling in Case 7 (material b (granule) 30% + material c (sand) 70%): (a) 0 min, (b) 5 min, (c) 10 min, and (d) 15 min after the start of wave generation.

point continued thereafter, and after 7 min from the start of wave generation, even the portion from the notch formed in the sand layer in the middle level to the crown of the filling collapsed. As the gradient of the upper part of the slope surface after the collapse was 1:0.43 (approximately 67°), it was steeper than in Case 6 (material b + material c; 40% granule, 60% sand). As shown above, in the case of beach nourishment filling of alternating sand and granule layers of uneven thickness, as with Case 6, the sand and gravel layers below the wave run-up point got eroded, and the sand and granule layers above it consequently collapsed. However, the gradient of the upper part of the slope surface was steeper when compared with Case 6.

4.3.4.8 Case 8 (Material a + Material b + Material c; 13.3% Pebble + 13.3% Granule + 73.3% Sand)

Figure 4.3.19 shows the deformation process of beach nourishment filling in Case 8. Right after starting wave generation, at the timing when the waves that ran up the slope surface of the filling receded, the mixed material around the wave run-up point was moved to the offshore side, and as the result of this erosion, the sand layer and mixed material layer (same material as Case 5) above it partially collapsed and the slope surface rose perpendicularly. Consequently, after 5 min from the start of wave generation, while beach scarp with perpendicular slope surface was formed across the mixed material layer in the lower level and the mixed material layer in the upper level, the slope gradient of the portion on the seaside of the wave run-up point became approximately 1:4.8 (b). Additionally, unlike other cases, since the video from the same angle of view was not taken after 5 min from the start of wave generation, the state after this is not analyzed. As shown above, in the case of beach nourishment filling of the alternating sand layer and mixed material layer (composed of pebble, granule, and sand of the same ratio) of the same thickness, the sand layer and mixed material layer above the wave run-up point got collapsed because of erosion of mixed material layer at the wave run-up point, and such sand layer rose straight up, and this process was repeated.

4.3.5 Comparison between the Test Result and the Actual Phenomenon

4.3.5.1 Interpretation of the Test Result

It is assumed that the mechanisms of formation and collapse of beach scarp presented by van Bemmelen et al.[4.3.8] are assuming a situation in which irregular waves run-up to a beach composed of material of single particle size, and this is in contrast with the

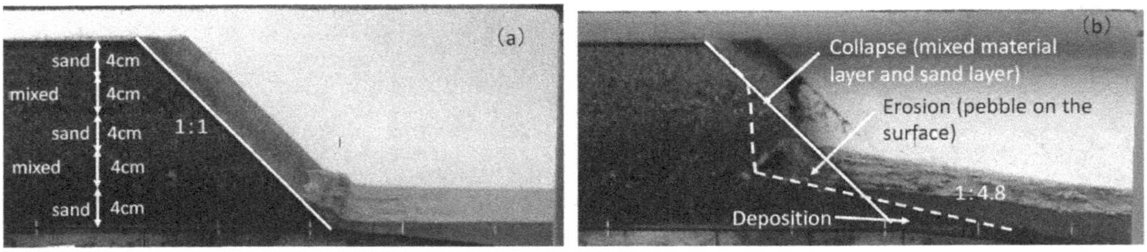

FIGURE 4.3.19
Deformation of beach nourishment filling in Case 8 (material a (pebble) 13.3% + material b (granule) 13.3% + material c (sand) 73.3%): (a) 0 min and (b) 5 min after the start of wave generation.

test in which regular waves run-up to beach nourishment filling composed materials of several particle sizes shown in Section 4.3.4. However, we will invoke the knowledge of van Bemmelen et al.[4.3.8] in an attempt to interpret the test result in this section.

Table 4.3.2 shows the height and gradient of the beach scarp formed in each case and their interpretations. In the test conducted with regular waves, the run-up height of waves did not go beyond the lower part of the slope surface of beach nourishment filling and was constant in all cases and throughout the whole period of the test. Although the equation to obtain the gradient of the upper part of the slope β_u presented by van Bemmelen (Section 4.3.2) cannot be used since the run-up height of the wave is constant, in the case that used sand as the material, it is assumed that beach scarp was formed because the equilibrium beach gradient was gentler than the initial gradient of the beach nourishment filling. Additionally, compared with Case 3 in which beach nourishment filling was composed of sand only, in Case 5 in which an even mixture of pebble, granule, and sand was used, notches were hardly formed, and a larger amount of the filling was eroded after 15 min from the start of wave generation. In Cases 1, 2, and 4 in which beach nourishment filling was composed of gravel (pebble and granule) only, a perpendicular slope surface was not formed. It is assumed that this is because the slope surface of the beach nourishment filling could not rise steeply because suction was small with beach nourishment filling composed of gravel only. Nevertheless, in Case 4 in which a mixture of pebble and granule was used, the gradient of the slope surface near the wave run-up point was steep, indicating that the interlocking effect of gravel of different particle sizes may have been produced. Conversely, in Cases 6 and 7 in which the layer of a single type of gravel and the layer of sand were installed in a layered structure, the form of erosion was different depending on the material or thickness of the sand layer near the wave run-up point. Specifically, in Case 6 (material b + material c; 40% granule, 60% sand), although the erosion near the run-up point was slightly

TABLE 4.3.2

Erosion Situation of Beach Nourishment Filling after the Test in Each Case

Case	Material	Height of Beach Scarp	Gradient of Beach Scarp	Explanation
1	Pebble	No scarp	–	Difficult to be eroded due to large particle size
2	Granule	No scarp	–	Eroded but no beach scarp formed
3	Sand	Medium	Almost upright	The upper part of the filling did not collapse due to suction
4	Pebble, granule	No scarp	–	The slope near the wave runup point became steeper due to the engagement of the pebble and the granule
5	Pebble, granule, sand	High	Steep	The upper part of the filling collapsed larger than case3
6	Granule, sand	High	Slightly mild	The slope of the granule layer with small suction became milder
7	Granule, sand	High	Steep	Beach scarp became steep because the granule layer was thin
8	Pebble, granule, sand	Medium	Steep	Beach scarp became steep due to the large amount of sand

* pebble (D_{50}:8.2 mm), granule (D_{50}:2.3 mm), sand (d_{50}:0.82 mm)

slower than in Case 7 (material b + material c; 30% granule, 70% sand) because the run-up point was on a gravel layer since the gravel layer above it could not rise steeply, the recession of the upper part of the filling was greater than in Case 7.

As shown above, it is assumed that the erosion of beach nourishment filling in tests presented in this section was caused through the processes of (1) the deformation of the slope of the beach nourishment filling because of the sand gravel on the slope surface of the beach nourishment filling moved by waves and (2) the collapse of the slope at the time it became unstable because of deformation of the slope.

4.3.5.2 Law of Similarity to Be Applied

Although Noguchi et al.[4.3.1] have not presented the scale of the test or law of similarity applied, considering the deformation processes of beach nourishing filling sorted out in Section 4.3.5, the law of similarity of a combined test regarding waves and ground discussed in Section 3.4.4 can be applied. However, as stated in Section 3.4.4, in tests using water in a gravitational force field, the similitude ratio of time for ground and fluid does not match. Thus, when applying the law of similarity related to the combined test of wave and ground, an appropriate time-related similitude ratio must be selected on the basis of the mechanism of each phenomenon that occurs during the deformation process of beach nourishment filling. Regarding the area in the beach nourishment filling where waves run-up since erosion occurs mainly because of bedload transport, it can be considered that the law of similarity of bedload transport specified in Section 3.4.1 can be applied. Conversely, the beach nourishment filling that is above the area where wave runs up, it is assumed that application of the law of similarity considering the characteristics of unsaturated ground is necessary since the effect of suction is observed as stated above. As shown in Section 3.3.5, the similitude ratios of unsaturated ground and saturated ground differ in coefficient of permeability and so on. In comparison, Honma et al.[4.3.6)] defined the particle size for the slope of the embankment on the basis of the law of similarity of permeability coefficient. Additionally, regarding the law of similarity considering the unsaturated capillary flow, Yoshida[4.3.11)] presented that the similarity of the flow rate of seepage flow increases by obtaining equal geometrical scale ratio and static capillary rise ratio and by reconciling the negative pressure–water content curve. In a miniature model in a gravitational field, by obtaining an equal geometrical scale ratio and static capillary rise ratio, you will be selecting a particle size that is larger than that of the site so that the static capillary rise will be smaller than that of the site. Additionally, it is assumed that material composition is adjusted to reconcile with the negative pressure–water content curve.

4.3.6 Conclusion

- In the case of beach nourishment filling that is composed of material that includes sand, vertical slope surface is formed easily when waves run-up partway of the slope surface of the embankment.
- In the case of beach nourishment filling of alternating sand layer and gravel layer, vertical slope surface is formed easily if there is a sand layer above the run-up point.

- The above trends are believed to be because the slope surface stabilizes easily since suction in the sand layer is greater than that in the gravel layer.
- Although the law of similarity that should be applied to erosion of beach nourishment filling is the law of similarity for combined tests regarding waves and ground, the law of similarity considering the characteristics of unsaturated ground must be applied to the portion above the run-up height of the waves.

References

4.2 Case (1) Scouring phenomenon under wave-dissipating blocks

4.2.1 Suzuki, K., Takahashi, K., Takano, T., Shimosako, K.: Settlement failure of wave dissipating blocks in front of caisson type breakwater due to scouring under the rubble mound: Field investigation and large scale experiment, Report of PARI, Vol.41, No.1, pp.51–90, 2002. (in Japanese)

4.2.2 Nishida, H., Yamaguchi, Y., Kondoh, T., Shimizu, K.: Koukan danseiha tansahou niyoru rigantei no maibotsu jyoukyou ni kansuru kousatsu [A measurement of the sinking status of wave dissipating blocks of detached breakwaters by seismic survey], Proceedings of the 32nd Japanese Conference on Coastal Engineering, pp.365–369, 1985. (in Japanese)

4.2.3 Irie, I., Kuriyama, Y., Asakura, H.: Study on scour in front of breakwaters by standing waves and protection method, Report of PHRI, Vol.25, No.1, pp.3–86, 1986. (in Japanese)

4.2.4 Suzuki, K., Takahashi, K.: An experiment on settlement of wave dissipating blocks in front of breakwaters, Proceedings of the 45th Japanese Conference on Coastal Engineering, pp.821–825, 1998. (in Japanese)

4.2.5 Nakamura, T. Muraoka, H., Cho, Y., Mizutani, N.: Experimental study on sand leakage from under composite breakwater covered with wave dissipating blocks, Journal of Japan Society of Civil Engineering, Ser. B3 (Ocean Engineering), Vol.74, No.2, p.I_258–I_263, 2018. (in Japanese)

4.2.6 Muraoka, H., Nakamura, T., Cho, Y., Mizutani, N.: Experimental study on erosion of sand seabed under rubble mound of composite breakwater covered with wave dissipating blocks and effects of gabions, Journal of Japan Society of Civil Engineering, Ser. B1 (Hydraulic Engineering), Vol.74, No.5, p.I_595–I_600, 2018. (in Japanese)

4.2.7 Rubey, W.W.: Settling velocities of gravel, sand and silt particles, American Journal of Science, Vol.25, pp.325–338, 1933.

4.2.8 Shigeo, T., Suzuki, K., Kang, Y., Tsunakazu, K.: An experiment on wave-induced liquefaction of a fine-grained sandbed, Proceedings of the 44th Japanese Conference on Coastal Engineering, pp.916–920, 1997. (in Japanese)

4.2.9 Suzuki, K., Tada, K., Shimosako, K., Yamazaki, H., Kang, Y.: An experiment on wave-induced liquefaction of sandbed in a large-scale wave flume, Proceedings of the 50th Japanese Conference on Coastal Engineering, pp.856–860, 2003. (in Japanese)

4.2.10 Chen, S., Suzuki, K., Arikawa, T.: Influence of scaling effect for sand leakage into the mound of wave dissipating blocks, Journal of Japan Society of Civil Engineering, Ser. B2 (Coastal Engineering), Vol.75, No.2, pp.I_547–I_552, 2019. (in Japanese)

4.2.11 Shimada, M., Fujimoto, T., Saito, S., Sakakiyama, T., Hiraguchi, H.: Shouha burokku no anteisei ni kansuru mokei shukushaku kouka ni tsuite [Model scale effects on the stability of wave dissipating blocks], Proceedings of the 33rd Japanese Conference on Coastal Engineering, pp.442–445, 1986. (in Japanese)

4.2.12 Sakakiyama, T., Kashima, R.: Shouha burokku ni sayousuru haryoku ni kansuru jikken suke-ru kouka [Experimental scale effects on wave forces acting on wave dissipating blocks], Proceedings of the 36th Japanese Conference on Coastal Engineering, pp.653–657,1989. (in Japanese)
4.2.13 Araki, S., Sawada, Y., Miyamoto, J., Ushiyama, H., Tanaka, Y., Kotake, Y.: Examination of sand scour under rubble mound of a composite breakwater with dissipating blocks in a drum centrifuge, Journal of Japan Society of Civil Engineering, Ser. B2 (Coastal Engineering), Vol, 74, No.2, p.I_1093–I_1098, 2018. (in Japanese)
4.2.14 Hughes, S.A.: Physical Models and Laboratory Techniques in Coastal Engineering, World Scientific, 568pp., 1993.
4.2.15 Hioki, K., Iwanaga, S., Nakamura, S. and Hongo, T.: Sairyuudo no busshitsu idou parameta to baria seinou ni kansuru kousatsu [Examination of the mass transfer parameters and functions as a barrier for fine grained soils], Journal of Japan Society of Civil Engineering, Ser. C (Geotechnics), Vol.67, No.2, pp.240–251, 2011. (in Japanese)
4.2.16 Uchida Y., Komine, H., Yasuhara, K., Murakami, S., and Endo, K.: Haikibutsu saishu shobunjyou fukudozai ni okeru metan sanka saikin no ikusei kanou na jyouken no teiji [Proposal of applicable conditions for methanotrophs activity in a final cover material of waste disposal site], Japanese Geotechnical Journal, 3(1), pp.85–93, 2008. (in Japanese)
4.2.17 Hideo Sekiguchi, Phillips, R.: Enshinryokuba ni okeru suimenha no zouha to sono tekiyou [Wave making in centrifugal force field and its application], Proceedings of Civil Engineering in the Ocean, Vol.6, pp.205–210, 1990. (in Japanese).

For the design of breakwaters discussed in the Section 4.2, it is advisable to refer to following literature:

4.2.18 The Overseas Coastal Area Development Institute of Japan: Technical Standards and Commentaries for Port and Harbour Facilities in Japan, OCDI, 2182pp, 2020.
4.2.19 Goda, Y.: Random Seas and Design of Maritime Structures, 3rd ed., World Scientific, 732pp., 2010.

4.3 Case (2) Erosion of Beach Nourishment Filling

4.3.1 Noguchi, K,, Kato, F. and Sato, S.: Sand-gravel mixed foreshore nourishment to improve erosion resistant performance, Journal of JSCE, B2, Vol.73, No.2, pp.I_799–I_804, 2017.
4.3.2 Serizawa, M., Uda, T. and Hoshigami, Y.: A model for predicting scarp formation in foreshore nourishment, Proceedings of Civil Engineering in the Ocean, Vol.21, pp.1011–1016, 2005.
4.3.3 Hattori, M. and Kakegawa, T.: Hamagake no keiseikatei to Hasseijoken, [Formation process and occurrence conditions of beach scarp], Proceedings of Coastal Engineering, JSCE, Vol.41, pp.546–550, 1994.
4.3.4 Uda, T. and Serizawa, M.: A Note on Formative Process of Scarp, Proceedings of Civil Engineering in the Ocean, Vol.12, pp.409–414, 1996.
4.3.5 Nishi, R. and Kraus, N.C.: Sakyushinshokukiko to moderuka ni tsuite, [On the mechanisms and modeling of sand dune erosion], Proceedings of Coastal Engineering, JSCE, Vol.43, pp.676–680, 1996.
4.3.6 Honma, D., Muyatake, M., Sassa, S., Kimura, K., Shirozu, H., and Ebiko, T.: An experiment to elucidate the mechanisms behind high-wave-related failure of backshore slopes on coastal roads in consideration of changes in ground properties, Journal of JSCE, B2, Vol.72, No.2, pp.I_1189–I_1194, 2016.
4.3.7 Erikson, L.H., Larson, M., and Hanson, H.: Laboratory investigation of beach scarp and dune recession due to notching and subsequent failure, Marine Geology, 245, pp.1–17, 2007.

4.3.8 van Bemmelen, C.W.T.: Beach Scarp Morphodynamics: Formation, Migration, and Destruction, Master thesis, Technical University of Delft, 2018.
4.3.9 van Bemmelen, C.W.T., de Schippera, M.A., Darnall, J., and Aarninkhofa, S.G.J., Beach scarp dynamics at nourished beaches, Coastal Engineering, vol.160, 2020. https://doi.org/10.1016/j.coastaleng.2020.103725
4.3.10 Horikawa, S., Sunamura, T. and Kondo, H.: Nami ni yoru nijigenkaihinhenkei ni kansuru jikkentekikenkyu, [Experimental studies on two-dimensional beach transformation induced by waves], Proceedings of the Japanese Conference on Coastal Engineering, Vol.21, pp.193–200, 1974.
4.3.11 Yoshida, S.: Fundamental studies of the flow through porous media, Bulletin of Yamagata University, Agric., Sci., Vol.5, No.3, pp.257–329, 1968.

5

Numerical Calculation Methodology for Wave–Seabed Interaction

Tomoaki Nakamura, Taro Arikawa, Kazunori Fujisawa, and Eiji Harada

5.1 Role of Numerical Calculation

As described in Chapters 1 to 4, it is useful to conduct model experiments for the understanding of phenomena since we can obtain much useful information through model experiments. Nevertheless, considering the economical efficiency such as labor, time, and expense in addition to constraints such as the scale and capability of experimental facilities, conducting model experiments by changing conditions in various ways can often be a burden. Additionally, there are problems regarding the acquisition of data since not all physical quantities you wish to know can be obtained through model experiments because of reasons such as the limitation of the measurement device.

Conversely, numerical calculations have the advantage of enabling evaluation from various viewpoints compared with model experiments since it can output spatiotemporally detailed information. Additionally, it is expected that various conditions can be modified easily once the setting for calculation is established, reducing labor.

Nevertheless, it is well-known that verification and validation are important when conducting numerical calculations. Although we would leave the explanation of details to Shiratori et al.,[5.1.1] verifying the appropriateness of the discretization of the given fundamental equation, the correctness of the coding, grid convergence properties, and the influence of the time step is indispensable. For example, when discretizing the advection term of Navier–Stokes equation, introducing excessive upwind difference can produce excessive numerical viscosity, leading to the inappropriate calculation result. Conversely, regarding validation, in direct numerical simulation (DNS) that conduct numerical calculations without utilizing a turbulence model in the fundamental equation, all of the fluctuations of turbulence from the whole area of the flow field to the turbulence of the smallest size (Kolmogorov scale) that bear energy dissipation will be calculated with the Navier–Stokes equation. Thus, since enormous computational capacity is necessary, it is very unlikely that this can be applied to practical problems.[5.1.2] Additionally, with phenomena that are related to the seabed, all of the particles of the materials of the seabed must be considered in the calculation, making it further unfeasible. Thus, modeling of turbulence, hydrodynamic force, constitutive equation, and so on are essential, making the validation of the model necessary. Also, for example, in the grid method discussed later, the feedback from the model of the seabed to the model

DOI: 10.1201/9781003424758-5

of the wavefield is hardly provided. As you can see, some phenomena are left without being modeled.

Additionally, as with model experiments, the numerical calculation is also affected by scale effects. When the grain size D of the permeable material and the mean flow velocity U are selected as a representative scale of the flow field, the Reynolds number Re will be $Re = UD/v$ when the kinematic viscosity of the fluid is v. In this case, if the scale of energy dissipation due to viscosity is η, the ratio of D and η (D/η) will be roughly proportional to $Re^{3/4}$.[5.1.3] Here, we would like to consider the situation in which the dimensional scale of the model is $1/N$ of the prototype on the basis of Froude's law. If the law of similarity of settling velocity (Dean's law) discussed in Chapter 3 is followed, when Rubey's equation is used to calculate settling velocity, D of the model will become asymptotically closer to $1/N$ of the prototype when D is relatively large. Assuming that fluid whose ν is equal in the model and prototype is used, Re of the model will be $1/N^{3/2}$ of the prototype. In other words, D/η of the model will be $1/N^{9/8}$ of the prototype. Since D of the model becomes asymptotically closer to $1/N$ of the prototype, η of the model is $N^{1/8}$ times that of the prototype, meaning the scale of the energy dissipation is larger in the experiments. This corresponds to the fact that the influence of the viscous force relatively increases under Froude's law. Since the experiments on the prototype are difficult due to constraints of experimental facilities, validation is performed on the model in general. Considering that this numerical calculation is expanded to the prototype which is N times the model and given that the scale in which energy dissipation occurs is $1/N^{1/8}$ times the model, if the grid of the prototype is assumed by simply multiplying the model by N based on Froude's law, the width of the scale that can be resolved will be different between the model and the prototype. Thus, it is vital to perform verification and validation that are appropriate for the phenomenon to be studied or the scale and to use them within their application range.

As stated above, since model experiments and numerical calculations both have merits and demerits, proceeding with the examination by combining model experiments and numerical calculations can be considered to be effective measures in solving problems related to interactions between waves and seabed. Among the above, the numerical calculations that can deal with both wave and seabed are covered in this chapter. First, in Section 5.2, we will review the flow of development and advancement of numerical calculation models introduced so far. Then, to give examples of numerical calculation models, we will introduce the following models in the next sections: a model that uniformly solves the fluid flow outside and inside the porous media in Section 5.3, a model that is classified as particle method (which will be explained later) in Section 5.4, and a model that is classified as grid method in Section 5.5. Details on turbulence models can be found in Kajishima[5.1.3] and Kajishima and Taira.[5.1.4]

5.2 Development of Numerical Calculation for Wave–Seabed Interaction

Research Subcommittee on Numerical Wave Flume of JSCE Coastal Engineering Committee[5.2.1] classified the numerical calculations that are related to both wave and seabed into four categories as shown in Table 5.2.1.

TABLE 5.2.1

Classification of Numerical Calculations for Wave–Seabed Interaction[5.2.1]

Classification	Wave	Ground
Type 1	Given by theories such as the small-amplitude wave theory	Analyzed by grid-based models such as the finite element method (FEM)
Type 2	Analyzed by numerical models	Analyzed by grid-based models such as FEM
Type 3	Given by theories such as the small-amplitude wave theory	Analyzed by particle-based models such as the distinct element method (DEM)
Type 4	Analyzed by numerical models	Analyzed by particle-based models such as DEM

5.2.1 Grid Method

Type 1 and Type 2 are methods that are based on a grid or element that is based on finite difference method (FDM), finite volume method (FVM), finite element method (FEM), and boundary element method (BEM) and referred to as the grid method. Although the model of Type 1 that gives the wavefield by theoretical value has been developed, since it cannot handle the interactions between waves and seabed, this section will cover Type 2, which gives the wavefield by numerical analysis. Since the research introduced hereafter is summarized in Tables 5.2.2 and 5.2.3, please refer to them as necessary. Please note that Type A and Type W, which are methods for coupling waves and seabed, are described later.

Park et al.[5.2.2] developed a model that is based on FEM that applies potential theory to wavefield and applies u–w formulation of Biot's equation to seabed. Here, u indicates the displacement of the soil skeleton, and w indicates the relative displacement of pore water regarding the soil skeleton. Although this model calculates in a frequency domain unlike the models introduced below, it is a two-way coupling of strong coupling that solves both wavefield and seabed by combining them in a single simultaneous equation. Thus, since it can solve a wavefield considering the displacement of the soil skeleton, the damping of wave height over soft seabed has been reproduced using this model.

Mizutani and Mostafa[5.2.3] developed a two-dimensional model that combines BEM that solves wavefield; FEM that analyzes the seepage flow of the seabed; and FEM based on Biot's consolidation equation that analyzes the coupling between pore water and soil of the seabed, and the constitutive equation of linear elastic solid. In this model, the one-way coupling is utilized. The flow field of the whole area, including the inside of the seabed, is solved first via BEM and FEM that analyze the seepage flow, and then the pressure at the seabed surface obtained from the result is given to the FEM that analyzes the coupling between water and soil. This coupling is hereinafter referred to as Type A (All). The developed model has been applied to the caisson-type composite breakwater shown in Figure 5.2.1.[5.2.4]

CADMAS-SURF (SUper Roller Flume for Computer-aided Design of MAritime Structure) was developed by "Research group for application of numerical wave flume to maritime structure design" to develop a numerical wave flume that replaces the existing hydraulic model experiment using two-dimensional wave flume on the basis of the idea that numerical calculation method that is based on the volume-of-fluid (VOF) method is a potential option of numerical wave flume.[5.2.5] This research group comprises research institutes of industry, government, and academia. It aims to develop

TABLE 5.2.2

List of Type A Models (the Flow Field of the Whole Area Including the inside of the Seabed Is Solved, and the Pressure and Flow Velocity at the Seabed Surface Obtained from the Result Is Given to the FEM)

Wave	Potential theory				
	BEM	•			
	VOF		•		
	CADMAS				
	OpenFOAM				
	RANS			•	•
Seepage		•			
Ground	Consolidation equation	•			
	u–*p* formulation		•	•	•
	u–*w* formulation				
	u–*w*–*p* formulation				
Constitutive equations	Linear elastic	•	•	•	
	Elastic-plastic				•
Year	Before 2000	Mizutani and Mostafa (1997)[5.2.3] [2D]			
	2000–2010		Nakamura et al. (2005)[5.2.18] [2D]	Hsu et al. (2002)[5.2.25] [2D]	
	2010–2020		Nakamura et al. (2013)[5.2.20] [3D]	Ye et al. (2013)[5.2.27] [3D]	Ye et al. (2015)[5.2.28] [3D]
Remarks	Coupling	One-way	One-way	One-way	One-way

the first numerical flume in Japan to create a numerical wave flume that can be applied to the practical task of design of maritime structures and its code was published in 2001 by Coastal Development Institute of Technology.[5.2.6] Based on the porous model of Sakakiyama and Imai,[5.2.7] its fundamental equations are the two-dimensional continuity and Navier–Stokes equations for incompressible fluid.

Additionally, Jiang et al.[5.2.8] developed a coupled calculation of CADMAS-SURF and soil deformation in *u*–*w* formulation as they thought it is important to consider the interactions with the seabed in breakwaters and seawalls from back then. The model of Jiang et al.[5.2.8] is a pseudo-two-way coupling that first solves the flow field of the area except for the inside of the seabed by FDM and then inputs the pressure at the seabed surface obtained by FDM to FEM and then returns the flow velocity at the seabed surface obtained by FEM to FDM. This coupling is hereinafter referred to as Type W (Wave). Takahashi et al.[5.2.9] then upgraded it to a coupled method with seabed calculation in *u*–*p* formulation to improve the stability of the numerical calculation and established the coupled calculation method with the seabed when wave force was subjected to the breakwater, applying it to various conditions, including caisson-type seawall shown in Figure 5.2.2. Here, p represents pore water pressure and since variables of u and p overlap with flow velocity and pressure in the wavefield, it is referred to as U-π formulation by Takahashi et al.[5.2.9] Arikawa et al.[5.2.10]

TABLE 5.2.3

List of the Type W Models (the Flow Field of the Area Except for the Inside of the Seabed Is Solved Assuming Impermeable or Permeable, and the Pressure at the Seabed Surface Obtained from the Result Is Given to the FEM)

Wave	Potential theory	•				
	BEM					
	VOF					
	CADMAS		•	•	•	•
	OpenFOAM					
	RANS					
Seepage						•
Ground	Consolidation equation			•		
	u–p formulation				•	•
	u–w formulation	•	•			
	u–w–p formulation					
Constitutive equation	Linear elastic	•	•	•	•	
	Elastic-plastic					•
Year	Before 2000	Park et al. (1996)[5.2.2] [2D]				
	2000–2010		Jiang et al. (2000)[5.2.8] [2D]	Maeno and Fujita (2001)[5.2.17] [2D]	Takahashi et al. (2002)[5.2.9] [2D]	Arikawa et al. (2009)[5.2.11] [3D]
	2010–2020					
Remarks	Coupling	Two-way	Pseudo two-way	One-way	One-way	Pseudo-two-way giving the deformation as the porosity. Seepage flow analysis is performed separately

(Continued)

TABLE 5.2.3 ***(Continued)***

List of the Type W Models (the Flow Field of the Area Except for the Inside of the Seabed Is Solved Assuming Impermeable or Permeable, and the Pressure at the Seabed Surface Obtained from the Result Is Given to the FEM)

Wave	Potential theory				
	BEM				
	VOF				
	CADMAS				
	OpenFOAM				•
	RANS	•	•	•	
Seepage					
Ground	Consolidation equation				
	u–*p* formulation	•		•	•
	u–*w* formulation				
	u–*w*–*p* formulation		•		
Constitutive equation	Linear elastic		•	•	•
	Elastic–plastic	•			
Year	Before 2000				
	2000–2010	Kumagai (2009)[5.2.21] [2D]			
	2010–2020		Zhang et al. (2015)[5.2.29] [3D]	Zhang et al. (2020)[5.2.31] [3D] Lin et al. (2020)[5.2.32] [3D]	Li et al. (2020)[5.2.34] [3D]
Remarks	Coupling	One-way	One-way	One-way	One-way

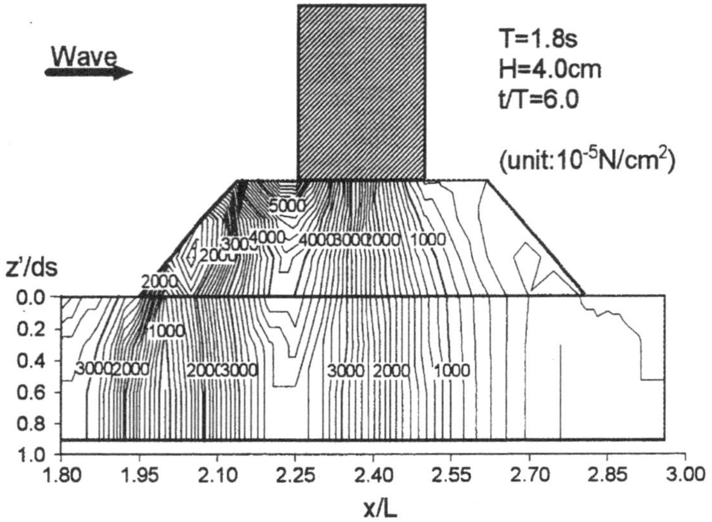

FIGURE 5.2.1
Pressure around the composite-type breakwater.[5.2.3]

developed a three-dimensional CADMAS-SURF/3D on the basis of the two-dimensional CADMAS-SURF. Then, the research aimed at putting CADMAS-SURF/3D into practical use was launched by the member and institution of "Research group for application of numerical wave flume to maritime structure design" and the code was published in 2010 by Coastal Development Institute of Technology.[5.2.11] Arikawa et al.[5.2.12] developed a coupled calculation with structure by FEM, and in 2011, they developed a gas–liquid–solid three-phase model that is also coupled with DEM.[5.2.13] Then, they enabled the coupled calculation of gas, liquid, solid, and seabed by utilizing the same method as coupling with soil deformation calculation regarding two-dimensional CADMAS-SURF, and furthermore, they developed a coupled calculation method between a nonlinear shallow water equation and CADMAS-SURF/3D as well as the gas–liquid–solid coupled model.[5.2.14]

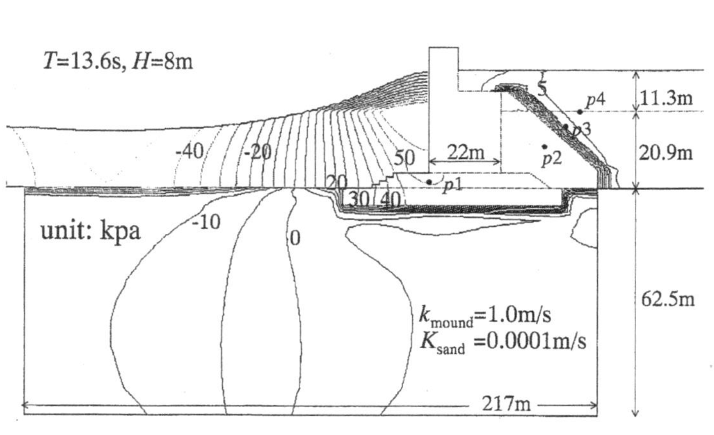

FIGURE 5.2.2
Pressure around the caisson-type seawall.[5.2.9]

Consequently, a series of calculations regarding the generation and propagation of tsunamis and the destruction of structures became possible. Additionally, Ohki et al.[5.2.15] conducted a verification of the validity regarding the calculation of seepage. This model was published in 2021.[5.2.16]

Maeno and Fujita[5.2.17] developed a two-dimensional VOF-FEM model that applied FEM based on CADMAS-SURF for wavefield and FEM based on Biot's consolidation equation and constitutive equation of linear elastic solid for seabed. The model of Maeno and Fujita[5.2.17] uses one-way coupling of Type W; the calculation of the wavefield by CADMAS-SURF assuming the seabed is impermeable is done first, and then the pressure at the seabed surface obtained from the result is input in FEM. Nakamura et al.[5.2.18] developed a two-dimensional model that combines FDM that is based on the VOF method applied to wavefield and FEM that is based on *u–p* formulation of Biot's equation and constitutive equation of linear elastic solid applied to the seabed. By using the model of Mizutani and Mostafa[5.2.3] as a reference, the model of Nakamura et al.[5.2.18] adopts a one-way coupling of Type A that solves the flow field of the whole area, including inside of seabed by FDM, and then passes the obtained pressure and flow velocity at the seabed surface to FEM. Additionally, Nakamura et al.[5.2.19] developed it into a three-dimensional model by replacing the model of the wavefield with large eddy simulation (LES) that is based on multi-interface advection and reconstruction solver (MARS). Moreover, Nakamura et al.[5.2.20] proposed a method of considering the effect of topographic changes due to sediment transport in a simple way, and by integrating this to the previously mentioned three-dimensional model, they developed the three-dimensional coupled fluid–structure–sediment–seabed interaction model (FS3M). Kumagai et al.[5.2.21, 5.2.22] developed FEM that is based on *u–p* formulation of Biot's equation that applies the constitutive equation of elastoplastic solid and constructed a two-dimensional model connected by VOF method that solves wavefield and one-way coupling of Type W. Then, they applied the model to the composite breakwater, etc. as shown in Figure 5.2.3.

Jeng et al.[5.2.23] and Ye et al.[5.2.24] developed a two-dimensional model called PORO-WSSI II (porous models for wave–seabed–structure interactions, version II) or FSSI-CAS 2D (FSSI: fluid–structure–seabed interaction, CAS: Chinese Academy of Sciences). This is a model that couples FDM of wavefield of Hsu et al.[5.2.25] that is based on volume-averaged Reynolds-averaged Navier–Stokes (VARANS) equation that applies VOF method to

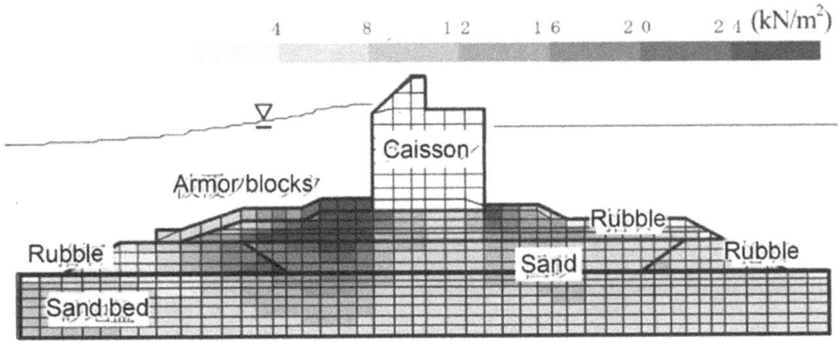

FIGURE 5.2.3
Maximum shear stress around the composite-type breakwater.[5.2.22]

track free surface and FEM of seabed that consist of *u–p* formulation of Biot's equation and constitutive equation of linear elastic solid. Adopting the same coupling method as the model of Mizutani and Mostafa,[5.2.3] this model is a one-way coupling of Type A that solves the flow field of the whole area, including inside of seabed, by FDM and then inputs the obtained pressure at the seabed surface to FEM. With FSSI-CAS 2D, Ye et al.[5.2.26] evaluated the response of the seabed in the vicinity of the composite breakwater when breaking wave is applied, as shown in Figure 5.2.4. Additionally, Ye et al.[5.2.27] made the model three dimensional and developed it into FSSI-CAS 3D. Ye et al.[5.2.28] also introduced the constitutive equation of elastoplastic solid to FSSI-CAS 2D to assess the probability of occurrence of liquefaction associated with an increase in excess pore water pressure. Zhang et al.[5.2.29] developed a three-dimensional FDM that utilizes FDM that is based on RANS equation and VOF method for the wavefield and utilizes FDM that is based on *u–w–p* formulation of Biot's equation and constitutive equation of linear elastic solid for the seabed. The same one-way coupling of Type W as the model of Maeno and Fujita[5.2.17] is used in the model of Zhang et al.[5.2.29] FDM of the wavefield is first solved by imposing nonslip conditions on the seabed surface, and then the obtained pressure of the seabed surface is input to FDM of the ground. Then, wave response analysis of the seabed in the vicinity of the monopile foundation was conducted with the model developed, as shown in Figure 5.2.5. Additionally, Zhang et al.[5.2.30] developed it to inhomogeneous seabed. Zhang et al.[5.2.31] developed a two-dimensional model that combines FVM that is based on RANS and VOF method and FEM that is based on *u–p* formulation of Biot's equation and constitutive equation of linear elastic solid. The one-way coupling of Type W similar to that of the model of Maeno and Fujita[5.2.17] is also used in the model

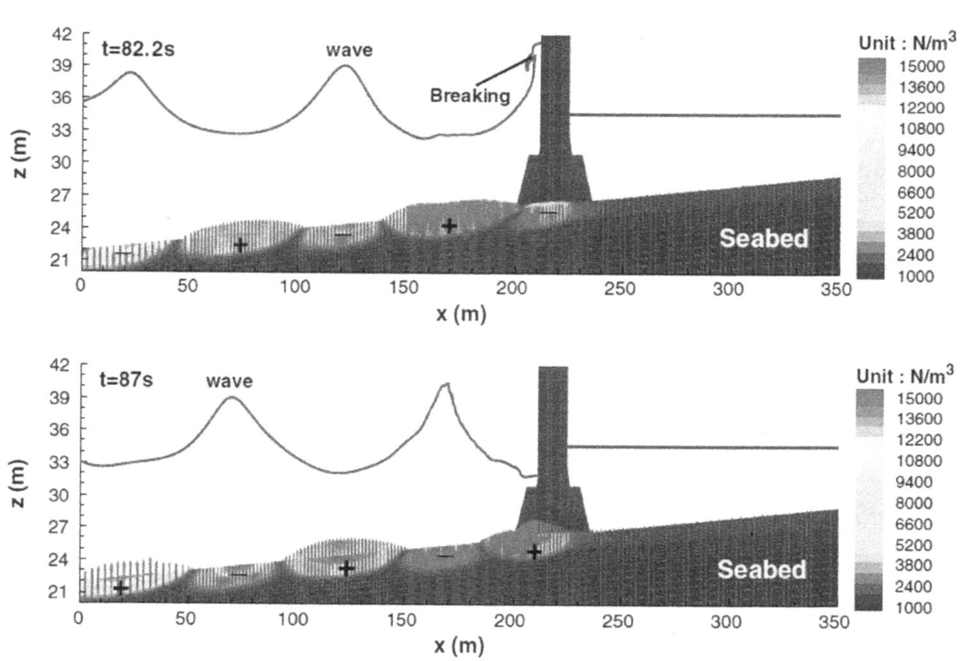

FIGURE 5.2.4
Seepage force around the composite-type breakwater during wave breaking.[5.2.26]

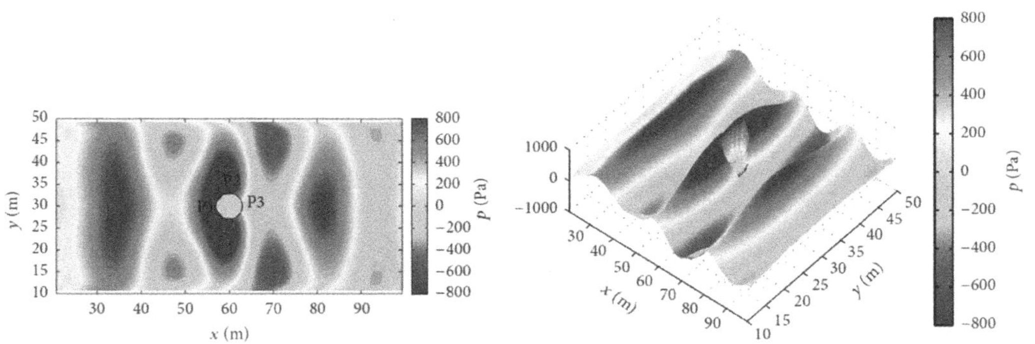

FIGURE 5.2.5
Pore water pressure inside the ground around the monopile (© 2015 Chi Zhang et al./CC BY 3.0).[5.2.29]

of Zhang et al.[5.2.31] FVM of the wavefield is first solved by imposing a boundary condition that is based on logarithmic law to the seabed surface, and then the obtained pressure and shearing force at the seabed surface are given to FEM. Other than this, Lin et al.[5.2.32] developed a three-dimensional model that combines FDM that is based on RANS and VOF method and FEM that is based on Biot's consolidation equation and constitutive equation of linear elastic solid using the one-way coupling of Type W similar to that of the model of Maeno and Fujita.[5.2.17]

Models that utilize commercial software or open-source program have been developed in recent years; Chang and Jeng[5.2.33] constructed a three-dimensional model that is similar to the model of Ye et al.[5.2.24] with COSMOL Multiphysics. Li et al.[5.2.34] introduced FVM that is based on *u–p* formulation of Biot's equation and constitutive equation of linear elastic solid to OpenFOAM and applied it to the wave response phenomena in the vicinity of the gravity-based foundation of offshore wind power generation, as shown in Figure 5.2.6. One-way coupling of Type W similar to that of the model of Maeno and Fujita[5.2.17] is also used in this model.

In the grid methods mentioned above, coupling calculation is performed in either of the following methods: Type A in which the model of the wavefield is extended and applied to the seepage flow within the seabed, and the pressure at the seabed surface obtained by interpolating the result is input to coupled water–soil analysis; or Type W in which calculation of the wavefield is performed by specifying infiltration/exfiltration flow velocity across the seabed surface, and the pressure at the seabed surface obtained by extrapolating the result is input to coupled water–soil analysis. Nevertheless, both methods have problems to be solved. In the case of Type A, applicability of the wavefield model to the inside of the seabed is in question. Therefore, it is preferred to be solved by applying coupled analysis of Darcy and Navier–Stokes that can solve the fluid flow outside and inside the porous media integrally by applying Navier–Stokes equation to the outside of the porous domain and Darcy–Brinkman equation to the inside of the porous domain.[5.2.35] Additionally, as pressure and flow velocity inside the seabed are obtained from both wavefield analysis and water–soil coupled analysis, the fact that the pressure and flow velocity obtained from both analytics do not necessarily match is recognized as a problem. Conversely, in the case of Type W, since only the pressure at the seabed surface is input to the water–soil coupled analysis, the fact that the establishment of the equation of continuity inside of the seabed at the vicinity of the surface is not ensured is recognized as a problem.

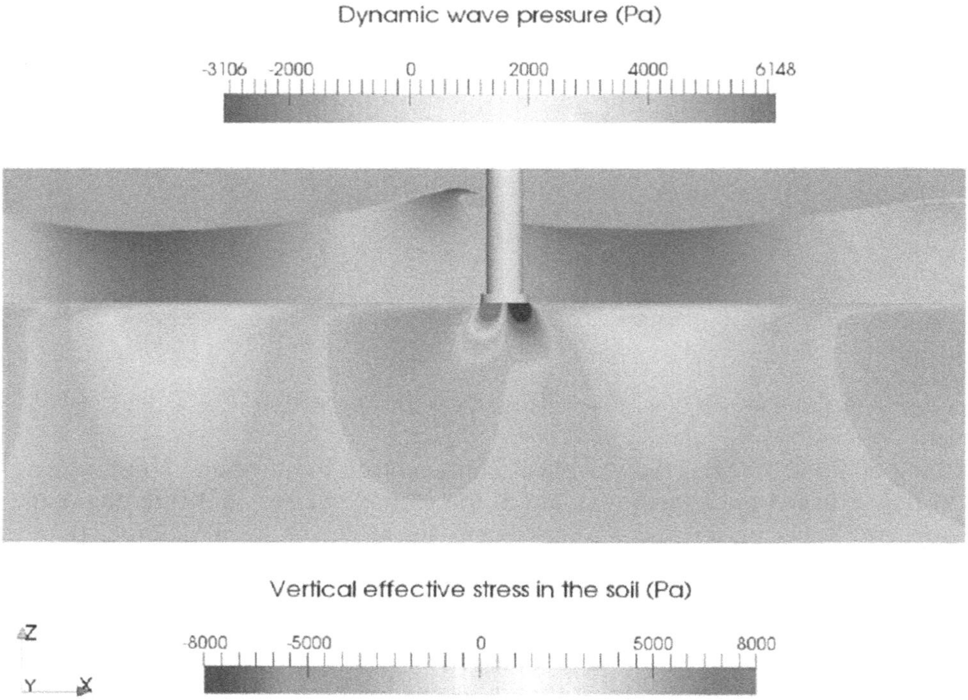

FIGURE 5.2.6
Vertical effective stress in the gravity-based foundation of the offshore wind power generation.[5.2.34]

5.2.2 Particle Method

Types 3 and 4 are based on distinct element method (DEM), smoothed particle hydrodynamics (SPH) method, moving particle semi-implicit (MPS) method, discontinuous deformation analysis (DDA), etc. and classified as particle method. Figure 5.2.7 shows a flow of a model of Gotoh[5.2.36] classified as particle method.

Gotoh et al.[5.2.37] developed a two-dimensional DEM model and assessed the occurrence region of liquefaction when time variation of the pore water pressure distribution inside the seabed is applied with an analytic solution. Additionally, Gotoh and Sakai[5.2.38] developed a two-dimensional numerical movable bed that is based on DEM. Sakai et al.[5.2.39] expanded the numerical movable bed of Gotoh and Sakai[5.2.38] so that the effect of additional lift force due to pore water pressure can be considered and the influence of pore water pressure on the sediment transport be investigated. Chen et al.[5.2.40] developed a two-dimensional model that is based on manifold method MM (manifold model), which is a combination of FEM and DDA, and examined the liquefaction process of the seabed when time variation of the pore water pressure distribution is applied. Gotoh et al.[5.2.41] expanded the numerical movable bed that is based on DEM developed by Gotoh and Sakai,[5.2.38] making it three dimensional. Additionally, Harada et al.[5.2.42] integrated the effect of suction to the three-dimensional DEM developed by Gotoh et al.[5.2.41] and revealed the transient expansion process of a cave due to sand leakage, as shown in Figure 5.2.8. Honda et al.[5.2.43] analyzed the deformation of the mound and seabed when an eccentric load is subjected to the rubble mound with two-dimensional DEM. Takayama et al.[5.2.44] developed a two-dimensional model that is

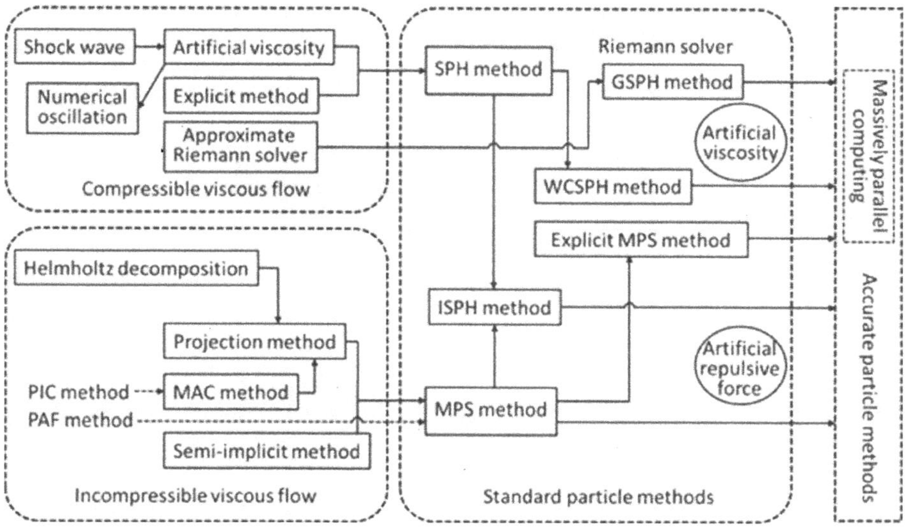

FIGURE 5.2.7
Advancement of the particle methods.[5.2.36] (The original figure can be found in Gotoh.[5.2.36] The author of this book is responsible for the translation.)

based on DEM whose ellipse element is mound rubble and assessed the stability of caisson of composite breakwater against waves. Additionally, Takayama and Takahashi[5.2.45] made an improvement to make caisson a single element. Additionally, Tsujio et al.[5.2.46] assessed the resilience of caisson-type composite breakwater against tsunami using the model of Takayama and Takahashi.[5.2.45]

Gotoh et al.[5.2.47] developed the two-dimensional solid–liquid two-phase MPS that solves both wavefield and movable bed using the MPS method. Model of erosion due to surface shear and water impact is incorporated in this model, and the model is applied to scouring phenomenon in front of the seawall due to return flow of tsunami, as shown in Figure 5.2.9. Ikari et al.[5.2.48] constructed an elastoplastic MPS method by incorporating the DEM-type spring–dashpot model and the constitutive equation of elastoplastic solid to solid–liquid two-phase MPS and simulated the generation process of a tsunami

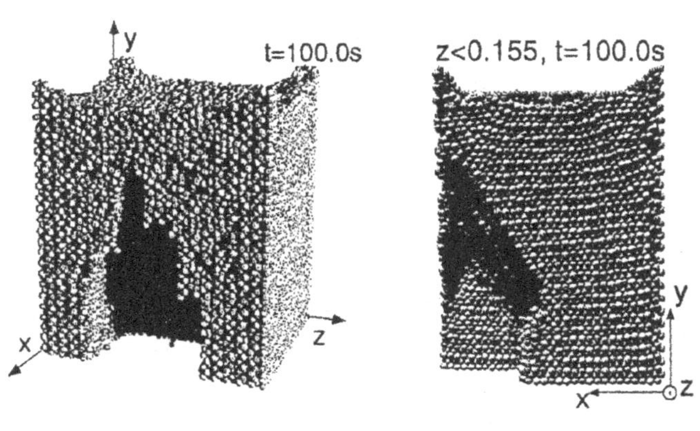

FIGURE 5.2.8
Stable form of the cave formed by sand leakage.[5.2.42]

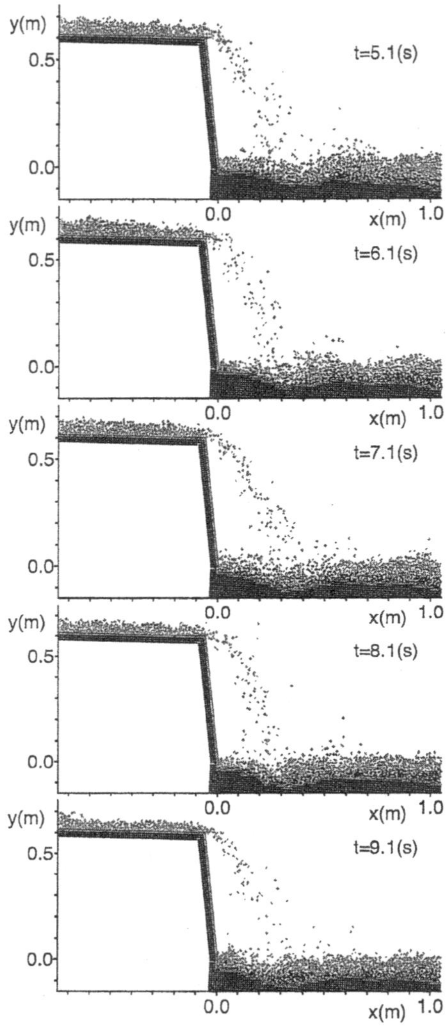

FIGURE 5.2.9
Local scour in front of the seawall induced by tsunami return flow.[5.2.47]

due to landslide. Additionally, Gotoh et al.[5.2.49] incorporated the model of erosion due to bottom shear to the elastoplastic MPS method of Ikari et al.[5.2.48] and analyzed the formation process of a cave due to sand leakage from the lower part of the seawall. Moreover, extending the elastoplastic MPS method to an effective stress base, Ikari et al.[5.2.50] developed the effective stress-type elastoplastic MPS method that can consider the coupling with pore water and analyzed the large deformation of the caisson-type composite breakwater under static load. Gotoh et al.[5.2.51] presented a method for conducting solid–liquid coupled analysis by overlapping the MPS method that solves the liquid phase and DEM that solves the solid phase and conducted a reproduction calculation for the dam-break flow that contains several solid phases. Gotoh et al.[5.2.52] incorporated the DEM-type spring–dashpot model, erosion model due to bottom shear,[5.2.49] and erosion model

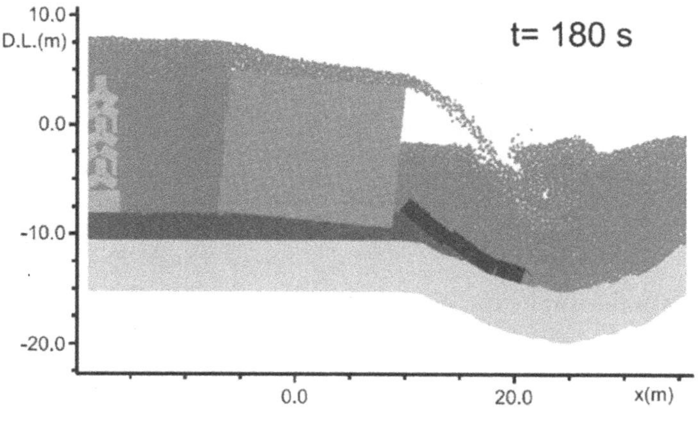

FIGURE 5.2.10
Damage of the composite-type breakwater induced by overflow tsunami.[5.2.52]

due to water impact[5.2.47] to the solid–liquid two-phase MPS and examined the failure process of caisson-type composite breakwater caused by tsunami overflow, as shown in Figure 5.2.10. Ikari et al.[5.2.53] enhanced the elastoplastic MPS method and applied it to the scouring phenomenon of the seabed by vertical water jet. Additionally, Ikari et al.[5.2.54] improved the seabed deformation model in the models developed by Ikari et al.[5.2.53] and incorporated a submodel that analyzes the pick-up, advection, diffusion, settlement, and deposition of fine sediment. Furthermore, Ikari et al.[5.2.55] improved it to analyze stress inside the seabed by considering the soil–water coupling and simulated the damage process of caisson-type composite breakwater during scouring by overflow, as shown in Figure 5.2.11. Additionally, Ikari et al.[5.2.56] introduced modification of stress. Shimizu et al.[5.2.57] constructed a three-dimensional DEM–MPS method by coupling the DEM and MPS method by using drag-type interaction force and elucidated the three-dimensional erosion process of the mound by a tsunami flowing over the caisson-type composite breakwater, as shown in Figure 5.2.12.

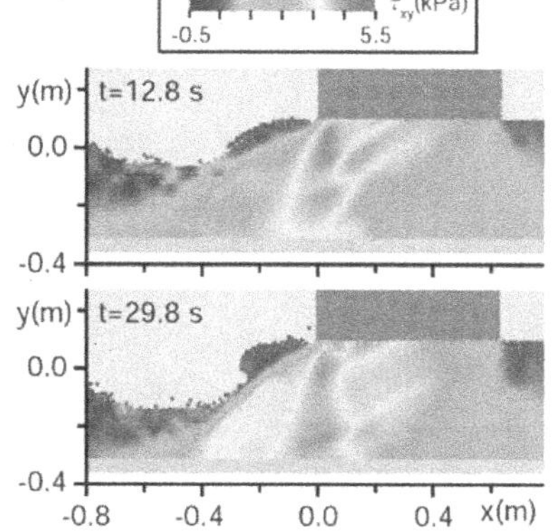

FIGURE 5.2.11
Shear stress around the composite-type breakwater during overflow-induced local scouring.[5.2.55]

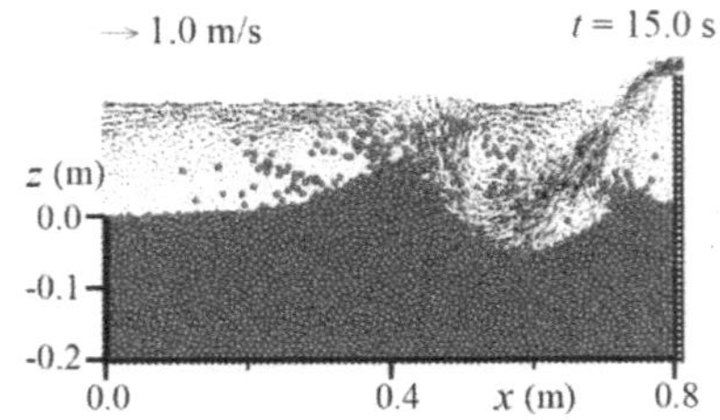

FIGURE 5.2.12
Erosion of the mound induced by the tsunami flowing over the caisson.[5.2.57]

Additionally, Imase et al.[5.2.58, 5.2.59] developed a two-dimensional model that enables soil–water coupled analysis using the SPH method and assessed the stability of the caisson-type composite breakwater against tsunami. Oie et al.[5.2.60] incorporated a model of erosion by surface shear and water impact to high-precision ISPH (incompressible SPH) method of Gotoh et al.[5.2.61] and performed numerical calculation of scouring behind the seawall that is due to overflow. Additionally, Mitourida and Arikawa[5.2.62] examined scouring in front of the seawall due to incoming waves using this model. Uejima and Sato[5.2.63] introduced the concept of redeposition to the high-precision ISPH method of Gotoh et al.[5.2.61] Gotoh et al.[5.2.64] developed the PARISPHERE (PARticle Implemented Simulator for PHysical and Engineering REsearch) that introduced the coupling model with DEM[5.2.51] and a model of erosion by surface shear and water impact[5.2.47] by utilizing the high-precision ISPH method of Gotoh et al.[5.2.61] as the base. Additionally, Tsuruta et al.[5.2.65] improved the PARISPHERE by incorporating the model of deposition of suspended sediment and verified that the accuracy of the deposition at the onshore side of scouring have been improved regarding the overflow tsunami-induced scouring phenomenon at the onshore side of caisson-type composite breakwater shown in Figure 5.2.13.

5.2.3 Coupling of Grid Method and Particle Method

Lastly, we would like to introduce examples of numerical calculations that combine the grid method and the particle method. Maeno et al.[5.2.66] developed a two-dimensional VOF–DEM–FEM model. This model applies FDM that is based on RANS and VOF method to wavefield, including submerged breakwater and the upper layer of seabed, and applies FEM that is based on Biot's consolidation equation and the constitutive

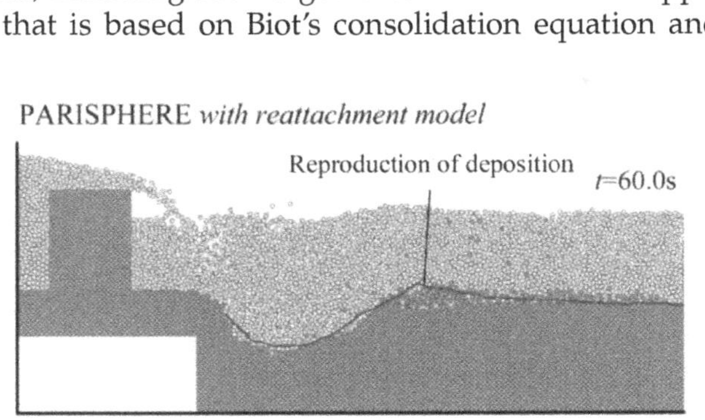

FIGURE 5.2.13
Local scour behind the composite-type breakwater induced by overflow tsunami.[5.2.65]

equation of linear elastic solid to the lower layer of the seabed that will not be deformed largely, and relatively large deformation of the submerged breakwater and the upper layer of the seabed is analyzed using DEM. In this model, the flow velocity and the pressure gradient obtained using FDM or FEM are given to DEM, and the porosity of the FDM is updated on the basis of the location of particles calculated using DEM. Additionally, Yamaguchi et al.[5.2.67] developed a two-dimensional DEM-CFD model. This model uses the computational fluid dynamics (CFD), which solves Navier–Stokes equation with FVM for the analysis of wavefield, including inside of the seabed, and applies DEM to the analysis of displacement of soil particles. Additionally, in this model, FVM and DEM are coupled by using interaction force that is proportional to the relative velocity of flow velocity and particle, and the porosity of FVM is updated on the basis of the location of particles calculated using DEM.

5.3 Treatment of Interface between Fluid and Porous Domains

Simultaneous computation of the seepage flow under the ground and the flow in the fluid domain, which is occupied by a single fluid, is required for predicting the mechanical behavior of the seabed that is affected by tidal waves. Moreover, when the soil structures, such as embankments, are considered in coastal and geotechnical engineering, it is necessary to understand both the fluid flow in cracks or soil pipes and the seepage flow in the intact soil in order to accurately predict their hydraulic effects on the stability of the soil structures. When it comes to the phenomenon of mass transfer, the transportation of substances due to both the overland flow, such as the river and lake, and the subsurface flow has been studied. Thus, methodologies which enable the simultaneous computation of two different flows in the fluid and porous domains provide a variety of applications to technical and practical problems.

The well-known Darcy's law has been usually applied to the seepage flow in porous media, while the Navier–Stokes equations describe the flow in the fluid domain occupied purely by a fluid such as water. Numerical analyses solving the Darcy flow coupled with the Navier–Stokes flow (hereinafter referred to as Darcy/Navier–Stokes coupled flow) has been actively conducted in recent years. For example, see studies by Çeşmelioğlu and Rivière,[5.3.1] Chidyagwai and Rivière,[5.3.2] Badea et al.,[5.3.3] Girault and Rivière,[5.3.4] Chidyagwai and Rivière,[5.3.5] Cai et al.,[5.3.6] Urquiza et al.,[5.3.7] and Mu and Xu,[5.3.8] in which various numerical methods based on FEM or FVM have been presented.

Numerical analysis of the Darcy/Navier–Stokes coupled flow requires the two different flows to be connected at the interface between the porous and fluid domains. How to connect the two different flows becomes the most important technique for accuracy and stability of the numerical computation. The beginning research for this problem was presented by Beavers and Joseph.[5.3.9] They conducted an experiment regarding a horizontal Hagen–Poiseuille flow (laminar flow) over a porous medium in which a seepage flow also occurs horizontally, as shown in Figure 5.3.1, and investigated proper boundary conditions imposed on the fluid–porous interface for the Hagen–Poiseuille flow. They concluded that the flow velocity tangential to the

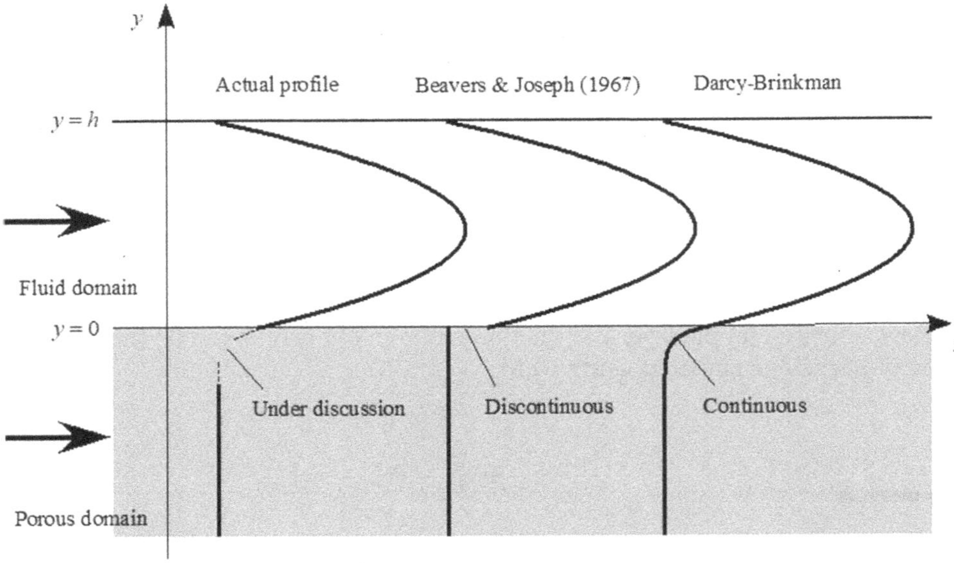

FIGURE 5.3.1
Hagen–Poiseuille flow over a porous medium.

interface in the fluid domain has a slip condition; hence, the different values of the flow velocity are given to the fluid domain and the porous medium at the interface, which means the flow velocity becomes discontinuous at the fluid–porous interface. The boundary condition proposed by Beavers and Joseph[5.3.9] is called the Beavers–Joseph condition. Saffman[5.3.10] modified the Beavers–Joseph condition, and Neale and Nader[5.3.11] explained the flow velocity and the stress around the interface obtained from the boundary conditions of the continuous flow velocity and stress. Bars and Worster[5.3.12] and Ochoa-Tapia and Whitaker[5.3.13–5.3.14] have provided further discussion on this subject; however, the best boundary condition at the fluid–porous interface is still under discussion. In order to solve this problem, accurate microscopic (at the scale of pore size of the porous medium) measurement and modeling of the fluid flow at the fluid–porous interface should be required.

The boundary conditions for pressure and flow velocity at the interface between the fluid and porous domains are called the interfacial conditions, and the above-mentioned studies focused on the flow velocity in tangential direction at the interface. Nevertheless, it should be noted that the flow velocity in the normal direction has a direct influence on the conservation of mass and has a significant influence on the stability of the numerical solution. The interfacial conditions proposed so far can be summarized as follows.

a. **Flow Velocity in the Tangential Direction:** In addition to the Beavers–Joseph condition,[5.3.9] other interfacial conditions are listed in Table 5.3.1.[5.3.15] The flow velocity in the tangential direction at the interface becomes discontinuous in the Beavers–Joseph condition. Some conditions give the continuous flow

TABLE 5.3.1

Interfacial Conditions for Tangential Flow Velocity[5.3.15]

Model	Velocity	Velocity Gradient	Reference
1		$\left.\frac{\partial u_x}{\partial y}\right\|_{\text{fluid}} = \frac{\alpha}{\sqrt{K}}\left(u_x\|_{\text{interface}} - \langle u\rangle_x\|_{\text{porous}}\right)$	Beavers and Joseph (1967)
2	$\langle u\rangle_x\|_{\text{porous}} = u_x\|_{\text{fluid}}$	$\left.\frac{\partial\langle u\rangle_x}{\partial y}\right\|_{\text{porous}} = \left.\frac{\partial u_x}{\partial y}\right\|_{\text{fluid}}$	Neale and Nader (1974) Vafai and Kim (1990) Kim & Choi (1996)
3	$\langle u\rangle_x\|_{\text{porous}} = u_x\|_{\text{fluid}}$	$\mu_{\text{eff}}\left.\frac{\partial\langle u\rangle_x}{\partial y}\right\|_{\text{porous}} - \mu\left.\frac{\partial u_x}{\partial y}\right\|_{\text{fluid}}$	Ochoa-Tapia and Whitaker (1995a)
4	$\langle u\rangle_x\|_{\text{porous}} = u_x\|_{\text{fluid}}$	$\frac{\mu}{\varepsilon}\left.\frac{\partial\langle u\rangle_x}{\partial y}\right\|_{\text{porous}} - \mu\left.\frac{\partial u_x}{\partial y}\right\|_{\text{fluid}} = \beta\frac{\mu}{\sqrt{K}}u_x\Big\|_{\text{interface}}$	Ochoa-Tapia and Whitaker (1995b)
5	$\langle u\rangle_x\|_{\text{porous}} = u_x\|_{\text{fluid}}$	$\frac{\mu}{\varepsilon}\left.\frac{\partial\langle u\rangle_x}{\partial y}\right\|_{\text{porous}} - \mu\left.\frac{\partial u_x}{\partial y}\right\|_{\text{fluid}} = \beta\frac{\mu}{\sqrt{K}}u_x\Big\|_{\text{interface}} + \beta_1 p u_x^2\Big\|_{\text{interface}}$	Ochoa-Tapia and Whitaker (1998)

velocity in the tangential direction, though the Brinkman's extension (the Brinkman term) is required for the description of the seepage flow in porous media.

b. **Flow Velocity in the Normal Direction:** In order to satisfy the conservation of mass transported through the interface between the fluid and porous domains, the flow velocity in the normal direction is required to be continuous at the interface. Therefore, the normal component of the flow velocity in the fluid domain must be the same as the Darcy velocity in the porous medium at the interface shared by the two domains.

c. **Pressure:** The condition of continuous pressure at the interface is commonly accepted.

The approaches for simultaneous computation of the flows in the fluid and porous domains are roughly classified into two types: the single-domain method and the two-domain method. As illustrated in Figure 5.3.2a, in the single-domain method, a resistance term is added to Navier–Stokes equation for reproducing the seepage flow in the porous medium, and then the flows in the entire domain are computed by a single set of governing equations (usually mesoscopic volume-averaged equations). The interfacial conditions, as mentioned above, are not necessary and the variables (pressure and flow velocity) are assumed to be continuous at the fluid–porous interface.

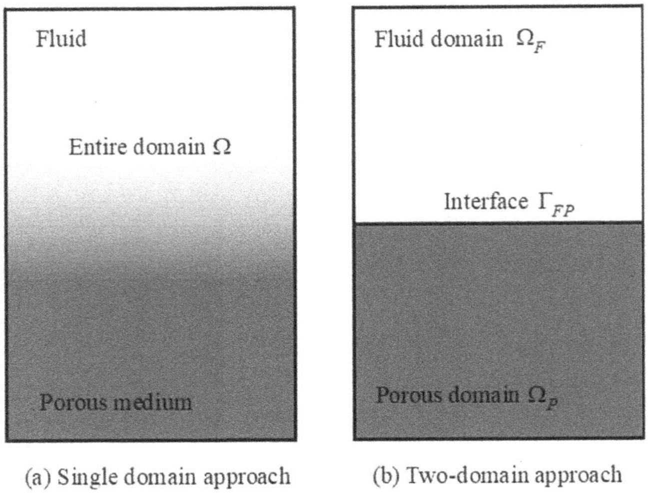

FIGURE 5.3.2
(a) Single-domain method and (b) two-domain method.

The single-domain method does not define the explicit interface between the fluid and the porous domains; however, the material properties, such as porosity and hydraulic conductivity, are averaged in the neighborhood of the interface and are allowed to change gradually in the transition zone. Therefore, the fluid–porous interface becomes unclear, which makes it difficult to obtain accurate numerical solution of the pressure and the flow velocity near the interface.

On the other hand, the two-domain method defines the fluid–porous interface and divides the entire computational domain into the fluid and porous domains clearly. Since a different set of governing equations is solved in each domain, the boundary conditions are required on the boundary and the interfacial conditions also need to be imposed on the interface which is the boundary shared by both the fluid and the porous domains. The interfacial conditions have the role of connecting the flow velocity and the pressure of the two domains. The two-domain method does not have the artificial transition zone as seen in the single domain method. Hence, accurate numerical solution near the interface, which satisfies the imposed interfacial conditions, can be obtained. Moreover, the clear fluid–porous interface is advantageous for computing the deformation of the porous medium because it enables precise boundary conditions of stress or displacement to be imposed onto the interface.

In Type A (All) analysis shown in the previous section, the fluid flow field (or wave field) in the fluid domain and the seepage flow field in the porous medium are simultaneously solved by the single-domain method, and then the pressure at the fluid–porous interface is given to the subsequent soil–water coupled deformation analysis as a boundary condition. Since the fluid flow field in the entire domain is solved by the single-domain method, both the flow velocity and the pressure are continuous at the fluid–porous interface. Hence, the conservation of mass is satisfied through the interface when the flow velocity in the fluid domain is continuously connected to the Darcy flow velocity in the porous medium. However, it should be noted that the seepage flow field in the porous medium is computed by the fluid flow analysis and the soil–water coupled analysis, which results in the two different sets of the seepage flow field. In Type W (Wave) analysis, the fluid flow field (or wave field) is solved only in the fluid domain and the computed pressure at the fluid–porous interface is inputted to the subsequent soil–water coupled deformation analysis for the porous medium. Thus, although the pressure in the fluid domain is continuously connected to that in the porous medium, the flow velocity generally becomes discontinuous at the interface in the normal direction as well as in the tangential direction, which implies that the conservation of mass is not strictly satisfied.

5.3.1 Role of Interfacial Condition

How the interfacial conditions connect the flow velocity in the fluid domain to that in the porous medium is explained through an example of the Hagen–Poiseuille flow (laminar flow) over the porous medium shown in Figure 5.3.1. Only the two-domain method is considered here because the interfacial conditions are not required for the single-domain approach.

The flow in the fluid domain is governed by the Navier–Stokes equations written below:

$$\frac{\partial u_i}{\partial x_i}=0,\ \frac{\partial u_i}{\partial t}+\frac{\partial u_i u_j}{\partial x_j}=-\frac{1}{\rho}\frac{\partial p}{\partial x_i}+\nu\frac{\partial^2 u_i}{\partial x_j \partial x_j} \tag{5.3.1}$$

where u_i, p, ρ, and n denote the flow velocity, the pressure (piezometric pressure), the fluid density and the kinematic viscosity, respectively, and x_i and t are the Cartesian coordinates and time. The subscripts i and j are the free index or the dummy index that follow the summation convention. On the other hand, assuming the Darcy's law, the following equation is obtained with the continuity equation in the porous medium:

$$\frac{\partial\langle u_i\rangle}{\partial x_i}=0,\quad \langle u_i\rangle=-\frac{K}{\mu}\frac{\partial\langle p\rangle^*}{\partial x_i} \tag{5.3.2}$$

where K and μ denote the intrinsic permeability and the viscosity, respectively. In Equation (5.3.2), $\langle\bullet\rangle$ and $\langle\bullet\rangle^*$ indicate the volume average and the phase average, respectively. Specifically, $<u_i>$ means the Darcy's flow velocity and * corresponds to the pressure (piezometric pressure) of the fluid occupying the pore of the porous medium.

Considering the Hagen–Poiseuille flow over the porous medium shown in Figure 5.3.1, the fluid is flowing only in the x direction (horizontal direction), and Equations (5.3.1) and (5.3.2) can be transformed into the following forms:

$$-p'+\mu\frac{d^2u}{dy^2}=0\quad (0<\mathrm{y}<\mathrm{h}) \tag{5.3.3}$$

$$-p'-\frac{\mu}{K}\langle u\rangle=0\quad (y<0) \tag{5.3.4}$$

where u and p' denote the horizontal flow velocity and the pressure gradient in the horizontal direction, respectively. The pressure gradient is assumed to be uniform in the entire domain. Then, the seepage flow velocity (Darcy flow velocity) can be easily obtained as

$$\langle u \rangle = -\frac{K}{\mu} p' \quad (y < 0) \tag{5.3.5}$$

As for the fluid domain, integration of Equation (5.3.3) gives the following equation:

$$u = \frac{p'}{2\mu} y^2 + C_1 y + C_2 \quad (0 < y < h) \tag{5.3.6}$$

where C_1 and C_2 are integration constants. By imposing the nonslip condition onto the top face ($y = h$) in the fluid domain,

$$u = 0 \quad \text{at} \quad y = h \tag{5.3.7}$$

and imposing the Beavers–Joseph condition (Model 1 of Table 5.3.1) onto the interface at the bottom,

$$\frac{du}{dy} = \frac{\alpha}{\sqrt{K}} \left(u - \langle u \rangle \right) \quad \text{at} \quad y = 0 \tag{5.3.8}$$

The above two integration constants of Equation (5.3.6) are determined and the following equation can be derived:

$$u = \frac{p'}{2\mu} \left(y + h + \frac{-h^2 + 2K}{h + \dfrac{\sqrt{K}}{\alpha}} \right) (y - h) \quad (0 < y < h) \tag{5.3.9}$$

It can be easily understood that the flow velocity is discontinuous at the fluid–porous interface (the flow velocity and the Darcy flow velocity at the interface is not the same).

The Darcy flow velocity in the porous medium can be continuously connected to the flow velocity in the fluid domain at the fluid–porous interface by introducing the viscous term called Brinkman term. The following Darcy–Brinkman equation is derived by volume-averaging of the Navier–Stokes equation in the porous domain. The partial differential equation includes the viscous and inertia terms in addition to the Darcy term.

$$\frac{\partial \langle u_i \rangle}{\partial t} + \frac{\partial}{\partial x_j} \left(\frac{\langle u_i \rangle \langle u_j \rangle}{\lambda} \right) = -\frac{\lambda}{\rho} \frac{\partial \langle p \rangle^*}{\partial x_i} + \nu \frac{\partial^2 \langle u_i \rangle}{\partial x_j \partial x_j} - \frac{\lambda \nu}{K} \langle u_i \rangle \tag{5.3.10}$$

where λ denotes the porosity of the porous medium. Assuming the steady flow in the x direction as shown in Figure 5.3.1, Equation (5.3.10) is transformed as follows:

$$-p' + \frac{\mu}{\lambda} \frac{d^2 \langle u \rangle}{dy^2} - \frac{\mu}{K} \langle u \rangle = 0 \quad (y < 0) \tag{5.3.11}$$

The general solution of Equation (5.3.11), which does not diverge at $y \to -\infty$, takes the following form:

$$\langle u \rangle = D \cdot e^{\sqrt{\frac{\lambda}{K}}y} - \frac{K}{\mu}p' \quad (y < 0) \tag{5.3.12}$$

where D is an integration constant. By imposing the interfacial condition (Model 2 of Table 5.3.1),

$$u = \langle u \rangle \text{ and } \frac{du}{dy} = \frac{d\langle u \rangle}{dy} \quad \text{at} \quad y = 0 \tag{5.3.13}$$

and the nonslip condition (Equation (5.3.7)) on the top face ($y = h$), three integration constants (C_1 and C_2 of Equation (5.3.6) and D of Equation (5.3.12)) are determined and the following solutions of the flow velocities in the fluid and the porous domains are obtained:

$$u = \frac{p'}{2\mu}\left(y + h + \frac{-h^2 + 2K}{h + \sqrt{\frac{K}{\lambda}}} \right)(y - h) \quad (0 < y < h) \tag{5.3.14}$$

$$\langle u \rangle = \frac{p'}{2\mu} \cdot \frac{-h^2 + 2K}{h + \sqrt{\frac{K}{\lambda}}} \cdot \sqrt{\frac{K}{\lambda}} e^{\sqrt{\frac{\lambda}{K}}y} - \frac{K}{\mu}p' \quad (y < 0) \tag{5.3.15}$$

The introduction of the Brinkman term allows the continuous flow velocity at the interface. The interfacial condition is turned into the boundary condition for each domain and connects the flow velocity in the porous medium to that in the fluid domain. It is noteworthy that Equation (5.3.9) becomes identical to Equation (5.3.14) when $\alpha = \sqrt{\lambda}$.

5.3.2 Examples of Numerical Analyses

In order to exemplify numerical analyses of Darcy/Navier–Stokes coupled flow, two-dimensional cavity flow, a well-known benchmark problem, over a porous medium is presented herein. The Navier–Stokes equations, Equation (5.3.1), are solved in the fluid domain and the Darcy–Brinkman equations, Equation (5.3.1), are applied to the seepage flow in the porous medium. The continuous flow velocity and its gradient is imposed onto the fluid–porous interface as the interfacial conditions for the flow velocity (see Model 2 of Table 5.3.1). The pressure is also assumed to be continuous at the interface. The computational domain is a square with the dimension of 0.1 m in height and width. The lower part of 0.02 m in height is the porous domain with the hydraulic conductivity of 2.0×10^{-3} m/s and porosity of 0.5. As for the boundary conditions, the flow velocity of 0.05 m/s in the horizontal direction was imposed to the top face, and the nonslip condition was imposed to the other boundaries. Taking the length of the square domain as the representative length and the flow velocity on the top face as representative velocity, Reynolds number Re is calculated as $Re = 5{,}000$, given the kinematic viscosity of water, 1.0×10^{-6} m^2/s.

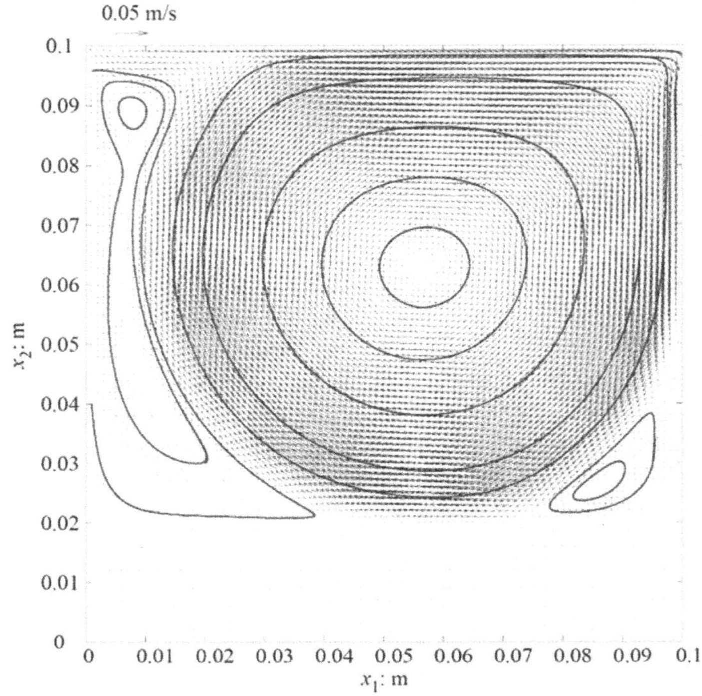

FIGURE 5.3.3
Flow velocity and stream lines in the fluid domain.[5.3.16]

The spatial distributions of the computed flow velocity and pressure are shown in Figures 5.3.3–5.3.5. A large vortex appears in the fluid domain over the porous medium and the pressure at the center of the vortex becomes lower. Since the seepage flow velocity in the porous medium is much smaller than that in the fluid domain, Figure 5.3.4 only shows the seepage flow velocity and the streamlines in the porous domain located at the bottom of the computational square (this result is obtained by

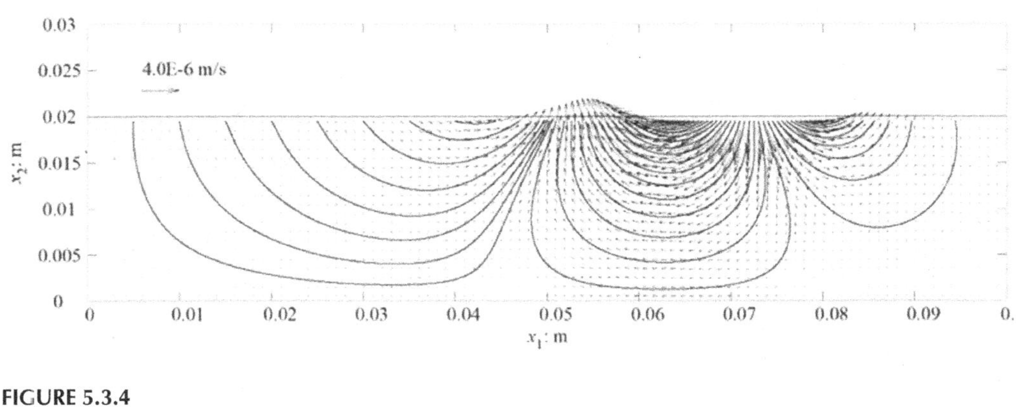

FIGURE 5.3.4
Flow velocity and stream lines in the porous domain.[5.3.16]

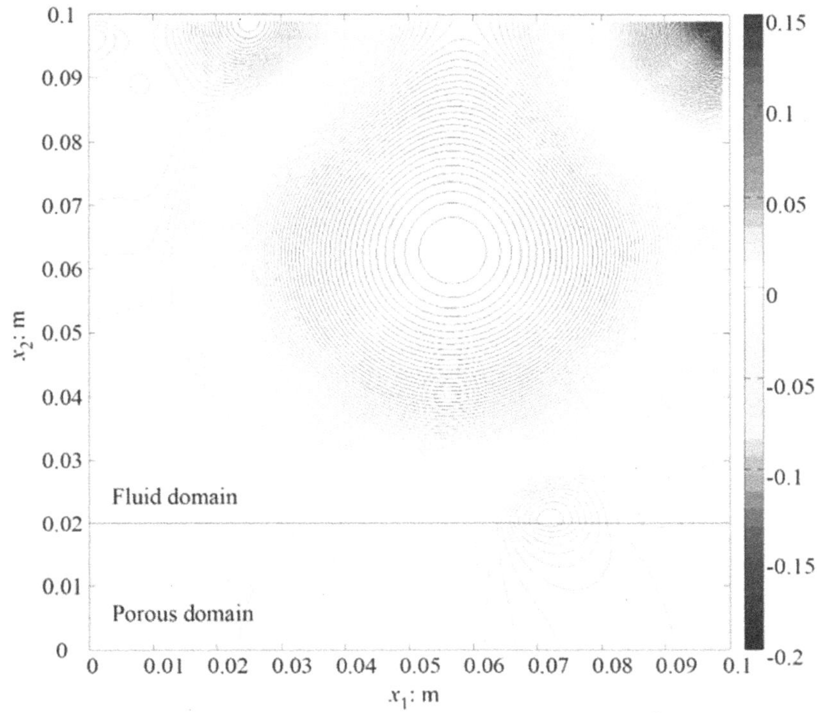

FIGURE 5.3.5
Computed pressure distribution in fluid and porous domains.[5.3.16]

solving the Darcy–Brinkman equations). It can be seen from the figure that the upward seepage flow, heading to the main vortex in the fluid domain, is produced in the central region of the porous medium, which results from the lower pressure at the vortex. As shown in Figure 5.3.5, the pressure is continuously changing at the interface as the interfacial conditions impose the continuous pressure.

By introducing turbulence models to the computation of flow in the fluid domain, the fluid flows under higher Reynolds numbers can be achieved. Numerical results of the large eddy simulation (LES), which employs the coherent structure model[5.3.17] as a turbulence model of the fluid domain, are explained here. The cavity flow with the same geometry as described above is considered, and the lower part of the computational domain is also assumed to be the porous medium with the porosity of 0.5 and the hydraulic conductivity of 2.0×10^{-3} m/s. Regarding the boundary conditions, the flow velocity of 1.0 m/s in the horizontal direction is imposed onto the top face, and the nonslip condition is imposed to the side walls and the bottom face. The Reynolds number of this example corresponds to $Re = 100{,}000$. The computed distributions of the fluid velocity and the vorticity in the fluid domain are shown in Figures 5.3.6 and 5.3.7. A single vortex is generated on the right side in the early stage ($t = 0.25$ s), and the vortex moves downward along the right side wall. After that, several vortices develop along the fluid–porous interface ($t = 2.75$ s) as the vortex approaches to the porous domain. The development of vortices can be observed in the vorticity distribution in Figure 5.3.7

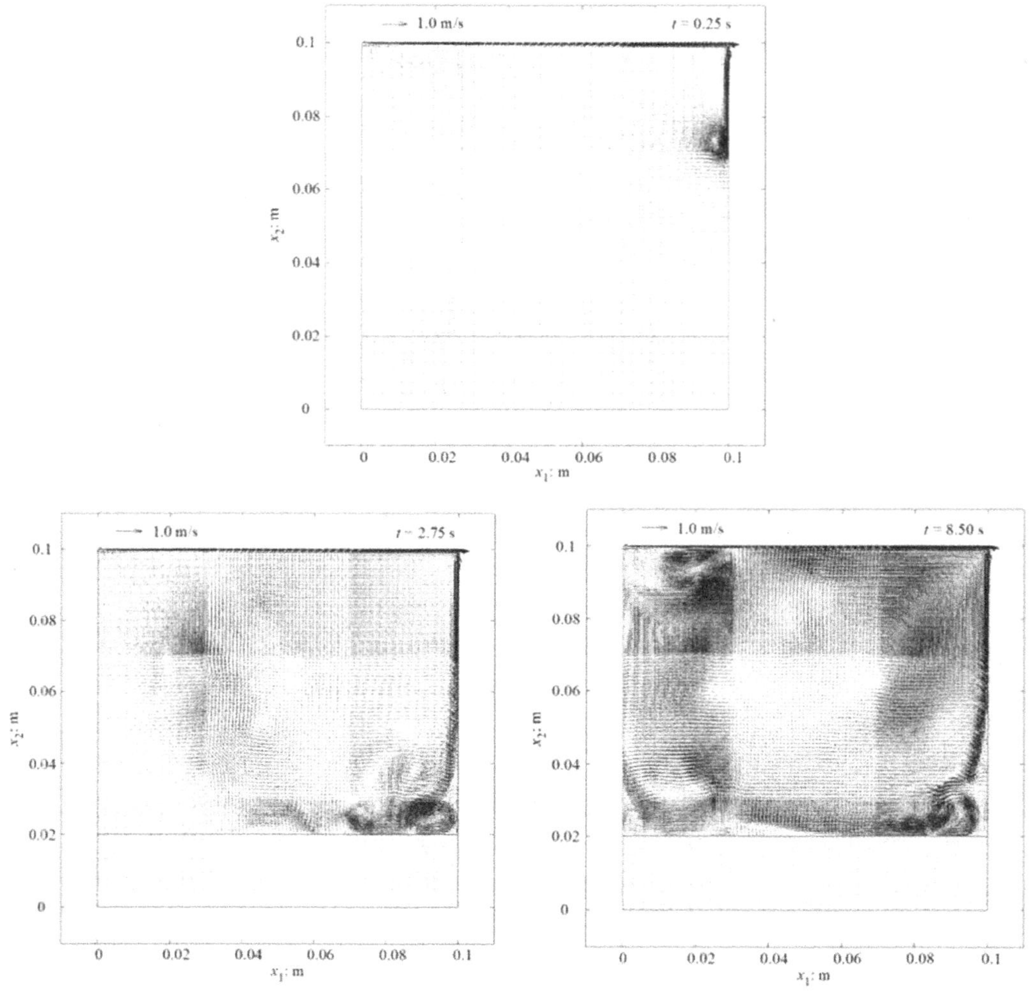

FIGURE 5.3.6
Flow velocity at $Re = 100{,}000$.

more easily than the fluid velocity distribution in Figure 5.3.6. The fluid flow does not settle at a steady state, and vortices repeatedly appear or disappear ($t = 8.50$ s) at the high Reynolds number ($Re = 100{,}000$). It should be noted that the seepage flow in the porous medium is still laminar as described by the Darcy's law and the vorticity in the porous medium is significantly small, as shown in Figure 5.3.7. Figure 5.3.8 shows the flow velocity in the porous medium after 2.75 and 8.50 s. Only the flow velocity in the porous medium is illustrated in these figures because the flow velocity is even smaller than that of the fluid domain. There are inflows and outflows around the surface of the porous medium where the vortices are generated in the vicinity of the fluid–porous interface. Simultaneously, analysis of the seepage flow in the porous medium and the turbulent flow in the fluid domain is also possible by introducing turbulence models.

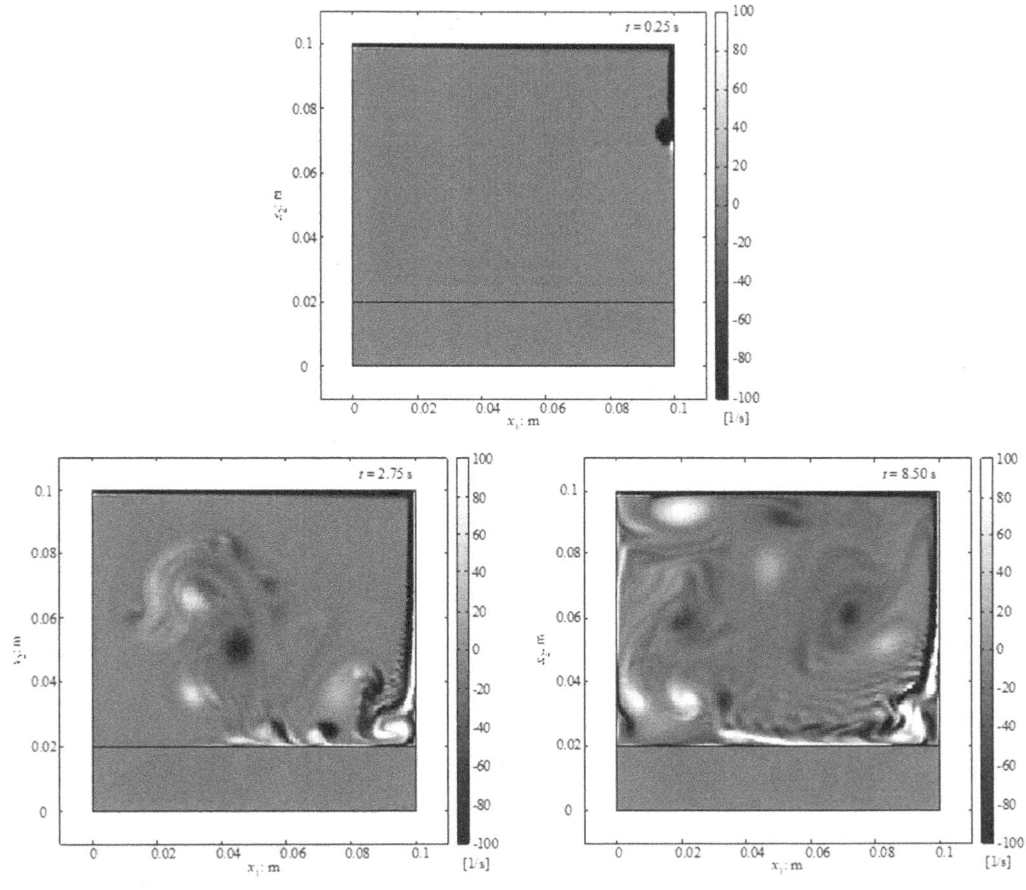

FIGURE 5.3.7
Vorticity at Re = 100,000.

5.4 Bottom Sediment Transport by Particle Method

The ripple on the coastal area formed by bottom sediment transport during numerical simulation of coastal process is an important shore side boundary condition because it affects the nearshore flow field as bed roughness. To thoroughly understand the formation mechanism of ripples by bottom sediment transport, it is reasonable to comprehend the ripple from the scale of sand particles and to describe it as an aggregate of sand particle motion. The distinct element method (DEM) presented by Cundall and Strack[5.4.1] is a numerical simulation method of the Lagrange type that enables the tracking of individual sand particle motion. Appropriately coupling with the flow field and constructing a solid–liquid two-phase flow model can help develop an understanding of the transportation mechanism of bottom sediment. In the bottom boundary layer, bottom sediment transport surfaces by active momentum exchange with sand particles

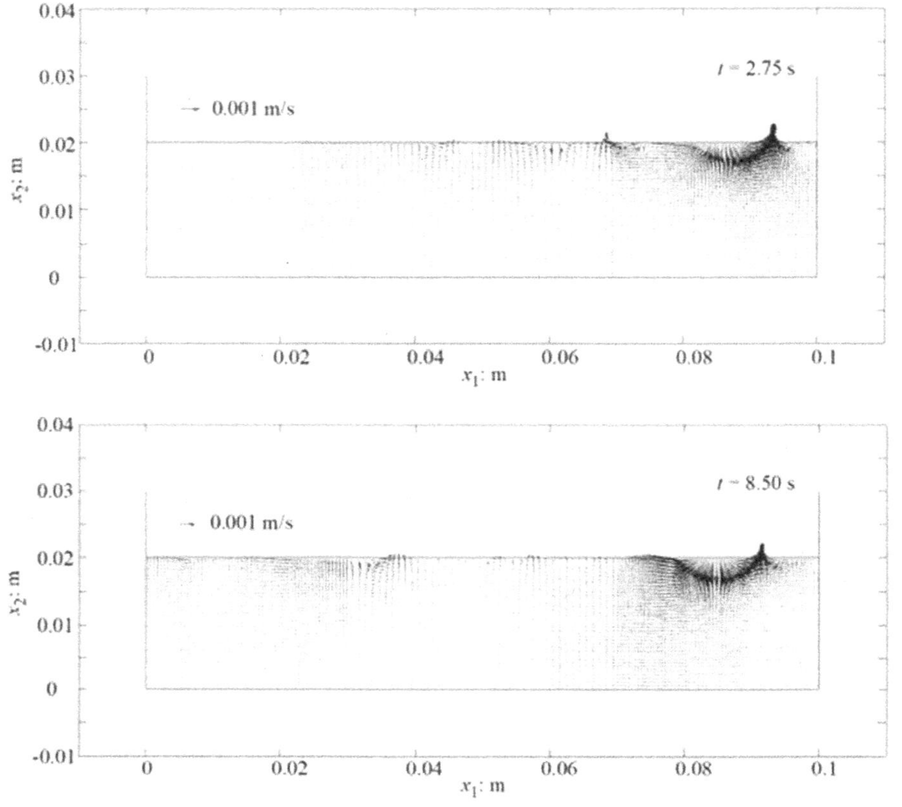

FIGURE 5.3.8
Flow velocity in porous medium.

at the surface layer of the sedimentary layer due to bottom shear. Besides this, at the surf zone and swash zone, bottom sediment transport is affected largely by momentum inflow from the water surface due to breaking waves. Thus, to understand the bottom sediment transport mechanism in a shallow water area, it is important to select a fluid simulation that can accurately handle the free water surface behavior.

Here, we would like to take a general view of the numerical simulation method intended for bottom sediment transport that applies particle-tracking-type multiphase flow model that uses a numerical movable bed of DEM type for tracking sand particle motion. First, we would like to show the basic framework of the multiphase flow model method related to solid–liquid two-phase flow that uses DEM and then introduce a numerical simulation case for ripple formation process that uses the DEM–MPS method.

5.4.1 Multiphase Flow Analysis

The execution of DNS is considered to be an effective option in comprehending the turbulence modulation caused by particle incorporation. The DNS of single-phase turbulence or solid–liquid phase turbulence under limited conditions has been performed for decades. Nevertheless, the computation load of DNS is extremely high, and the execution of DNS having high Reynolds number conditions is still difficult even with today's computing environment. Because of this condition, although the scale of turbulence that can be resolved will be rough, numerical simulation of solid–liquid phase flow that reduces computation load by utilizing governing equations that have average processing applied has been conducted. Average processing is applied by considering the scales of time and length regarding the target phenomenon to focus on. Assessment of particle interaction force is essential for multiphase flow field that contains sand particles of nonnegligible amount considering the impact of the contact or collision between sand particles on the phenomenon. With the particle-tracking method using DEM, Lagrangian tracking of sand particle motion is performed while assessing the particle interaction force. Additionally, a numerical calculation method of either Euler type or Lagrange type can be chosen for the fluid calculation method that is coupled with DEM. If sand drift process in the surf zone or swash zone that accompanies intense water-level fluctuation is targeted, numerical simulation adopting Lagrange type for fluid calculation is effective since it can eliminate the unclearness of the interface due to numerical diffusion. The following is a model example of solid–liquid two-phase flow of particle-tracking type using DEM for solid-phase sand particle tracking.

5.4.1.1 Euler–Lagrange Coupling

Although it is written in detail in publications related to CFD (Kobayashi,[5.4.2] for example), the fundamental equation of fluid motion in Euler–Lagrange coupling is fundamentally the same as the two-fluid model and is composed of the equation of continuity and equation of motion that are obtained by various averaging, including mean volume and time average. Thus, the solution obtained by numerical analysis will be the flow properties that are averaged at a range larger than the scale of fluctuation at the interface caused by particle motion.

$$\frac{\partial}{\partial t}\alpha_f \rho + \nabla \cdot \left(\alpha_f \rho \boldsymbol{u}\right) = 0 \tag{5.4.1}$$

$$\frac{\partial}{\partial t}\alpha_f \rho \boldsymbol{u} + \nabla \cdot \left(\alpha_f \rho \boldsymbol{u}\boldsymbol{u}\right) = -\nabla\left(\alpha_f p\right) + \nabla \cdot \left(\alpha_f \boldsymbol{\tau}\right) + \alpha_f \rho \boldsymbol{g} + \boldsymbol{M}_{\text{int}} \tag{5.4.2}$$

Here, α_f is the volume fraction of fluid phase, ρ is the fluid density, $\boldsymbol{u}$ is the flow vector, p is the pressure, $\boldsymbol{\tau}$ is the viscosity stress, $\boldsymbol{g}$ is the gravitational acceleration, and $\boldsymbol{M}_{\text{int}}$ is the interaction force of solid–liquid phase correlation. Conversely, translational motion equation and rotational motion equation that are based on Basset–Boussinesq–Oseen (BBO) equation are used for point mass movement of sand particles.

$$m\frac{d\boldsymbol{u}_p}{dt} = -V_p \nabla p + V_p \nabla \cdot \boldsymbol{\tau} + \boldsymbol{F}_f + \boldsymbol{F}_c + m\boldsymbol{g} \tag{5.4.3}$$

$$I\frac{d\boldsymbol{\omega}_p}{dt} = \boldsymbol{T}_f + \boldsymbol{T}_c ; \boldsymbol{T}_c = \boldsymbol{r} \times \boldsymbol{F}_c \tag{5.4.4}$$

Here, m is the particle mass, $\boldsymbol{u}_p$ is the particle migration velocity vector, V_p is the particle volume, $\boldsymbol{F}_f$ is the fluid force that is due to the effect of the right-hand side of Equation (5.4.2) except for terms 1, 2, and 3, $\boldsymbol{F}_c$ is the particle interaction force, $\boldsymbol{r}$ is the relative position vector based on the center of gravity of the particle, $\boldsymbol{T}_f$ is the torque that is due to fluid force, and $\boldsymbol{T}_c$ is the torque that is due to particle interaction force. In the Euler–Lagrange method, a suitable modeling is required for the interaction force of solid–liquid phase because the representative scale for the flow field is larger than that of sand particles. The interaction force of solid–liquid phase $\boldsymbol{F}_f$ is given by the sum of fluid forces such as drag and virtual mass force.

$$\boldsymbol{F}_f = \boldsymbol{F}_d + \boldsymbol{F}_v + \boldsymbol{F}_b + \boldsymbol{F}_l + \boldsymbol{F}_m \tag{5.4.5}$$

Here, $\boldsymbol{F}_d$ is the fluid drag, $\boldsymbol{F}_v$ is the added mass force, $\boldsymbol{F}_b$ is the Basset force, $\boldsymbol{F}_l$ is the Saffman lift force that is due to the velocity gradient of the fluid, and $\boldsymbol{F}_m$ is the Magnus lift force that is due to the particle rotation. Since the models for the fluid force of each of these terms have not been fully developed, special care must be taken regarding their application range when using these terms. Additionally, in some cases, only the dominant fluid forces are considered and simplified in the actual numerical simulation. For example, in a single-particle tracking model of numerical movable bed, the sand particle motion is tracked by translational motion equation that considers drag $\boldsymbol{F}_d$ and virtual mass force $\boldsymbol{F}_v$ as fluid force, but the impact of other fluid force terms are ignored, as shown in the following equation (for example, Gotoh[5.4.3]):

$$\begin{aligned}(\sigma + \rho C_M) V_p \frac{d\boldsymbol{u}_p}{dt} &= \frac{1}{2} C_D \rho \frac{\pi}{4} d^2 \left|\boldsymbol{u} - \boldsymbol{u}_p\right| \left(\boldsymbol{u} - \boldsymbol{u}_p\right) \\ &\quad + \rho(1 + C_M) V_p \frac{d\boldsymbol{u}}{dt} + V_p(\sigma - \rho)\boldsymbol{g}\end{aligned} \tag{5.4.6}$$

Here, σ is the particle density, C_D is the drag coefficient, π is the number pi, and C_M is the added mass coefficient. Note that particle interaction force can be excluded if the density of collision of particles is low enough to be ignored, and Equation (5.4.6) is handled like this. In the numerical simulation of the movable bed phenomenon that includes sedimentary layer and dispersed bed of sand particles, the term $\boldsymbol{F}_c$ that is due to particle interaction force is included in Equation (5.4.3) and the motion equation of sand particles is calculated with the spring–dashpot model of DEM. Note that if the impact of turbulence is to be considered, turbulent flow stress $\boldsymbol{\tau}_{Re}$ must be included in the stress of the motion (Equation (5.4.2)). Although the calculation of solid–liquid phase turbulence has been attempted with various turbulence models so far, the effect of particle incorporation on the structure of turbulence has not been fully understood. Additionally, regarding the correlation term related to the variation component of density and flow velocity, the fundamental Equations (5.4.1) and (5.4.2) must be considered if the effect of their variation on the phenomenon cannot be ignored. However, because of the difficulty of its formulation, only a few cases among previous studies have considered the correlation term of variation.

Numerical analysis of Euler–Lagrange method that conducts Lagrangian tracking on particle motion by adopting fundamental equation of two-fluid model to describe fluid movement and by considering particle interaction force is suitable for describing the particle movement that indicates discrete behavior. Hence, it has been conducted many times to explain the sediment transportation phenomenon. Yeganeh et al.[5.4.4] conducted a numerical simulation using Euler–Lagrange method for steady bedload transport process and presented that assessment of the particle interaction force is important for reproducing multilayer structure of movable bed composed of high concentration particle–fluidized bed and saltation layer, which are observed under high bottom surface shear force condition. Additionally, Drake and Calantoni[5.4.5] as well as Calantoni and Thaxton[5.4.6] conducted numerical simulations of sheet-flow sediment transport under oscillatory flow using coupling of one-dimensional fluid model and three-dimensional discrete particle model (DPM) and proposed a sediment transport rate equation (power law) that focuses on practical use. Note that particle motion is tracked using a fundamental equation that is in the form of Equation (5.4.6) with particle interaction force added. In the field of geotechnical engineering, it can be confirmed that research conducted by El Shamy and Zeghal[5.4.7] and Suzuki et al.[5.4.8] use this type of method.

5.4.1.2 Particle–fluid DNS and LES

Euler–Lagrange method is useful in predicting the average structure of the phenomenon. However, because the equation of continuous phase is described assuming an averaged field and since the flow field around the particle such as trailing vortex involved associated with particle motion cannot be fully described if the sand particle motion is tracked using a point mass model, the interaction of the sand particle motion and turbulence field cannot be examined thoroughly. To understand the essence of solid–liquid two-phase turbulence mechanism, including the understanding of turbulence modulation due to sand particle incorporation, numerical analysis using a fluid computational grid that can resolve the minimum vortex (Kolmogorov scale) that bear the dissipation of energy is required. Models such as turbulence model and constitutive equation are not included in the analyses of these types, and since they directly execute the numerical calculation of fundamental equation, they are referred to as DNS.

Regarding the calculation method for resolving the flow field around the particle in high accuracy, there is a method of applying a boundary-fitted grid or unstructured grid along the boundary of particles (see Figure 5.4.1). Although methods of these types can give precise boundary conditions with ease and their calculation accuracy is higher than methods that adopt a fixed rectangular grid, their computation load is higher because grid generation associated with particle migration is required. With the calculation method using a fixed rectangular grid, analysis accuracy at stepwise boundary must be considered. Although a method for subdividing the computational grid at the boundary or cut cell method that divides the rectangular grid along the boundary has been proposed, the process of redefining the periphery of the boundary for particle migration remains required as with boundary fitted grid and unstructured grid. The immersed boundary (IB) method proposed by Peskin[5.4.9] is available as a method that uses a rectangular grid that does not require the process of reconstruction of interface boundary via particle migration. IB method is a method of approximating the boundary by adding compelling force to the Navier–Stokes equation. Although various calculation methods that are based on the IB method have been proposed so far, in all of them, compelling force on the computational grid is given by considering the velocity near the interface given by interpolation. Kajishima and Takiguchi[5.4.10] and Tsuji et al.[5.4.11] conducted DNS of particle assemblage behavior by using a numerical calculation method that is based on the IB method that calculates the compelling force using the volume average velocity at the computational grid that includes particles. The approach taken by Kajishima

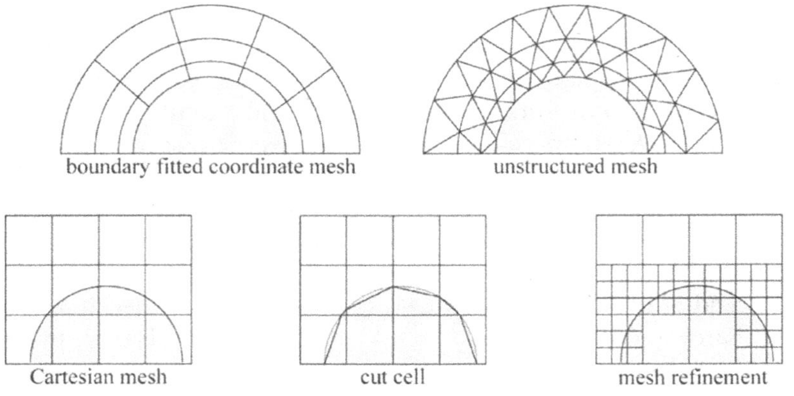

FIGURE 5.4.1
Computational grid.

and Takiguchi[5.4.10] regarding DNS for particle–fluid is as follows: the fundamental equations of fluid are the continuity equation of the incompressible Newtonian fluid and Navier–Stokes equation. In the numerical calculation, the following equations are used for average velocity $\boldsymbol{u}$ that consider the volume fraction of each phase of computation cell (see Figure 5.4.2):

$$\nabla \cdot \boldsymbol{u} = 0 \tag{5.4.7}$$

$$\frac{\partial \boldsymbol{u}}{\partial t} + \boldsymbol{u} \cdot (\nabla \boldsymbol{u}) = -\frac{1}{\rho} \nabla p + \nu \nabla \cdot \left[\nabla \boldsymbol{u} + (\nabla \boldsymbol{u})^T \right] + \boldsymbol{g} + \boldsymbol{f}_p \tag{5.4.8}$$

$$\boldsymbol{u} = (1 - \alpha_p)\tilde{\boldsymbol{u}} + \alpha_p \boldsymbol{v}_p; \; \boldsymbol{v}_p = \boldsymbol{u}_p + \boldsymbol{\omega}_p \times \boldsymbol{r} \tag{5.4.9}$$

Here, ν is the coefficient of kinematic viscosity, $\boldsymbol{f}_p$ is the compelling force, α_p is the solid-phase particle volume fraction, $\boldsymbol{v}_p$ is the velocity within the particle, $\boldsymbol{r}$ is the relative position vector based on the center of gravity of the particle, and w_p is the angular velocity. The temporary velocity $\tilde{u}$ is solved first, assuming that the whole flow field is filled with fluid.

$$\tilde{\boldsymbol{u}} = \boldsymbol{u}^n + \Delta t \left(-\boldsymbol{u} \cdot \nabla \boldsymbol{u} - \frac{1}{\rho} \nabla p + \nu \nabla \cdot \left[\nabla \boldsymbol{u} + (\nabla \boldsymbol{u})^{\mathrm{T}} \right] + \boldsymbol{g} \right) \tag{5.4.10}$$

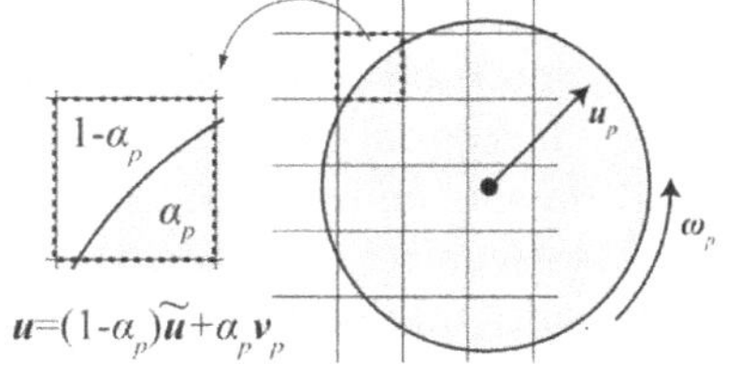

FIGURE 5.4.2
Velocity, including solid–liquid interface.

Here, superscript n indicates calculation steps. The computation cell that does not include particles will be velocity $\boldsymbol{u}^{n+1}$ when this $\tilde{\boldsymbol{u}}$ is at the time $n+1$ step. For computation cell that includes particles, the time evolution completes by correcting the velocity using compelling force $\boldsymbol{f}_p$ that is defined by the following equation:

$$\tilde{\boldsymbol{u}} = \boldsymbol{u}^n + \Delta t\left(-\boldsymbol{u}\cdot\nabla\boldsymbol{u} - \frac{1}{\rho}\nabla p + \nu\nabla\cdot\left[\nabla\boldsymbol{u} + (\nabla\boldsymbol{u})^{\mathrm{T}}\right] + \boldsymbol{g} + \boldsymbol{f}_p\right) \tag{5.4.11}$$

$$\boldsymbol{f}_p = \frac{\alpha_p\left(\boldsymbol{v}_p - \tilde{\boldsymbol{u}}\right)}{\Delta t} \tag{5.4.12}$$

On the other hand, individual particle motion is tracked from the volume integral of compelling force $\boldsymbol{f}_p$ that is solved by Equation (5.4.12), as shown in the following equations:

$$m\frac{d\boldsymbol{u}_p}{dt} = -\rho\int_{V_p}\boldsymbol{f}_p dV + \boldsymbol{G}_p \tag{5.4.13}$$

$$I\frac{d\boldsymbol{\omega}_p}{dt} = -\rho\int_{V_p}\boldsymbol{r}\times\boldsymbol{f}_p dV + \boldsymbol{N}_p \tag{5.4.14}$$

Here, $\boldsymbol{G}_p$ and $\boldsymbol{N}_p$ are external force and external force moment, respectively, and V_p is the particle volume. Note that since the same computation cell as the fluid calculation is used for the integral computation of fluid force that is applied to the particle, the balance of momentum will be conserved.

Apart from the above, some methods use fluid density considering the volume fraction of each phase within the computational grid. In the numerical calculation using the one-field model seen in the numerical analysis of gas–liquid two-phase flow, the fluid motion with density weighted with volume fraction within the computational grid is calculated (for example, Akiyama and Aritomi[5.4.12]).

$$\nabla\cdot\boldsymbol{u} = 0 \tag{5.4.15}$$

$$\frac{\partial\boldsymbol{u}}{\partial t} + \nabla\cdot(\boldsymbol{u}\boldsymbol{u}) = -\frac{1}{\bar{\rho}}\nabla p + \frac{1}{\bar{\rho}}\nabla\cdot\mu\left[\nabla\boldsymbol{u} + (\nabla\boldsymbol{u})^{\mathrm{T}}\right] + \boldsymbol{g} + \boldsymbol{f}_s \tag{5.4.16}$$

$$\bar{\rho} = \sum_k \alpha_k\rho_k \tag{5.4.17}$$

Here, μ is the coefficient of kinematic viscosity, $\boldsymbol{f}_s$ is the body force from the surface tension effect, subscript k is a symbol indicating k phase, and α_k is the volume fraction of k phase within the computational grid. Note that the transportation equation related to the volume fraction of the following equation is used for capturing gas–liquid interface:

$$\frac{D\alpha_k}{Dt} = 0 \tag{5.4.18}$$

Since Lagrangian tracking of the rigid particle is performed in Euler–Lagrange method for solid–liquid two-phase flow, Equation (5.4.18) related to volume fraction seen in the numerical analysis of gas–liquid two-phase flow is not required for capturing the interface of the solid–liquid phase, and consideration on suppression of numerical diffusion or interface gradient is also not required. A numerical analysis method for particle–fluid mixture considering volume fraction of each phase has been developed in the multiphase incompressible flow solver with collocated grid system (MICS) developed by Ushijima et al.[5.4.13] Yamada et al.[5.4.14] conducted a numerical simulation (a computational grid width of 1/8 of solid-phase cylindrical element diameter is used, and it is confirmed that drag coefficient is properly reproduced) of incipient motion of bottom elements using MICS and presented that MICS produces satisfactory reproducibility by comparing with experiments. Although modeling of fluid force applied to particles is unnecessary as flow field around particles can be well resolved with DNS, because of its high computation load, it is limited to analysis of phenomenon with low Reynolds number. Execution of DNS for high Reynolds number (such as examination of active sediment transportation mechanism under high Shields number condition) is not realistic since a large number of computational grids is required; examination using Reynolds-averaged Navier–Stokes simulation (RANS) or large eddy simulation (LES) is convenient. Turbulence analysis of solid–liquid two-phase flow using LES is a reasonable option to comprehend the unsteady flow around individual sand particles with as much detail and to examine the sediment transportation mechanism. In LES, the subgrid-scale (SGS) term will be introduced in the fundamental equation of fluid. Since it adopts a fluid computational grid that is finer than the scale of the particle, especially when there is active particle motion from the collision of particles, particle occupancy is contained in the computation grid and the fluctuation of the relative velocity of fluid and particle will become apparent. Thus, in the coupling of the solid–liquid phase, a model for correlation amount of fluctuation amount that is an effect of turbulence modulation induced by sand particle motion must be considered, and it is important to consider the subgrid-scale (SGS) term that considers the effect of particle laden in predicting the phenomenon. Although it appears that Smagorinsiky model of the continuous phase is applied in the SGS term used in solid–liquid two-phase turbulence analysis by LES regarding movable bed so far, vertical sorting process of sheet-flow sediment transport under oscillatory flow (Harada et al.[5.4.15]) and stone gravel movable bed calculation (Fukuoka[5.4.16]) can be mentioned as examples.

5.4.1.3 Lagrange–Lagrange Coupling

Regarding the examination of the sediment transportation mechanism in the surf zone and swash zone, the degree to which the wave breaking interface with intense water-level fluctuation is captured with high accuracy will impact the success or failure of the result of the simulation. The particle method which is a fluid solver of Lagrange type is confirmed to have high reproducibility in violent flow analysis, and smoothed particle hydrodynamics (SPH) method of Ginglod and Monaghan[5.4.17] and moving particle semi-implicit (MPS) method of Koshizuka and Oka[5.4.18] are well-known particle methods. Additionally, a similar concept developed using fluid calculation of DEM and Euler type is adopted to these methods and solid–liquid phase coupling with DEM, and there is a model that assesses the momentum exchange of solid–liquid phase using drag-type equation and model that gives it from the integral of hydraulic stress around the DEM particles. Also, a fundamental equation of fluid motion that gives a single-phase or particle–fluid mixture is proposed. Specifically, the fluid motion equation for single-phase is as follows:

$$\rho \frac{Du_l}{Dt} = -\nabla p + \mu \nabla \cdot \left[\nabla u_l + (\nabla u_l)^{\mathrm{T}} \right] + \rho g + M_{\text{int}}; M_{\text{int}} = \frac{\sigma v_p - \rho u_l}{\Delta t} \tag{5.4.19}$$

The equation for particle–fluid mixture is based on the following:

$$\bar{\rho} \frac{D\bar{u}}{Dt} = -\nabla \bar{p} + \nabla \cdot \bar{\tau} + \bar{\rho} g + M_{\text{int}}; M_{\text{int}} = \frac{\sigma v_p - \bar{\rho}\bar{u}}{\Delta t} \tag{5.4.20}$$

Here, $\bar{u}$ is the flow velocity vector of particle–fluid mixture, $\bar{\tau}$ is the viscosity stress of particle–fluid mixture, and $\bar{p}$ is the pressure of particle–fluid mixture. Note that Equations (5.4.19) and (5.4.20) are equations that couple fluid and DEM particle motion by using the difference between the momentum of fluid particle and DEM particle. Additionally, solid–liquid phase model of Lagrange–Lagrange type that couples DEM and particle method is explained by Ohta et al.[5.4.19] and Gotoh[5.4.20] along with study cases.

5.4.2 Example of Numerical Analysis

As discussed in Section 5.2, several numerical simulations regarding solid–liquid multiphase flow have been performed in the field of coastal engineering to predict and understand various phenomena related to sediment transportation. In this section, we would like to show recent calculation examples for ripple, which is a seabed microtopography, that uses Lagrange–Lagrange coupling.

The scale of the unevenness of ripples at shallow water areas is an essential geometric quantity that gives bottom boundary conditions for beach process prediction. Thus far, the data regarding wavelength and wave steepness of ripple has been accumulated through experiments using oscillatory flow device and field observation (for example, Inman,[5.4.21] Bagnold,[5.4.22] Mogridge and Kamphuis,[5.4.23] Sleath,[5.4.24] and Dingler and Inman[5.4.25]). However, since intense water-level fluctuation that is accompanied by air bubbles due to uprush and backwash waves is exhibited at a shallow swash zone, it is difficult to measure the flow field around the ripple or geometry of the ripple. Additionally, since the numerical simulation for gas–liquid–solid-phase turbulence phenomenon that is attendant with wave breaking is also difficult to perform, sufficient knowledge regarding the formation mechanism of ripples in a swash zone has not been obtained.

Ripples are generated by the interaction between bottom sediment and water particle motion from waves. Regarding sand particle motion near the threshold of sediment movement, in analysis or numerical simulation, it is acceptable not to solve the water surface since the impact of waves to the seabed is insignificant and it is acceptable to focus on the interaction between the flow of the bottom boundary layer and sand particles. Additionally, it is considered to be fine to focus on measurement near the boundary layer in experiments and observations. Nevertheless, since the supply of kinetic energy due to breaking wave coming from the water surface or water surface gradient of traveling wave strongly affect the formation of ripples in areas that are shallower than the swash zone, measurements including water surface fluctuation and analysis must be considered. Thus, a fluid analysis method that can accurately handle water surface in numerical simulation should be selected. Since ripple is a phenomenon of sand particle scale, to thoroughly understand the formation mechanism of ripples, it is important to solve the motion of individual sand particles, which is the fundamental process of the phenomenon. From this standpoint, there is research aimed at examining the formation mechanism of ripples from the viewpoint of computational mechanics of sediment transport and there some researches are based on the framework of the DEM–MPS method. In the DEM–MPS method, it adopts the MPS method, which is one of the Lagrange-type solvers that have an established reputation for analyzing violent flows and adopts DEM which enables Lagrangian tracking of individual sand particle motion to sand particle tracking, and couples them.

The following are examples of a hydraulic experiment intended for ripples generated on the surface of a movable bed in a forced oscillation water tank and a numerical simulation using the three-dimensional DEM–MPS method, both performed as basic research aimed at gaining an understanding of the early development mechanism of ripples generated in a swash zone. We would like to give an outline of the content of the hydraulic experiment, a fundamental equation of the numerical simulation model, and an examination of the development mechanism of ripples in a swash zone. Note that a similar experiment and calculation has been introduced by Harada et al.[5.4.26]

5.4.2.1 Hydraulic Model Experiment

The experimental setup is shown in Figure 5.4.3 (left). The oscillating condition is inputted from a computer using waveform control software and the glass water tank installed on a small oscillating device is vibrated through the controller unit. The oscillating condition was a sinusoidal signal with a period of 0.5 s and half amplitude of 1.0 cm. Additionally, a rubber mat was placed between the oscillating device and the glass water tank to prevent the grass water tank on the oscillating device from sliding. The glass water tank has 25 cm width, 17 cm depth, and 21 cm height (outer dimensions). A sedimentary layer (movable bed part) that is 1 cm thick (case 1) or 2 cm thick (case 2) was formed inside the tank using alumina particles with a particle size of $d = 1.0$ mm and specific gravity of 2.65.

Additionally, the thickness of the glass water tank is 0.5 cm. The surface of the sedimentary layer was flattened horizontally as possible and water was gently filled up to a water depth of 1.0 cm on the movable bed in a manner not to disarrange the surface

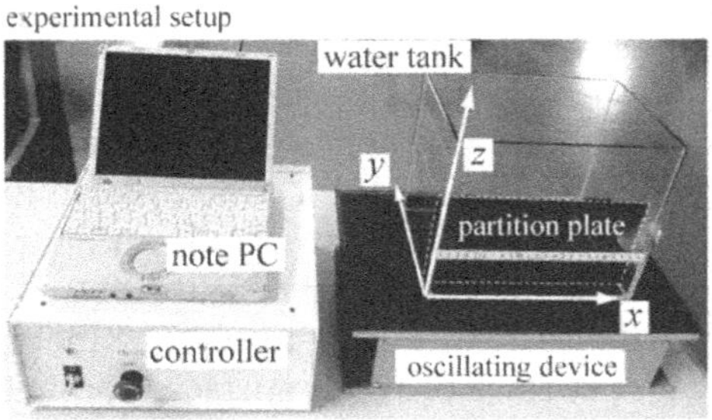

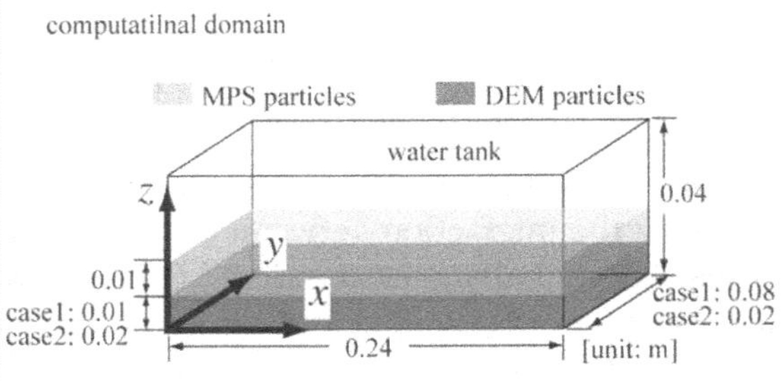

FIGURE 5.4.3
Experimental setup (*left*) and numerical domain (*right*).

layer of the sedimentary layer. The development process of ripples at the surface of the movable bed that is due to oscillation is filmed with a video camera mounted on the oscillating device. In the experiment of case 2 in which the sedimentary layer is 2 cm thick, red and blue dyes were poured inside the movable bed using a syringe, and the behavior of dyes from oscillation was recorded along with the development process of ripples.

5.4.2.2 Numerical Simulation Method and Computational Condition

The solid phase (DEM method) and the liquid phase (MPS method) are solved in different computational spaces. For coupling of both phases, momentum is exchanged through drag with the overlap-type solid–liquid phase-type DEM–MPS method (Gotoh[5.4.27]). Although small breaking waves are generated at the water surface from oscillation, to accurately capture the fluctuation of water surface that has strong nonlinearity, enhanced MPS method (MPS–HS–HL–ECS–GC–DS method: Khayyer and Gotoh,[5.4.28, 5.4.29, 5.4.30] Tsuruta et al.[5.4.31]), which introduced the accurate scheme to the MPS method (Koshizuka and Oka[5.4.18]), was applied to the discretization of the governing equation of fluid analysis.

The governing equations of fluid are the following equation of continuity and equation of motion:

$$\frac{D\rho_l}{Dt} + \rho_l \nabla \cdot \boldsymbol{u}_l = 0 \tag{5.4.21}$$

$$\rho_l \frac{D\boldsymbol{u}_l}{Dt} = -\nabla p + \mu \nabla^2 \boldsymbol{u}_l + \rho_l \boldsymbol{g} - \frac{1}{\varepsilon} \boldsymbol{F}_{\text{drag}} + \boldsymbol{F}_{\text{e}} \tag{5.4.22}$$

Here, ρ_l is the fluid density, $\boldsymbol{u}_l$ is the velocity vector, p is the pressure, μ is the coefficient of viscosity, $\boldsymbol{g}$ is the gravitational acceleration vector, $\boldsymbol{F}_{\text{e}}$ is the external force, ε is the porosity of the movable bed, and subscript l indicates the fluid phase. $\boldsymbol{F}_{\text{drag}}$ is the interaction force vector between the solid phase and the liquid phase. Solid-phase particle motion is described with translational and rotational motion equations:

$$\rho_s A_3 d_s^{\,3} \frac{d\boldsymbol{u}_s}{dt} = \frac{A_3 d_s^{\,3}}{1-\varepsilon} \boldsymbol{F}_{\text{drag}} - A_3 d_s^{\,3} \nabla p + \rho_s A_3 d_s^{\,3} \boldsymbol{g} + \boldsymbol{F}_{\text{pcol}} \tag{5.4.23}$$

$$I_s \frac{D\boldsymbol{\omega}_s}{Dt} = \boldsymbol{T}_{\text{pcol}} \tag{5.4.24}$$

Here, A_3 is the three-dimensional geometrical coefficient of the solid-phase particle, d_s is the diameter, $\boldsymbol{F}_{\text{pcol}}$ is the vector of force between solid-phase particles, I_s is the inertia tensor, $\boldsymbol{T}_{\text{pcol}}$ is the torque that is due to force between solid-phase particles, and the subscript s indicates solid-phase particle. The second term on the right-hand side of Equation (5.4.23) is equivalent to the inertia force of the flow and includes the buoyant force that acts upon solid-phase particles. Additionally, $\boldsymbol{F}_{\text{pcol}}$ and $\boldsymbol{T}_{\text{pcol}}$ that are due to particle interaction force are calculated using the spring–dashpot system installed between particles that are in contact (Harada and Gotoh[5.4.32]). The contact of solid-phase particles and the wall is handled as if the wall surface is a plain face without roughness, and contact force is calculated with the spring–dashpot system installed between particles and the wall as with contact between particles.

The force of interaction of fluid and solid-phase particles that overlap was assessed with the following equation related to porosity ε of the movable bed (Ergun,[5.4.33] Wen and Yu[5.4.34]):

$$\boldsymbol{F}_{\mathrm{drag}} = \beta(\boldsymbol{u}_l - \boldsymbol{u}_s) \tag{5.4.25}$$

$$\beta = \begin{cases} \dfrac{\varepsilon^2 \mu}{K} + F_{ch}\dfrac{\varepsilon^3 \rho_l}{\sqrt{K}}\left|\boldsymbol{u}_l - \boldsymbol{u}_s\right| & (\varepsilon \le 0.8) \\ \dfrac{3}{4} C_D (1-\varepsilon)\varepsilon^{-1.65}\dfrac{\rho_l}{d_s}\left|\boldsymbol{u}_l - \boldsymbol{u}_s\right| & (\varepsilon > 0.8) \end{cases} \tag{5.4.26}$$

$$K = \frac{\varepsilon^3 d_s^{\,2}}{\alpha(1-\varepsilon)^2}; F_{ch} = \frac{1.75}{\sqrt{150\varepsilon^3}} \tag{5.4.27}$$

Here, K indicates the permeability of the movable bed and C_D indicates the drag coefficient. The drag coefficient was given in the following equations as a function of particle Reynolds number (Re_p):

$$C_D = \begin{cases} \dfrac{24}{Re_p}\left(1 + 0.15 Re_p^{\,0.687}\right) & \left(Re_p \le 1000\right) \\ 0.4 & \left(Re_p > 1000\right) \end{cases} \tag{5.4.28}$$

$$Re_p \equiv \frac{\left|\boldsymbol{u}_l - \boldsymbol{u}_s\right| \varepsilon \rho_l d_s}{\mu} \tag{5.4.29}$$

To compare with the results of hydraulic experiments of case 1 and case 2 shown in the previous section, numerical simulation is performed using DEM–MPS method, in which DEM particles with the same particle size and specific gravity as with the alumina sphere used in the experiment is used. Additionally, to guarantee the resolution of the DEM particle scale that compose the numerical movable bed, fluid calculation using MPS particles of 1.0 mm in diameter was performed in the comparison with the hydraulic experiment of case 1. Additionally, an almost uniform ripple shape along the depth direction of the water tank was observed from the result of several experiments. Since the three-dimensionality was small, the width of the depth was set to be half of the experimental condition in consideration of computation cost (see the right image of Figure 5.4.3). On the other hand, in the comparison with the hydraulic experiment of case 2 in which the behavior of dyes in the movable bed is observed as well, the resolution of MPS particles was set to two times the resolution of DEM (0.5 mm in diameter) to capture the behavior of the dyes in detail. Since the computation load increased because the MPS particle size was decreased to half the size of DEM particles (two times higher resolution), to reduce computation load, the depth of the computation area was limited to 2 cm (see the right image of Figure 5.4.3). Note that it was confirmed that almost uniform ripple as with case 1 was formed in the experiment.

5.4.2.3 Comparison between Numerical Simulation and Hydraulic Model Experiment

a. Regarding case 1

Figure 5.4.4 shows the instantaneous images of the development process of ripples from the oscillation experiment. The movable bed development process in which unevenness is formed by oscillation on the movable bed surface that was flattened horizontally was tracked, and the initial formation of the ripple was observed near the center of the water tank ($x/d = 120$) at the time $t = 2.0$ s. A situation in which the wave height of ripples at the center of the water tank develops over time and the wave steepness increases can be observed. Note that after time $t = 4.0$ s, the outstanding developmental tendency of ripples was not observed at the movable bed surface within the observation area and the movable bed was developed sufficiently for the oscillating condition, indicating that it is nearly in an equilibrium state. With the oscillating condition of this experiment, besides the ripples near the center of the flume ($x/d = 120$), ripples having a wave height of approximately half of that at the center of the water tank was formed at positions around $x/d = 60$ and $x/d = 180$, showing bilateral symmetry to the center part of the water tank. This developmental tendency of ripples with the same type of experiment was observed in previous experiments (Ayrton and Ayrton[5.4.35]) as well.

Figure 5.4.5 shows the comparison between the movable bed surface shapes of the experiment and the numerical simulation. In the numerical simulation, although slight fluctuation in the unevenness of the movable bed surface is observed after time $t = 4.0$ s, no clear change was observed, indicating that the development of movable bed has settled for the most part. The same tendency was observed in the experimental result. Additionally, although the wave height of the ripples near the center of the flume ($x/d = 120$) in the numerical

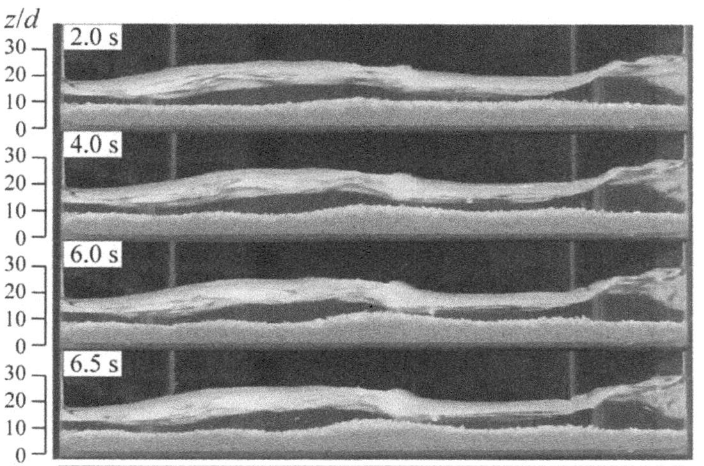

FIGURE 5.4.4
Developing process of ripple in the oscillating water tank experiment.

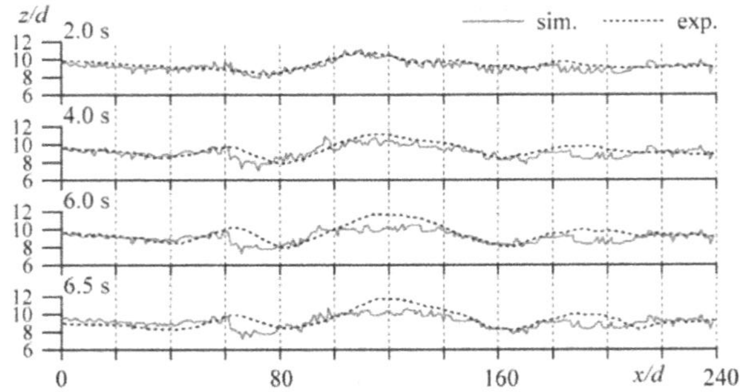

FIGURE 5.4.5
Comparison of movable bed surface form (case 1).

simulation is smaller than that of the experiment by one to two particles, the appearance position of crest and trough formed on the movable bed surface indicates that the numerical simulation well reproduced the tendency observed in the experiment. The comparison of water surface profiles of the experiment and the numerical simulation regarding the representative phase of a single period (0.5 s) from time t = 6.0 s is shown in Figure 5.4.6. The general form of water surface created by forced oscillation in the numerical simulation well reproduced the experimental result.

The magnitude of DEM particle migration velocity, as well as spatial distributions of drag in the x-axis (horizontal) direction and pressure gradient force that is applied to DEM particles at the timing when the acceleration of the oscillating device becomes zero (t = 6.12 s and t = 6.36 s), is shown in Figure 5.4.7 to examine the formation mechanism of ripples. At time t = 6.12 s and t = 6.36 s,

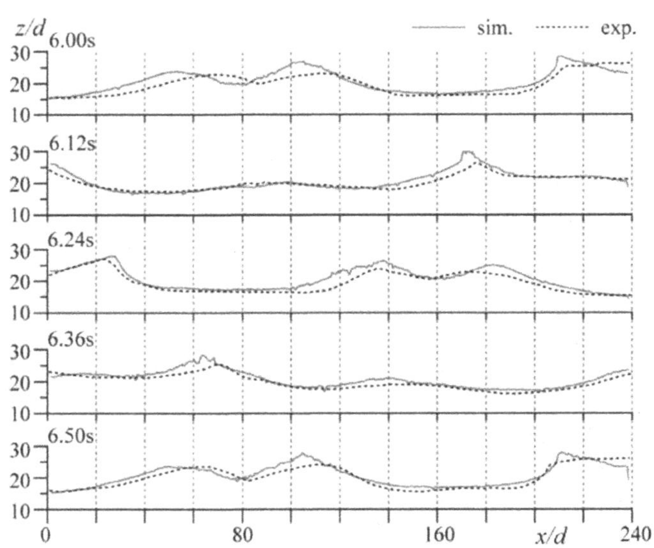

FIGURE 5.4.6
Comparison of water surface level (case 1).

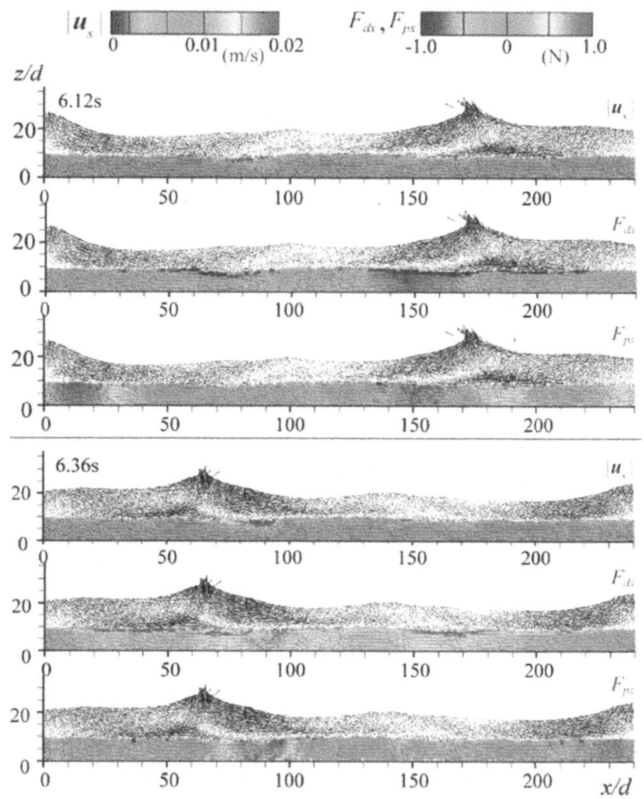

FIGURE 5.4.7
Spatial distributions for magnitude of DEM particle velocity, drag force acting on DEM particles and pressure gradient force.

the migration velocity of DEM particles near the movable bed surface at x/d = 70–90 and x/d = 140–160 shows relatively high-velocity distribution compared with other areas. Additionally, high value was confirmed in the horizontal component of drag and pressure gradient force that indicate the fluidization of DEM particles in the area at the same timing, suggesting that the formation of protrusions of ripple at the center of the movable bed is due to the DEM particle motion at the surface from the fluid force.

b. Regarding case 2

As with case 1, the comparison between the development process of the unevenness of the movable bed surface and water surface profile at the representative phase is shown to confirm the reproducibility of numerical simulation results regarding the experimental result. Figure 5.4.8 shows the development process of the movable bed surface. Although the wave steepness of the movable bed in the numerical simulation is relatively small compared with the experiment, the position on the surface of the movable bed where unevenness was generated almost matches. Additionally, at and after time t = 15.0 s, it can be seen that the development of ripples on the movable bed has settled and

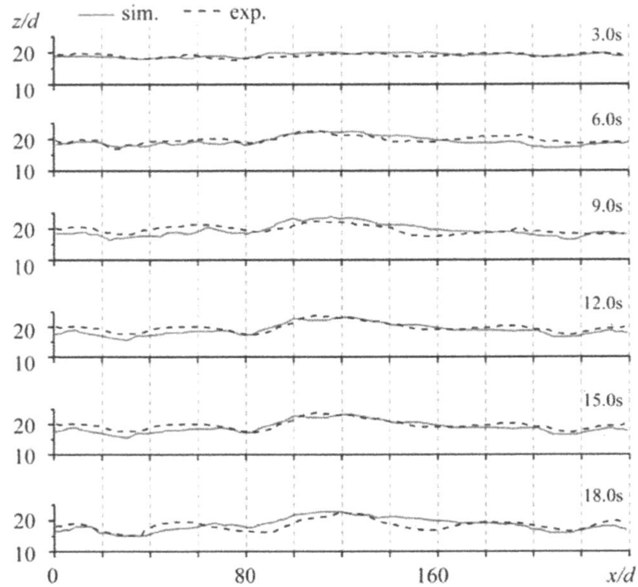

FIGURE 5.4.8
Comparison of movable bed surface form (case 2).

reached its equilibrium process. Figure 5.4.9 shows the water surface profile of the representative phase at the 37th period (time $t = 18.0$ s and after), and it indicates that the result of numerical simulation has well reproduced the experiment regarding water surface profile as well (Kawano et al.[5.4.36]).

We would like to examine the behavior of pore water within the movable bed regarding the development process of ripples. The comparison of the behavior

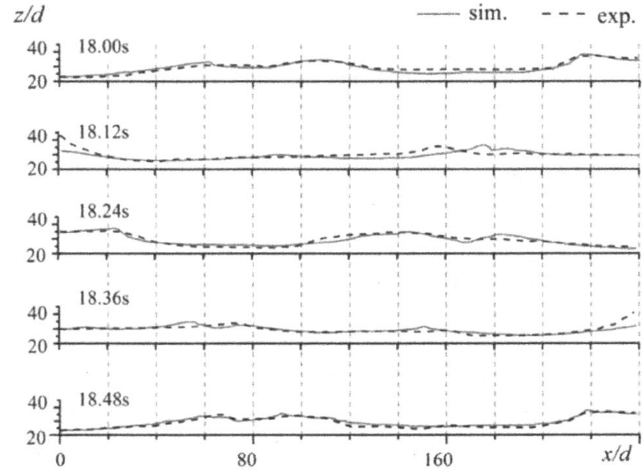

FIGURE 5.4.9
Comparison of water surface level (case 2).

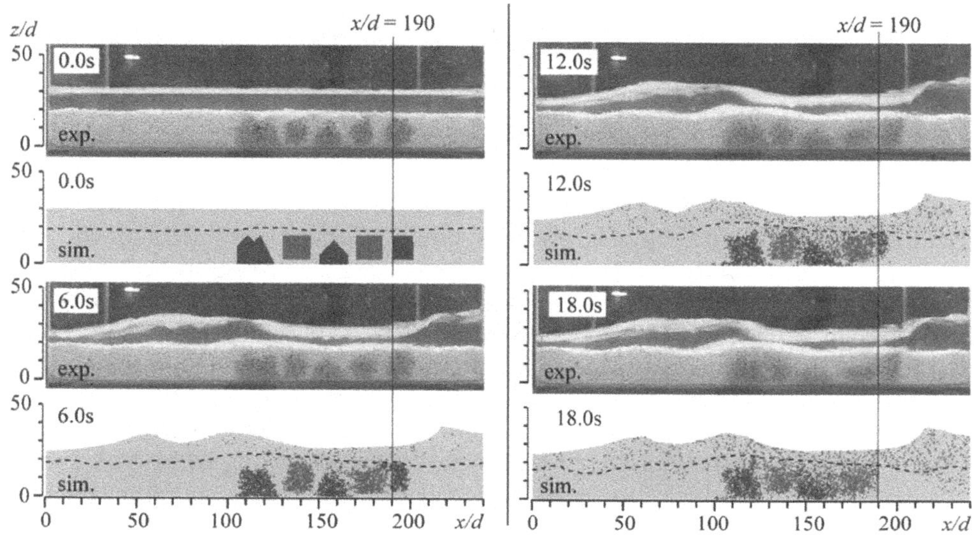

FIGURE 5.4.10
Comparison of dye behavior injected into movable bed.

of dyes injected inside the movable bed is shown in Figure 5.4.10 (Kawano et al.[5.4.36]). The instantaneous image shows the MPS particles in a cross section at $y = 1$ cm, and the height of the surface layer of the DEM sedimentary layer is indicated by the dotted lines. Note that to get closer to the same distribution state as that of when dyes were initially injected in the experiment, in the numerical simulation, the colored area of MPS particles was specified to the extent possible and Lagrangian tracking was performed. At the initial time of $t = 0.0$ s, the position near $x/d = 190$–200 is filled with the colored area of blue dye. In the experiment, distortion from the initial colored area was observed (see time $t = 18.0$ s) as the dyes at the upper area of the blue area near the movable bed surface were transported to the right. Meanwhile, the transportation rate of the colored particles was higher in the numerical simulation than in the experiment; at least half of the colored particles initially installed have exuded from the movable bed surface at time $t = 12.0$ s, and little or no blue particles were observed within the movable bed near $x/d = 190$–200 at time $t = 18.0$ s. Although spatial distribution of interstitial velocity at the vicinity of the end portion of the water tank differed between the experiment and the numerical simulation, regarding the pore water behavior near the ripples formed at the vicinity of the center of the water tank ($x/d = 120$), the tendency of exudation of dye near the crest of the ripples was confirmed in both the experiment and the numerical simulation. Regarding the dye behavior at the vicinity of the end portion of the water tank, the reason for the low reproducibility of the numerical simulation regarding the experiment is assumed to be the difference in the experimental area and the area for the numerical simulation. In the numerical simulation, the depth was limited to 2 cm, which is 1/8 the size of the experiment, from the viewpoint of computation cost. Thus, the end portion of the water tank was susceptible to the effect of binding from the sidewall compared with the experiment, causing a difference in the distribution of pore water flow at the end of the water tank. This must be considered in the future.

5.4.3 Conclusion

To examine the sediment transport mechanism in the swash zone, it is necessary to comprehend the water surface, which is a gas–liquid interface that fluctuates complicatedly, and to comprehend the solid–liquid interface that is due to bottom sediment transport. As (1) high reproducibility of water surface computation and (2) accurate tracking of the motion of particles that compose the movable bed are pointed as strengths of the DEM–MPS method, numerical simulation using the DEM–MPS method is believed to be effective in examining the sediment transport mechanism in the swash zone. Section 5.4.2 shows an example of application to numerical simulation that targets the examination of the microscopic structure of ripple formation process using DEM–MPS method. We confirmed the reproducibility of the result of the numerical simulation regarding the experiment through a comparison of the ripple development process and pore water behavior within the movable bed. Additionally, we presented a case study that examined the ripple formation mechanism from the viewpoint of acting force against movable bed particles, which is difficult to measure through experiments. The computation load of these types of calculation is high, and although there are constraints on area and resolution due to computer performance or computer resource, it demonstrates its true value when attempting to understand the internal structure. Understanding the microscopic structure will lead to deep knowledge regarding sand drift mechanism, and it is expected to contribute to the inducement of constitutive law as expansion from microscale to mesoscale.

5.5 Wave-induced Seabed Response by Grid Method

To solve the phenomena associated with waves and seabed by the grid method, different governing equations are used for a wavefield and seabed as mentioned in the following. Thus, in addition to the models of a wavefield and seabed, the appropriate coupling method of both is indispensable. This section covers CADMAS GEO-SURF,[5.5.1] CADMAS-SURF/3D,[5.5.2] and FS3M[5.5.3] as examples of the grid method and describes the wavefield model, the seabed model, and the coupling method of both models. The simulation results of FS3M are also introduced as application examples of the model.

5.5.1 Wave Model

The governing equations generally used in a wavefield model include the continuity equation and the Navier–Stokes equation for incompressible viscous fluids. As mentioned at the beginning of Chapter 5, since a DNS is not realistic, it is essential to apply Reynolds-averaged numerical simulation (RANS) based on the Reynolds average and large eddy simulation (LES) based on the spatial average. Two types of volume-of-fluid (VOF) methods, namely, the simple linear interface calculation (SLIC) type and the piecewise linear interface calculation (PLIC) type; level set (LS) method; the coupled level set and volume of fluid (CLSVOF) method, which is a combination of the VOF method and LS method; and other methods are also used for tracking complicated gas–liquid interfaces observed in coastal areas. The CADMAS GEO-SURF and CADMAS-SURF/3D mentioned here are RANS based on high Reynolds number two-equation k–ε model and FS3M is LES based on a coherent structure model. Both use the VOF method to track the gas–liquid interface. The following explains the outline of the governing equations and discretization method of a wavefield in CADMAS-SURF/3D and FS3M: for details, refer to Arikawa et al.[5.5.2] and Nakamura et al.[5.5.3]

The governing equations of CADMAS-SURF/3D are given as in the following equations.

Continuity equation:

$$\frac{\partial \gamma_x u}{\partial x}+\frac{\partial \gamma_y v}{\partial y}+\frac{\partial \gamma_z w}{\partial z}=\gamma_v S_\rho \tag{5.5.1}$$

Navier–Stokes equation:

$$\begin{aligned}\lambda_v \frac{\partial u}{\partial t}+\frac{\partial \lambda_x uu}{\partial x}+\frac{\partial \lambda_y vu}{\partial y}+\frac{\partial \lambda_z wu}{\partial z}=-\frac{\gamma_v}{\rho}\frac{\partial p}{\partial x}+\frac{\partial}{\partial x}\left\{\gamma_x \nu_e\left(2\frac{\partial u}{\partial x}\right)\right\}\\ +\frac{\partial}{\partial y}\left\{\gamma_y \nu_e\left(\frac{\partial u}{\partial y}+\frac{\partial v}{\partial x}\right)\right\}+\frac{\partial}{\partial z}\left\{\gamma_z \nu_e\left(\frac{\partial u}{\partial z}+\frac{\partial w}{\partial x}\right)\right\}-\gamma_v D_x u-R_x+\gamma_v S_u\end{aligned} \tag{5.5.2}$$

$$\begin{aligned}\lambda_v \frac{\partial v}{\partial t}+\frac{\partial \lambda_x uv}{\partial x}+\frac{\partial \lambda_y vv}{\partial y}+\frac{\partial \lambda_z wv}{\partial z}=-\frac{\gamma_v}{\rho}\frac{\partial p}{\partial y}+\frac{\partial}{\partial x}\left\{\gamma_x \nu_e\left(\frac{\partial v}{\partial x}+\frac{\partial u}{\partial y}\right)\right\}\\ +\frac{\partial}{\partial y}\left\{\gamma_y \nu_e\left(2\frac{\partial v}{\partial y}\right)\right\}+\frac{\partial}{\partial z}\left\{\gamma_z \nu_e\left(\frac{\partial v}{\partial z}+\frac{\partial w}{\partial y}\right)\right\}-\gamma_v D_y v-R_y+\gamma_v S_y\end{aligned} \tag{5.5.3}$$

$$\begin{aligned}\lambda_v \frac{\partial w}{\partial t}+\frac{\partial \lambda_x uw}{\partial x}+\frac{\partial \lambda_y vw}{\partial y}+\frac{\partial \lambda_z ww}{\partial z}=-\frac{\gamma_v}{\rho}\frac{\partial p}{\partial z}+\frac{\partial}{\partial x}\left\{\gamma_x \nu_e\left(\frac{\partial w}{\partial x}+\frac{\partial u}{\partial z}\right)\right\}\\ +\frac{\partial}{\partial y}\left\{\gamma_y \nu_e\left(\frac{\partial w}{\partial y}+\frac{\partial v}{\partial z}\right)\right\}+\frac{\partial}{\partial z}\left\{\gamma_z \nu_e\left(2\frac{\partial w}{\partial z}\right)\right\}-\gamma_v D_z w-R_z+\gamma_v S_w-\frac{\gamma_v \rho^* g}{\rho}\end{aligned} \tag{5.5.4}$$

Advection equation of VOF function F:

$$\gamma_v \frac{\partial F}{\partial t}+\frac{\partial \gamma_x uF}{\partial x}+\frac{\partial \gamma_y vF}{\partial y}+\frac{\partial \gamma_z wF}{\partial z}=\gamma_v S_F \tag{5.5.5}$$

Here, x and y are the horizontal coordinates; z is the vertical coordinate; t is the time; u, ν, and w are the flow velocity in the x, y, and z directions, respectively; p is the pressure; F is the VOF function (expresses the liquid phase when $F = 1$, gas–liquid interface when $0 < F < 1$, and gas phase when $F = 0$); ρ is the reference density; ρ^* is the density considering the buoyancy; ν_e is the sum of molecular kinematic viscosity and kinematic eddy viscosity; g is the gravitational acceleration; γ_v is the volume porosity; γ_x, γ_y, and γ_z are the surface porosity in the x, y, and z directions, respectively; $\lambda_v = \gamma_v + (1-\gamma_v) C_M$, $\lambda_x = \gamma_x + (1-\gamma_x) C_M$, $\lambda_y = \gamma_y + (1-\gamma_y) C_M$, $\lambda_z = \gamma_z + (1-\gamma_z) C_M$, C_M are the inertia coefficient; D_x, D_y, and D_z are the coefficient for the energy damping zones in the x, y, and z directions, respectively; and S_ρ, S_u, S_v, S_w, and S_F are the source terms for the wave generation source. Moreover, when using the Dupuit–Forchheimer resistance law, the drag terms R_x, R_y, and R_z received from a porous medium are expressed as follows:

$$R_x = \gamma_v (\gamma_x u)\left\{\alpha + \beta\sqrt{(\gamma_x u)^2 + (\gamma_y v)^2 + (\gamma_z w)^2}\right\} \tag{5.5.6}$$

$$R_y = \gamma_v (\gamma_y v)\left\{\alpha + \beta\sqrt{(\gamma_x u)^2 + (\gamma_y v)^2 + (\gamma_z w)^2}\right\} \tag{5.5.7}$$

$$R_x = \gamma_v (\gamma_z w)\left\{\alpha + \beta\sqrt{(\gamma_x u)^2 + (\gamma_y v)^2 + (\gamma_z w)^2}\right\} \tag{5.5.8}$$

Here, α and β are the parameters given by Engelund's expression:

$$\alpha = \alpha_0 \frac{(1-\gamma_v)^3}{\gamma_v^2} \frac{\nu}{d^2} \tag{5.5.9}$$

$$\beta = \beta_0 \frac{(1-\gamma_v)}{\gamma_v^3} \frac{1}{d} \tag{5.5.10}$$

Here, ν is the kinematic viscosity, d is the representative diameter of the porous media, and α_0 and β_0 are parameters with values depending on the type and stacking pattern of the porous media. The staggered grid is used for the computational grid. The simplified marker-and-cell (SMAC) method is applied to the coupled calculation of the flow velocity field and pressure field; first-order upwind difference scheme, second-order finite difference scheme, or a hybrid scheme of both to the discretization of the convection term of the Navier–Stokes equation; MILU-BiCGSTAB method to the Poisson equation; and the donor–acceptor method or a method that considers the inclination of the interface to the advection term of the advection equation of the VOF function.

The governing equations of FS3M are given as in the following equations.

Continuity equation:

$$\beta = \beta_0 \frac{(1-\gamma_v)}{\gamma_v^3} \frac{1}{d} \tag{5.5.11}$$

Navier–Stokes equation:

$$\begin{aligned} &\frac{\partial}{\partial t}\left[\left\{m + (1-m)C_A\right\}\bar{v}_i\right] + \frac{\partial}{\partial x_j}\left(m\bar{v}_i\bar{v}_j\right) = -\frac{m}{\hat{\rho}}\frac{\partial \bar{p}}{\partial x_i} + mg_i \\ &+ \frac{m}{\hat{\rho}}\left(f_i^s + R_i + f_i^{ob}\right) + \frac{1}{\hat{\rho}}\frac{\partial}{\partial x_j}\left(2m\hat{\mu}\bar{D}_{ij}\right) + \frac{\partial}{\partial x_j}\left(-m\tau_{ij}^a\right) + Q_i + m\beta_i \end{aligned} \tag{5.5.12}$$

Advection equation of VOF function F:

$$\frac{\partial(mF)}{\partial t}+\frac{\partial(m\overline{v}_jF)}{\partial x_j}=Fq^* \tag{5.5.13}$$

Here, x_i is the position vector, t is the time, v_i is the grid-scale (GS) component of the flow velocity, p is the GS component of the pressure, g_i is the gravitational acceleration vector, $\hat{\rho}$ is the apparent density of the fluid, $\hat{\mu}$ is the coefficient of the apparent viscosity of the fluid, m is the porosity of the porous media, C_A is the added mass coefficient of the porous media, f_i^s is the surface tension vector based on the continuum surface force (CSF) model, f_i^{ob} is the interaction force vector between the fluid and structure, $\overline{D}_{ij}$ is the GS component ($=(\partial\overline{v}_j/\partial x_i+\partial\overline{v}_i/\partial x_j)/2$) of the strain velocity tensor, τ_{ij} is the eddy stress tensor based on the coherent structure model, Q_i is the wave generation source/sink vector, q^* is the wave generation source/sink strength per unit time, and β_i is the damping function vector in the artificial damping zones. Superscript a expresses the anisotropic component of the tensor. Moreover, R_i is the linear and nonlinear drag force vector received from the porous media and is expressed as follows:

$$R_i=-\frac{m\hat{\rho}g}{k^w}\overline{v}_i-\frac{C_{D1}\hat{\rho}(1-m)}{2md_{50}}\overline{v}_i\sqrt{\overline{v}_j\overline{v}_j} \tag{5.5.14}$$

$$k^w=k_s^wS_e^{1/2}\left\{1-\left(1-S_e^{1/m^*}\right)^{m^*}\right\}^2 \tag{5.5.15}$$

$$k_s^w=\frac{1}{12C_{D2}}\frac{m^2}{1-m}\frac{gd_{50}^2}{\nu_w} \tag{5.5.16}$$

Here, C_{D2} and C_{D1} are the coefficients of the linear and nonlinear drag force, respectively, g is the gravitational acceleration, ν_w is the kinematic viscosity of water, d_{50} is the median particle size of the porous media, S_e is the effective degree of saturation of the porous media, and m^* is the shape parameter of the van Genuchten method. Unlike CADMAS-SURF/3D, the volume porosity γ_v and the surface porosity γ_x, γ_y, and γ_z are not distinguished in FS3M. Thus, porosity m is set for both. Conversely, the temporal change of porosity m is considered in FS3M. The difference in the modeling of the drag force received from the porous media is explained later. Additionally, there are differences in the modeling of the damping force in the energy damping zones and artificial damping zones, etc. As with CADMAS-SURF/3D, the staggered grid is used for the computational grid. Moreover, the SMAC method is applied to the coupled calculation of the flow velocity field and pressure field; the fifth-order accuracy MUSCL TVD (monotone upstream-centered schemes for conservation laws, total variation diminishing) scheme to the discretization of the convection term of the Navier–Stokes equation; the modified incomplete Cholesky conjugate gradient (MICCG) method to the Poisson equation; and the multi-interface advection and reconstruction solver (MARS) to the advection term of the advection equation of the VOF function.

Here, the modeling of the drag force received from the porous media associated with both the waves and the seabed is focused on. For simplification, the volume porosity γ_v and the surface porosity γ_x, γ_y, and γ_z are not distinguished in CADMAS-SURF/3D, and m is set for both. In FS3M, $\partial m/\partial t=0$ is assumed neglecting the temporal change of m. It is also assumed that the median grain size d_{50} of the porous media of FS3M is equal to the representative diameter d of CADMAS-SURF/3D. At this time, in the case of the x direction, for example, the general form of the Navier–Stokes equation is set as follows:

$$c\frac{\partial u}{\partial t}+\cdots=-\frac{m}{\rho}\frac{\partial p}{\partial x}-agm^2u-bgm^3u\sqrt{u^2+v^2+w^2}+\cdots \tag{5.5.17}$$

Here, a, b, and c are the coefficients, and a is equivalent to the inverse of the hydraulic conductivity k.

First, regarding the inertia force, when coefficient c of the time-derivative term provided in the first term on left-hand side of Equation (5.5.17) is focused on, it becomes $c=\gamma_v=\gamma_v+(1-\gamma_v)\,C_M=m+(1-m)\,C_M$ in CADMAS-SURF/3D, where C_M is the inertia coefficient. Nevertheless, it becomes $c=m+(1-m)\,C_A$ in FS3M, where C_A is an added mass coefficient. Herein, since there is a relationship of $C_M=1+C_A$ for C_M and C_A, it is obvious that the modeling of the inertia force differs between CADMAS-SURF/3D and FS3M.

Then, focusing on the linear components of the drag force, the coefficient a which is an inverse of hydraulic conductivity k can be summarized as in Table 5.5.1. In addition to CADMAS-SURF/3D and FS3M, Table 5.5.1 shows modeling by the Kozeny–Carman equation[5.5.4] and van Gent[5.5.5] for spherical particles. Here, α is the coefficient. According to Table 5.5.1, although the order of denominator m and numerator $(1-m)$ of a changes by the model, all models are proportional to d^{-2}. Here, when the linear component of the drag force becomes predominant, i.e., a one-dimensional space in the x direction in which Darcy's law holds, the hydraulic gradient i can be written as $i = a\ mu$ using the discharge velocity (Darcy velocity) mu. For example, consider a condition where the reduced scale of a model is $1/N$ of the prototype based on Froude's law. Although hydraulic gradient i is nondimensional and does not depend on the scale at this time, the flow velocity u of the model becomes $1/N^{0.5}$ of the prototype. Thus, a of the model should be set to $N^{0.5}$ times that of the prototype. As mentioned above, since the modeling of a is proportional to d^{-2}, particle size d of the model should be set to $1/N^{0.25}$ of the prototype. As mentioned in Chapter 3, according to the law of similarity (Dean number) of the settling velocity, when the particle size of the permeable material where the linear component of the drag force becomes predominant is relatively small, particle size d of

TABLE 5.5.1

Linear Component of the Resistance Force

	$a[s/m]$
CADMAS SURF/3D	$\alpha_0\dfrac{(1-m)^3}{m^2}\dfrac{\nu}{gd^2}$
FS3M	$12C_{D2}\dfrac{1-m}{m^2}\dfrac{\nu}{gd^2}$
Kozeny–Carman equation	$180\dfrac{(1-m)^2}{m^3}\dfrac{\nu}{gd^2}$
Van Gent[5.5.5]	$\alpha\dfrac{(1-m)^2}{m^3}\dfrac{\nu}{gd^2}$

the model asymptotically approaches $1/N^{0.25}$ of the prototype. Thus, it can be concluded that the modeling of Table 5.5.1 and the law of similarity of the settling velocity are in agreement. However, it has been confirmed that C_{D2} tends to become higher with the increase of the Keulegan–Carpenter (KC) number.[5.5.6] When the particle size of the model is $1/N^{0.25}$ of the prototype, the KC number of the model decreases to $1/N^{0.75}$ of the prototype. Thus, to match the permeability and the type of sediment transport between the model and prototype, it is necessary for the particle size of the model to be $1/N^{0.25}$ of the prototype and then to adjust the coefficient α, α_0, and C_{D2} for the decreased amount of the KC number.

The nonlinear component of the resistance force can be summarized as shown in Table 5.5.2. Table 5.5.2 also shows modeling by van Gent.[5.5.5] Here, β is the coefficient and KC is the KC number. Although a difference in the expression of the coefficient can be observed in Table 5.5.2, the model is inversely proportional to d because $(1-m)/m^3/(gd)$ is included in all the b. In this condition, when considering the one-dimensional space of the x direction in the condition where the nonlinear component of the drag force becomes predominant, the hydraulic gradient i can be written as $i = b\,mu\,|mu|$. The same as that previously mentioned, b of the model should be adjusted to N times the prototype when considering the condition where the reduced scale of a model is $1/N$ of the prototype based on Froude's law. Since b is the function form inversely proportional to d in either model, the particle size d of the model should be set to $1/N$ of the prototype. According to the law of similarity (Dean number) of the settling velocity, when the particle size of the permeable material where the nonlinear component of the drag force becomes predominant is relatively large, the particle size of the model asymptotically approaches $1/N$ of the prototype. Thus, the result indicates that all of the modeling of Table 5.5.2, the law of similarity of the settling velocity, and Froude's law are in agreement. However, it is known that the drag coefficient depends on the KC number and Reynolds number.[5.5.7] When the particle size of the model is $1/N$ of the prototype, the KC number of the model and prototype are in agreement. Conversely, when a fluid with an equal kinematic viscosity is used for the prototype and model, the Reynolds number of the model decreases to $1/N^{1.5}$ of the prototype. Thus, to match the permeability and the type of sediment transport in the model and prototype, it is necessary for the particle size of the model to be $1/N$ of the prototype and then to adjust the coefficient β_0, C_{D1}, and β according to the Reynolds number.

Such facts are confirmed from the point of view of the model experiments. Suppose that the reduced scale of the model is $1/N$ of the prototype and a fluid with an equal

TABLE 5.5.2

Nonlinear Component of the Resistance Force

	$b[s^2/m^2]$
CADMAS SURF/3D	$\beta_0 \frac{1-m}{m^3}\frac{1}{gd}$
FS3M	$\frac{1}{2}C_{D1}\frac{1-m}{m^2}\frac{1}{gd}$
Van Gent[5.5.5]	$\beta\left(1+\frac{7.5}{KC}\right)\frac{1-m}{m^3}\frac{1}{gd}$

kinematic viscosity is used for the prototype and model. At this time, since the Reynolds number of the model decreases to $1/N^{1.5}$ of the prototype as mentioned above, the drag coefficient of the model, i.e., the drag force that the fluid receives from the porous media, generally becomes larger than that of the prototype. This means that the drag force, which the porous media receives from a fluid body, also becomes larger when compared with that of the prototype and corresponds to the experimental results of Kajima et al.[5.5.8] and Sakakiyama and Ogasawara[5.5.9] who mentioned that the reflected and transmitted waves decrease as the scale of the experiment becomes smaller. Using the modeling of CADMAS-SURF/3D as an example, the model experiment may be brought closer to that of the prototype if ingenuity is used to increase porosity m only for the portion in which β_0 of the model has increased so that $\beta_0(1 - m)/m^3$ becomes equal in the model and the prototype. Nevertheless, when m is changed, the skeleton structure of the porous media becomes different in the prototype and the experiment. As β_0 is expected to change as well, it seems that adjustment by m is not so simple.

Finally, the case where both the linear component and nonlinear component of the drag force become predominant is considered. The same as that previously mentioned, considering a one-dimensional space in the x direction, the hydraulic gradient i becomes $i = a\, mu + b\, mu\, |mu|$. When the reduced scale of the model is $1/N$ of the prototype based on Froude's law, as mentioned above, it is necessary for particle size d of the model to be $1/N^{0.25}$ of the prototype based on the linear component and d of the model to be $1/N$ of the prototype based on the nonlinear component. Thus, d of the model cannot be uniquely determined. Nevertheless, according to the law of similarity (Dean number) of the settling velocity, the particle size of the model will take a value between $1/N$ and $1/N^{0.25}$ of the prototype. Thus, when focusing on a sediment transport phenomenon, for example, we believe that the permeability should be matched by adjusting the drag force coefficient after the particle size of the model is determined on the basis of the law of similarity of the settling velocity.

The above is summarized in Table 5.5.3. Here, Gu and Wang[5.5.10] was referred to for the classification of the permeable material and the predominant component of the drag force. As a case study, when this point of view is applied to the erosion problem of sandy seabed under rubble mound of breakwater covered with wave-dissipating blocks as mentioned in Chapter 4, it becomes as shown in Figure 5.5.1. First, regarding the wave-dissipating blocks and armor blocks, since both the prototype and model are equivalent to the size of large gravel or coarser in Table 5.5.3, the nonlinear component among the drag force is considered to be predominant. Thus, the size of the model shall be $1/N$ of the prototype. Moreover, regarding sandy seabed, since both the prototype and model are equivalent to the size of coarse sand or finer in Table 5.5.3, the linear component among the drag force is considered to be predominant. Hence, the particle size of the model shall be $1/N^{0.25}$ of the prototype. Nevertheless, because the drag force coefficient depends on the KC number and Reynolds number, an adjustment is required according to the change in the KC number and Reynolds number. Conversely, materials in the mound and gabion are equivalent to the size of small gravel in Table 5.5.3. Thus, the particle size needs to be determined depending on the sediment transport, linear component of the drag force, or the nonlinear component of the drag force. Moreover, in the phenomenon currently being focused on, leakage of the sandy seabed through the mound is occurring. An influence from the scale effect seems to be unavoidable since the particle size ratio of the sandy seabed and mound is involved.

TABLE 5.5.3

Predominant Components of the Resistance Force and Scale of the Particle Size under Each Similarity Law

			Particle Size of the Model Relative to the Prototype			
Description	**Size Range**	**Predominant Component of the Resistance Force**	**Froude's Law**	**Law of Similarity of the Settling Velocity**	**Law of Similarity of the Resistance Force**	**Coefficient of the Resistance Force**
Coarse sand or finer	<2 mm	Linear	$1/N$	$1/N^{0.25}$	$1/N^{0.25}$	Depends on the KC and Reynolds numbers
Pebble or small gravel	1 cm	Linear and nonlinear	$1/N$	$1/N$–$1/N^{0.25}$	Unable to determine uniquely	
Large gravel crusted stone	10 cm	Nonlinear	$1/N$	$1/N$	$1/N$	Depends on the Reynolds number
Boulder crusted stone	0.3–1.0 m					
Artificial blocks, large rocks	>1.0 m					

5.5.2 Seabed Model

Biot's equation in u–w–p formulation with the displacement of soil skeleton u, the relative displacement of the pore water to soil skeleton w, and pore water pressure p are derived from the equilibrium equation of the seabed, and equilibrium and continuity equations of the pore water. Assuming that the bulk elastic modulus of water K_w is finite, p can be eliminated; hence, u–w formulation of Biot's equation can be derived. Conversely, when w is eliminated from u–w–p formulation of Biot's equation assuming that the relative acceleration of the pore water is smaller than the acceleration of the soil skeleton, u–p formulation of Biot's equation can be derived. Furthermore, Biot's consolidation equation can be derived when all the acceleration terms are negligible. As

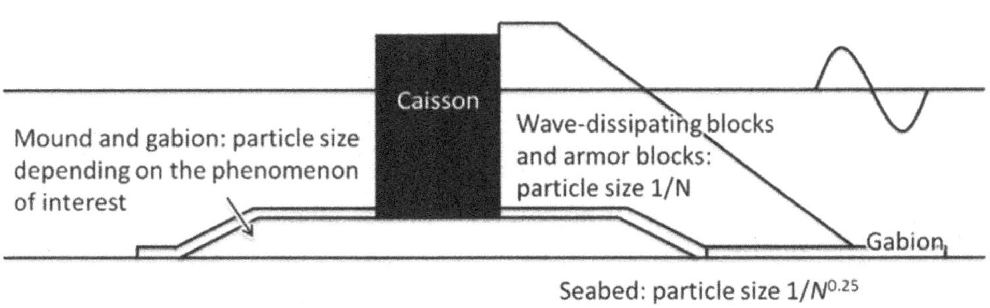

FIGURE 5.5.1
One method for determining the particle size based on Table 5.5.3 for the erosion phenomenon under the mound of the composite-type breakwater covered with wave-dissipating blocks.

mentioned in the previous section, one of these equations is used in the seabed model as the governing equation. Additionally, the overall governing equation will be formed by adding the principle of the effective stress, constitutive equation of the soil skeleton, and strain–displacement relations. Takahashi et al.[5.5.11] revealed that the accuracy of the pore water pressure deteriorated because slow compression waves could not be calculated appropriately using *u–w* formulation of Biot's equation. Therefore, *u–p* formulation of Biot's equation was applied in both CADMAS GEO-SURF and FS3M addressed here as the governing equation (called the *U-π* formulation in CADMAS GEO-SURF). The following explains an outline of the governing equation in FS3M: for details, refer to Takahashi et al.[5.5.11] and Nakamura et al.[5.5.3]

When the displacement vector of the soil skeleton is u_i and the relative displacement vector of the pore water is w_i, the governing equation of FS3M is given by *u–p* formulation of Biot's equation below, derived by assuming that the relative acceleration of the pore water $\ddot{w}_i$ is smaller than the acceleration of the soil skeleton $\ddot{u}_i$:

$$\rho \ddot{u}_i = -\sigma'_{ji,j} - p_{,i} + \rho g_i \tag{5.5.18}$$

$$-\frac{\partial \varepsilon_{kk}}{\partial t} + \frac{m}{K_w}\dot{p} + \left\{ \frac{k_s}{\rho_w g}\left(-\rho_w \ddot{u}_i - p_{,i} + \rho_w g_{si}\right) \right\}_{,i} = 0 \tag{5.5.19}$$

Here, p is the pore water pressure, σ'_{ij} is the effective stress tensor where the compression is positive, ε_{ij} is the strain tensor, m is the porosity of the seabed, ρ is the density of the seabed (= $(1-m)\rho_s + m\,\rho_w$), ρ_s is the density of soil particles, ρ_w is the density of the water, k_s is the hydraulic conductivity, K_w is the apparent bulk elastic modulus of water, g_{si} is the gravitational acceleration vector, and g is the gravitational acceleration. Subscript i expresses the partial differential $\partial/\partial x_i$.

The *u–p* formulation of Biot's equation is normally solved by FEM. At this time, there are two formulations: Sandhu's approach, which defines the pore water pressure by nodal points, and Christian's approach, which defines the pore water pressure by elements. Formulation of Sandhu's approach is applied in FS3M, considering that it is easier to connect to the wavefield if the definition point of the pore water pressure is located on the surface of the seabed. Moreover, the Newmark-β method is used for the displacement of the soil skeleton as time integration and the Crank–Nicolson method for the pore water pressure. As a computational grid, a 20-noded isoparametric element is used for displacement of the soil skeleton, and an 8-noded isoparametric element that is one order lower is used for the pore water pressure.

5.5.3 Coupling of Wave Model and Seabed Model

In the coupling of the wave model and seabed model, as mentioned in the previous section, there is Type A (All), which solves the flow field in an area including the inner part of the seabed by the wave model to give pressure acquired on the seabed surface to the seabed model, and Type W (Wave), which solves the flow field in an area except for the inner part of the seabed to give pressure acquired on the seabed surface to the seabed model. Coupling of Type W is adopted in CADMAS GEO-SURF addressed here and Type A in FS3M. The following introduces the coupling of Type A in FS3M.

FS3M consists of a main solver, which is a wave model; a VOF module based on the volume-of-fluid (VOF) method for tracking the gas–liquid interface; an IB (immersed boundary) module based on the IB method for computing fluid–structure interaction; a sediment transport (ST) module for computing the topographic changes of a movable bed by sediment transport and distribution of suspended sediment concentration; and an FEM module which is the seabed model. To perform coupled analysis of fluid, structure, and topographic change, weak two-way coupling is used between the main solver, VOF module, IB module, and ST module, and Type A one-way coupling is used between these and the FEM module as mentioned above. Figure 5.5.2 shows the calculation steps. Here, n is the number of calculation steps. Specifically, the following steps are taken to perform a calculation in FS3M:

1. The main solver is executed to calculate the flow velocity $\overline{v}_i$ and pressure $\overline{p}$.
2. The VOF module is executed using the flow velocity field acquired from the main solver, and VOF function F is calculated.
3. The ST module is executed using the value acquired from the main solver and the VOF module, and the change of seabed height z_s by sediment transport and distribution of suspended sediment concentration C is calculated.
4. The IB module is executed using the value acquired from the main solver, VOF module, and ST module, and the displacement of movable structures is calculated.
5. The FEM module is executed using the value acquired from the main solver and ST module, and the displacement of ground skeleton u_i and pore water pressure p are calculated.
6. The value acquired from the VOF module, ST module, and IB module is fed back to the main solver, which is then executed. The above steps are repeated until the time the calculation is completed.

Here, the continuity of the flow velocity and pressure on the seabed surface is ensured by interpolating the flow velocity and pressure on the seabed surface to calculate the

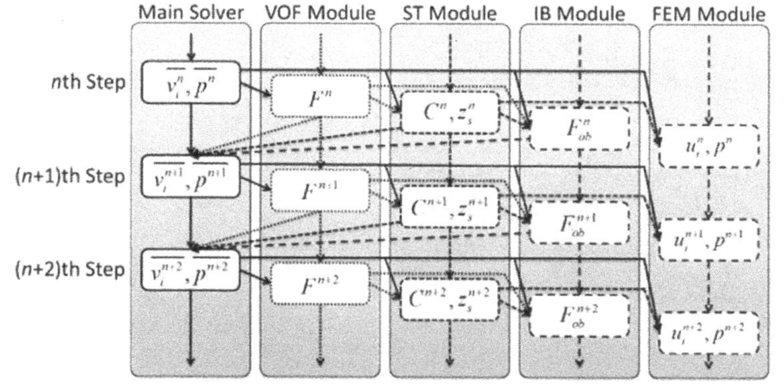

FIGURE 5.5.2
Coupling method of FS3M.

value from flow velocity $\overline{v}_i$ and pressure $\overline{p}$ acquired by the main solver in step 5 and entering the acquired value in the FEM module.

5.5.4 Examples of Numerical Analysis

As application examples of FS3M, the following introduces calculation examples of seepage phenomenon inside the sand column, seepage phenomenon inside dike, erosion phenomenon of dike by overflow, and scouring phenomenon around an inland structure by a run-up tsunami.

A calculation was implemented using the main solver and VOF module of FS3M for the unsaturated seepage test in the sand column by Yoshikawa et al.[5.5.13] The sand column with a length of 1.30 m and height of 0.45 m was set in a computational domain with a length of 2.20 m and height of 0.65 m, and the water level at the right end of the sand column was kept at 0.46 m. The sand column was composed of silica sand with the porosity m of 0.5, the initial saturation of 8.0%, and the shape parameter α and n^* of the van Genuchten equation of 0.28 kPa^{-1} and 12.898. Moreover, the median grain size d_{50} was set to 0.2 mm following the grain size accumulation curve of Sugii et al.[5.5.14] The residual saturation S_{rr} was set to 0.0% and the maximum saturation S_{rs} to 100% following Yoshikawa et al.[5.5.13] The linear drag force coefficient C_{D2} was set to $C_{D2} = 30.0$. Figure 5.5.3 shows examples of the results. Here, the value in which the water pressure measured at the bottom surface of the sand column converted to hydraulic head is indicated with

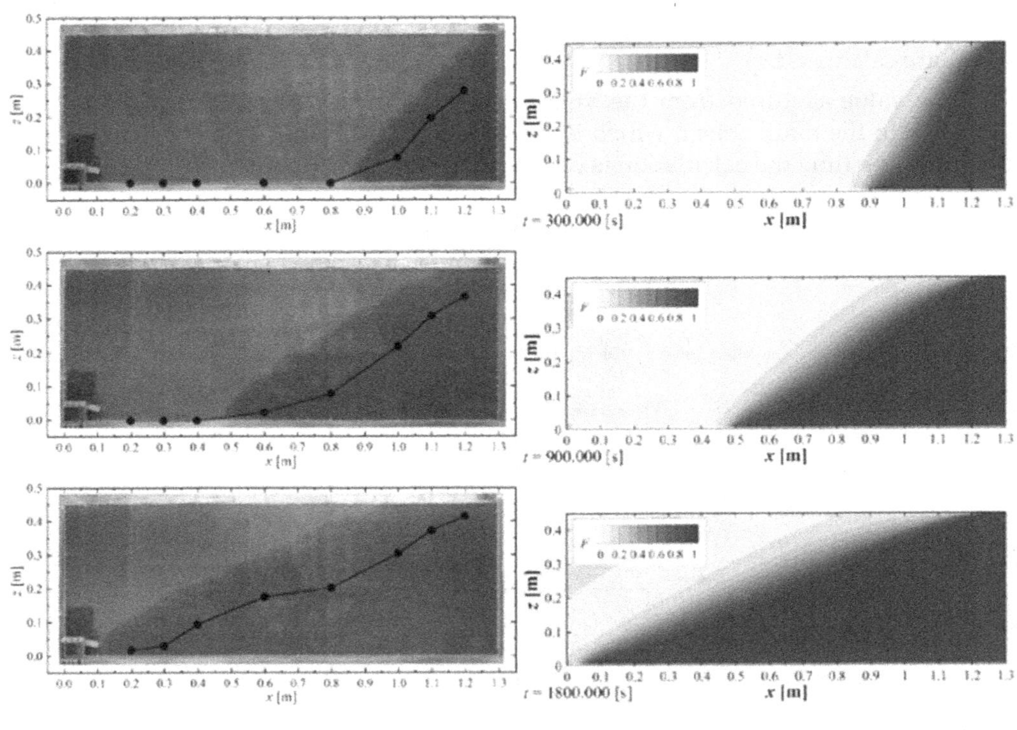

FIGURE 5.5.3
Numerical results for the unsaturated seepage experiments inside the sand column.[5.5.12]

dots and solid lines. As shown in the figure, seepage is progressing from the bottom right of the sand column, and the speed of the seepage surface of the calculation results is almost same as the test results. It can also be observed in the calculation results that there is almost no change in the degree of saturation on the upper side of the seepage surface. Thus, the unsaturated condition inside the sand is maintained by considering the effects of suction in the method based on VOF.

Next, the calculation of the unsaturated seepage test inside the dike by Yoden[5.5.15] is shown as follows: The main solver and VOF module in FS3M were also used in this calculation. The sandbed with a length of 1.20 m and height of 0.15 m and a dike with a crown width of 0.10 m, height of 0.15 m, and slope of 1/2 were set in a computational domain with a length of 2.80 m and height of 0.50 m, and the left side of the dike was flooded up to the crown. Silica sand with a median grain size d_{50} of 0.239 mm was used as the material of the bed and dike. The porosity m was set to 0.53, initial saturation to 9.17%, residual saturation S_{rr} to 11.5%, and shape parameter α and n^* of the van Genuchten equation to 0.3924 kPa^{-1} and 3.852, respectively. As in the case above, linear drag force coefficient C_{D2} was set to 30.0 and the maximum saturation S_{rs} to 100%. Figure 5.5.4 shows examples of the calculation results. The solid line in the figure indicates the position of the seepage surface obtained in the test. As in Figure 5.5.3, Figure 5.5.4 shows that the speed of the seepage surface in the test results and calculation results are in agreement reasonably well. According to the result of t = 1,800 s, which indicated the final state, the seepage point on the downstream slope of the dike is positioned close

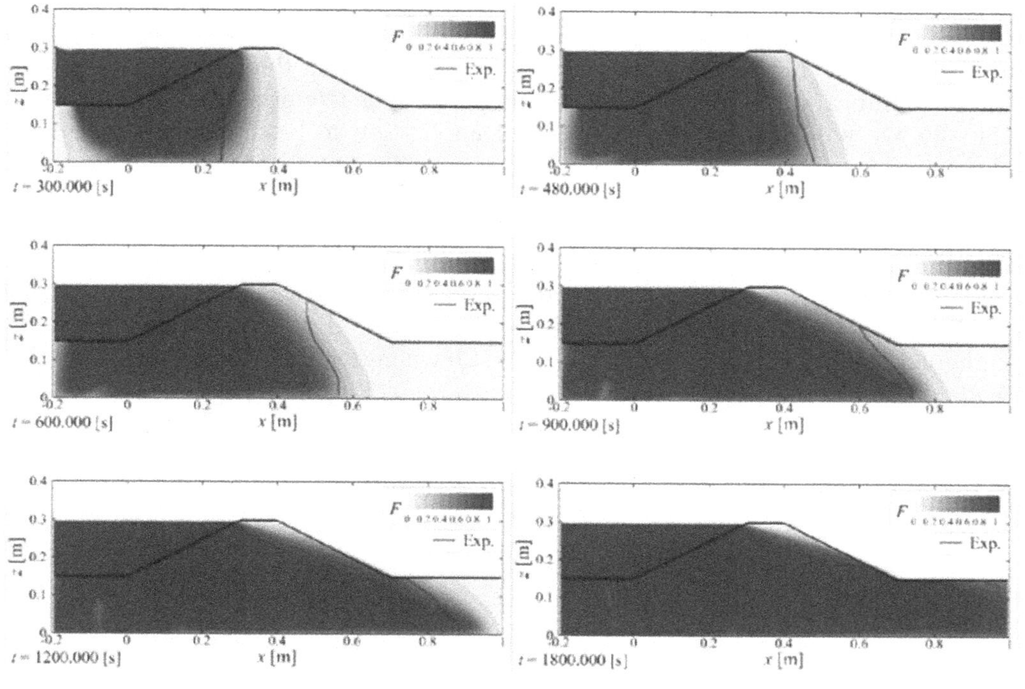

FIGURE 5.5.4
Numerical results for the unsaturated seepage experiments inside the dike.[5.5.12]

to $x = 0.568$ m and $z = 0.216$ m obtained by the Casagrande method, and computational capability of the model can also be confirmed from the viewpoint of the seepage point.

Next, the calculation of an overflow erosion test of an unsaturated dike by Yoden[5.5.15] is shown as follows: in addition to the main solver and VOF module in FS3M, the ST module, which performs the sediment transport calculation, is also used in this calculation. A dike made of silica sand with a crown width of 0.35 m and height of 0.40 m and slope of 1/2 was set in a computational domain with a length of 3.35 m and height of 0.50 m, and water with the same flow rate as the hydraulic experiment was flowed from the left end of the domain. The same as the test, porosity m of the dike was set to 0.51, initial saturation to 33.2%, and residual saturation S_{rr} to 12.0%. Other parameters were set the same as in the previous paragraph. Regarding the parameters of the sediment transport to be input into the ST module, the density of sand particles ρ_s was set to 2.65×10^3 kg/m^3, static friction angle and angle of repose to 45°, kinematic friction angle to 27°, and the coefficient of the pickup function of suspended sediment to 0.00055. The critical Shields number was obtained from Iwagaki's equation. To reduce the deposition on the horizontal bed at the downstream side of the dike, the settling velocity of the sand particles was set to 1/10 of that of Rubey's equation. Figure 5.5.5 shows a comparison of the erosion shape of the dike. As shown in the figure, the erosion shape of the dike agrees with the test results reasonably well, where the downstream slope is eroded with almost initial slope along as the crown gradually became lower in height. Figure 5.5.6 shows the distribution of VOF function F and suspended sediment concentration C during the erosion. The solid line in Figure 5.5.6a indicates the outer shape of the dike, and the solid line in Figure 5.5.6b indicates the water surface ($F = 0.5$) and the outer shape of the dike. As shown in the figure, the sand was picked up by the overflow, which increased suspended sediment concentration C as well as the gradual erosion of the dike from the downstream slope. At this time, it can be observed that the flow over the dike is extremely thin, and the erosion is progressing due to the interaction of this thin flow.

Finally, we will introduce an example of application to local scouring around an inland structure by a run-up tsunami. In this example, in addition to the main solver, VOF module, and ST module in FS3M, the FEM module, which performs the soil–water coupled analysis, was also used. Figure 5.5.7 shows a schematic diagram of the computational domain. As shown in the figure, a square cylinder where width B is 0.14 m was fixed in a position 0.28 m behind the seawall, and sand substrate in which median grain size d_{50} is 0.2 mm was set around the cylinder. A single leading long-period wave with a height of 0.064 m at $W1$, duration of 6.0 s, and still water depth of 0.455 m was generated.

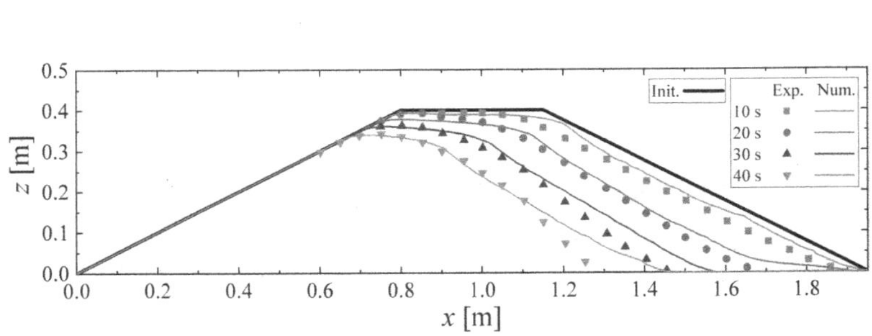

FIGURE 5.5.5
Numerical results for the experiments of overflow-induced erosion of the unsaturated dike.[5.5.12]

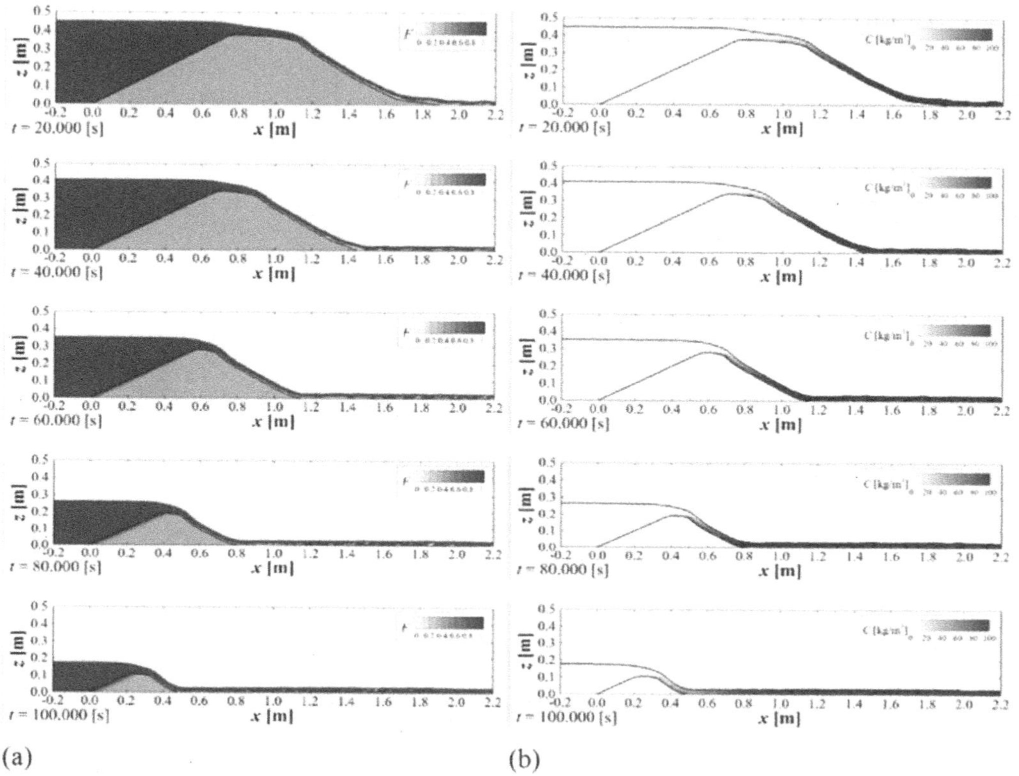

FIGURE 5.5.6
Flow fields during overflow-induced erosion of the unsaturated dike.[5.5.12] (a) VOF function F and (b) Suspended sediment concentration C.

Due to the lack of the experimental data of the sand parameter, porosity m was set to 0.4, density of sand particles ρ_s to 2.65×10^3 kg/m^3, shear modulus of elasticity G to 10^8 N/m^2, and Poisson ratio ν to 0.33, and hydraulic conductivity k_s was set to 4.34×10^{-4} m/s from the Kozeny–Carman equation. Moreover, the apparent bulk modulus of water K_w was changed to 2.2×10^5 N/m^2 and 2.2×10^7 N/m^2. Figure 5.5.8 shows a comparison of the excess pore water pressure in the sand. As shown in the figure, in the case of $K_w = 2.2 \times 10^5$ N/m^2, an increase in the excess pore water pressure in P2, P4, P6, and P8 deep in the sand is slightly small compared with the hydraulic experiments. However, the experimental results can be predicted satisfactorily for $K_w = 2.2 \times 10^7$ N/m^2. Figure 5.5.9 shows the distribution of excess pore water pressure P, mean effective stress σ'_m, and relative effective stress ratio RESR (=$1 - \sigma'_m/\sigma'_{m0}$; σ'_{m0}: initial value of σ'_m) when scouring is formed in the corner on the offshore side of the cylinder by the run-up tsunami. When focusing on $t = 7.2$ s when the tsunami reached the offshore surface of the cylinder, P close to the sand surface is slightly increased on the onshore side of the seawall and offshore side of the cylinder. At this time, σ'_m of the sand surface on the offshore side of the cylinder increased, and RESR decreased in response to this. Conversely, σ'_m of the sand surface on the onshore side of the seawall decreased, and RESR increased in response to this, which suggests the onset of liquification. Subsequently, whereas

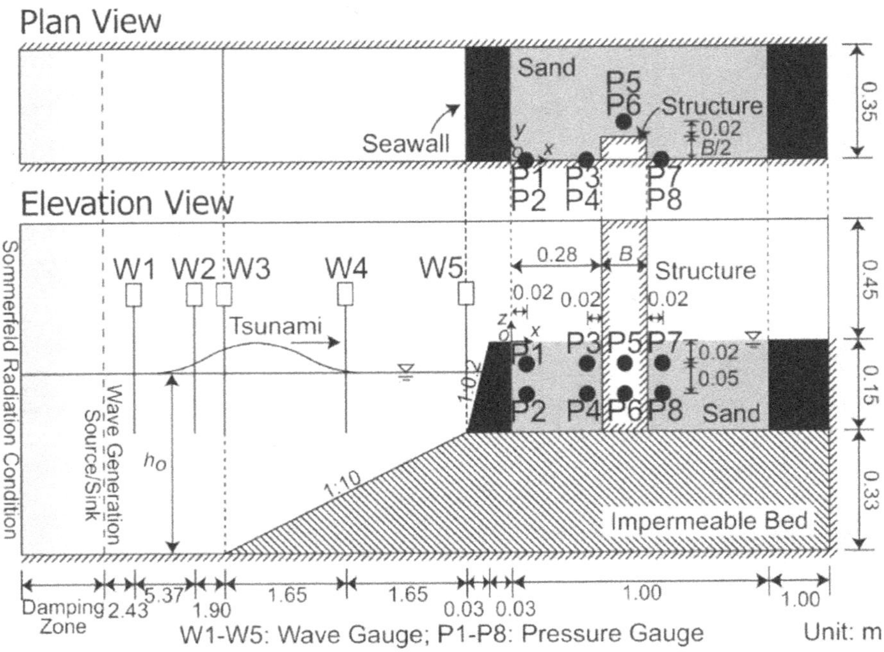

FIGURE 5.5.7
Computational domain for local scouring around the inland structure induced by run-up tsunami.[5.5.3]

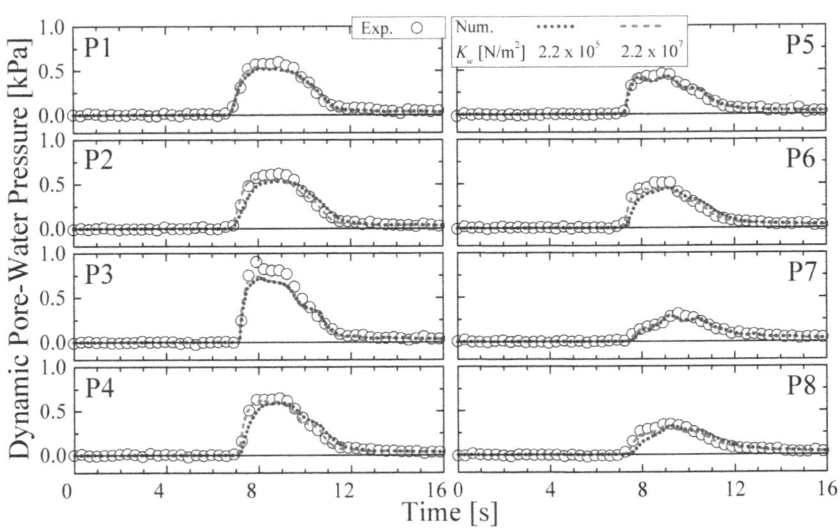

FIGURE 5.5.8
Comparison of the excess pore water pressure inside the ground P1 to P8.[5.5.3]

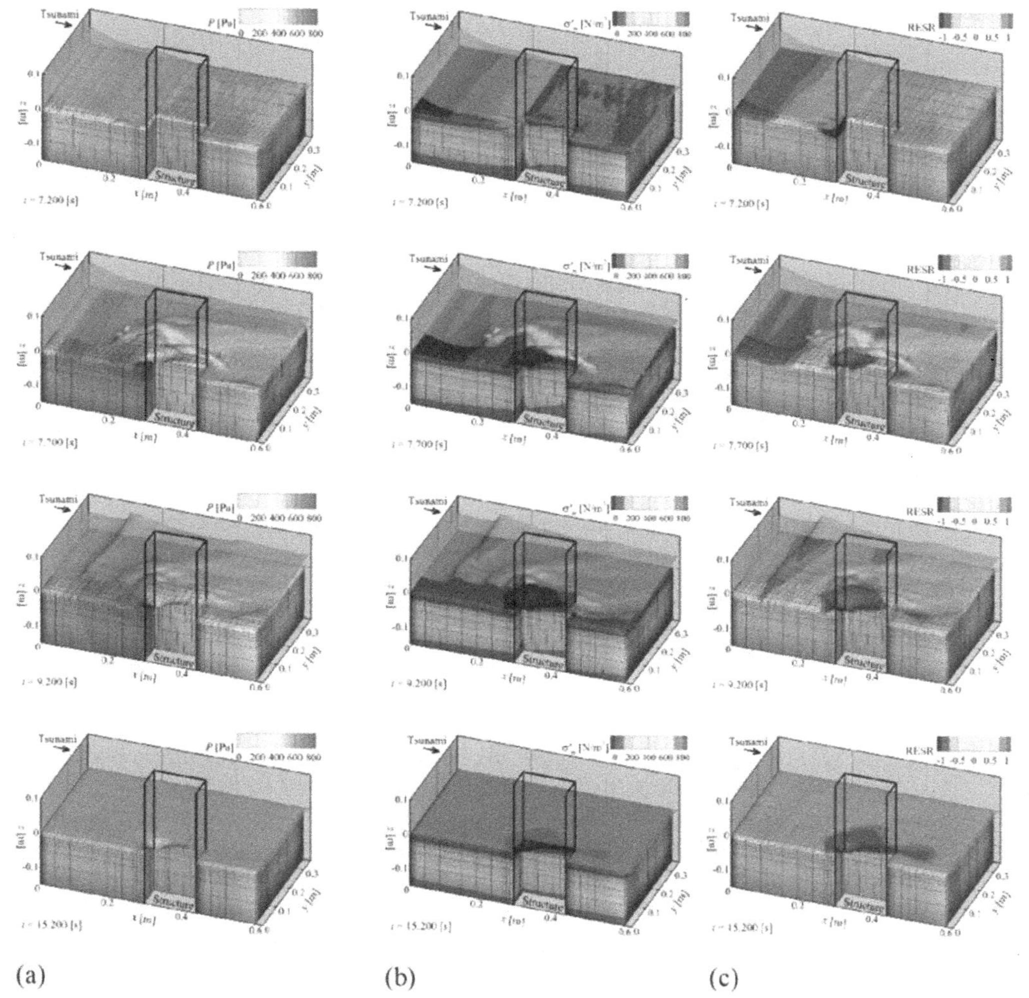

FIGURE 5.5.9
Response of the ground during wave-induced local scouring ($Kw = 2.2 \times 10^7$ N/m^2).[5.5.3] (a) Excess pore-water P, (b) Mean effectie stress σ'_m, and (c) RESR.

the tsunami is running up on the offshore surface of the cylinder, P increased further mainly on the sand surface below that. However, whereas an increase in P cannot be confirmed around the corner on the offshore side of the cylinder where the scouring started, σ'_m decreased because the weight of the sand decreased in connection with the scouring. Thus, RESR increased around the area, which suggests the onset of liquefaction. Furthermore, although the condition is approaching the initial state along with the decrease in the water level on the sand, it can be observed that σ'_m remained low on the corner of the offshore side of the cylinder where the scouring occurred.

As mentioned above, various phenomena associated with wave-seabed interaction can be calculated using FS3M. However, since there are issues that remain, such as the soil–water coupled analysis only simply taking account of the influence of topographic change due to sediment transport, further improvement and advancement are expected.

References

5.1 Role of numerical calculation

5.1.1 Shiratori, M., Koshizuka, S., Yoshida, Y., Nakamura, H., Hotta, A., and Takano, N.: Quality Assurance and Verification and Validation of Engineering Simulation, Maruzen Publishing Co., Ltd., 160 pp., 2013. (in Japanese)

5.1.2 Kajishima, T.: Dynamic SGS models for large eddy simulation of turbulent flow, Japanese Journal of Multiphase Flow, Vol.10, No.4, pp.372–378, 1996. (in Japanese)

5.1.3 Kajishima, T.: Numerical Simulation of Turbulent Flows, Revised Edition, Yokendo, 285 pp., 2014. (in Japanese)

5.1.4 Kajishima, T., and Taira, K.: Computational Fluid Dynamics: Incompressible Turbulent Flows, Springer, 385 pp., 2017.

5.2 Development of numerical calculation for wave–seabed interaction

5.2.1 Research Subcommittee on Numerical Wave Flume, JSCE Coastal Engineering Committee: Numerical Wave Flume: Toward Advanced Computation of Wave Breaking and Innovation of Coastal Structure Design, Maruzen Publishing Co., Ltd., 228 pp., 2012. (in Japanese)

5.2.2 Park, W.S., Takahashi, S., Suzuki, K., and Kang, Y.K.: Nami-jiban-kouzoubutsu no sougo sayo ni kansuru yuugenyousohou kaiseki [Finite element analysis of interaction between waves, seabed, and structures], Proceedings of Coastal Engineering, Vol.43, pp.1036–1040, 1996. (in Japanese)

5.2.3 Mizutani, N. and Mostafa, A.M.: Nami no hisenkeisei wo kouryoshita konseitei kiso jiban no harou outou ni kansuru kenkyu [Study on wave-induced response of foundation of composite breakwater considering wave nonlinearities], Proceedings of Coastal Engineering, Vol.44, pp.926–930, 1997. (in Japanese)

5.2.4 Mizutani, N. and Mostafa, A.M.: Nonlinear wave-induced seabed instability around coastal structures, Coastal Engineering Journal, Vol.40, No.2, pp.131–160, 1998.

5.2.5 Isobe, M., Takahashi, S., Yu, S.P., Sakakiyama, T., Fujima, K., Kawasaki, K., Jiang, Q., Akiyama, M., and Ohyama, H.: Interim development of a numerical wave flume for maritime structure design, Proceedings of Civil Engineering in the Ocean, Vol.15, pp.321–326, 1999. (in Japanese)

5.2.6 https://www.cdit.or.jp/program/cadmas.html

5.2.7 Sakakiyama, T. and Imai, S.: Syoha gogan no eppa ni kansuru suchi simyure-syon [Numerical simulation of overtopping of wave dissipating seawalls], Proceedings of Coastal Engineering, Vol.43, pp.696–700, 1996. (in Japanese)

5.2.8 Jiang, Q. Takahashi, S., Muranishi, Y., and Isobe, M.: A VOF-FEM model for the interaction among waves, soils and structures, Proceedings of Coastal Engineering, Vol.47, pp.51–55, 2000. (in Japanese)

5.2.9 Takahashi, S., Suzuki, K., Muranishi, Y., and Isobe, M.: U-π form VOF-FEM program simulating wave–soil interaction: CADMAS-GEO-SURF, Proceedings of Coastal Engineering, Vol.49, pp.881–885, 2002. (in Japanese)

5.2.10 Arikawa, T., Yamada, F., and Akiyama, M.: Study of the applicability of tsunami wave force in a three-dimensional numerical wave flume, Annual Journal of Coastal Engineering, Vol.52, pp.46–50, 2005. (in Japanese)

5.2.11 Coastal Development Institute of Technology: CADMAS-SURF/3D suchi hadou suisou no kenkyu-kaihatsu [Research and development of numerical wave flume CADMAS-SURF/3D], Coastal Technology Library No.39, https://www.cdit.or.jp/program/cadmas-3d.html, 2010.

5.2.12 Arikawa, T., Hamaguchi, K., Kitagawa, K., and Suzuki, T.: Development of numerical wave tank coupled with structure analysis based on FEM, Journal of Japan Society of Civil Engineers, Series B2 (Coastal Engineering), Vol.65, No.1, pp.866–870, 2009. (in Japanese)
5.2.13 Arikawa, T., Akiyama, M., and Yamazaki, N.: Development of solid–gas–liquid coupling system by using CADMAS-SURF/3D, Journal of Japan Society of Civil Engineers, Series B2 (Coastal Engineering), Vol.67, No.2, pp.I_21–I_25, 2011. (in Japanese)
5.2.14 Arikawa, T., Seki, K., Oki, Y., Hirano, H., Chida, Y., Araki, K., Ishii, K., Takagawa, T., and Shimosako, K.: Development of high precision tsunami runup calculation method coupled with structure analysis, Journal of Japan Society of Civil Engineers, Series B2 (Coastal Engineering), Vol.73, No.2, pp.I_325–I_330, 2017. (in Japanese)
5.2.15 Ohki, Y., Kusano, M., Seki, K., Taenaka, S., Moriyasu, S., Deji, T., Ueda, H., and Arikawa, T.: Study on verification of pore water pressure in state of raising up water level using fluid and soil weak coupling model, Journal of Japan Society of Civil Engineers, Series B2 (Coastal Engineering), Vo.73, No.2, pp.I_1111–I_1116, 2017. (in Japanese)
5.2.16 Research Group on CADMAS-SURF (tentative): https://github.com/CADMAS-SURF
5.2.17 Maeno, S. and Fujita, S.: VOF-FEM analysis of dynamic seabed behavior around revetment under wave motion, Proceedings of Coastal Engineering, Vol.48, pp.971–975, 2001. (in Japanese)
5.2.18 Nakamura, T., Hur, D.S., and Mizutani, N.: Study on suction mechanism of reclaimed sand based on a numerical simulation of wave, structure and seabed interaction, Annual Journal of Coastal Engineering, Vol.52, pp.836–840, 2005. (in Japanese)
5.2.19 Nakamura, T., Hur, D.S., and Mizutani, N.: Suction mechanism of reclaimed sand behind a rubble seawall, Doboku Gakkai Ronbunshuu B, Vol.62, No.1, pp.150–162, 2006. (in Japanese)
5.2.20 Nakamura, T. and Mizutani, N.: Development of coupled fluid–sediment–seabed interaction model and its application, Journal of Japan Society of Civil Engineers, Series B2 (Coastal Engineering), Vol.69, No.2, pp.I_1026–I_1030, 2013. (in Japanese)
5.2.21 Kumagai, T.: Deformation analysis for seabed and structure under wave action by VOF-elasto-plastic FEM model, Journal of Japan Society of Civil Engineers, Series B2 (Coastal Engineering), Vol. B2-65, No.1, pp.871–875, 2009. (in Japanese)
5.2.22 Kumagai, T.: Numerical analysis for dynamic behavior and deformation of composite breakwater and seabed under wave action by VOF-elasto-plastic FEM model, Journal of Japan Society of Civil Engineers, Series B2 (Coastal Engineering), Vol.66, No.1, pp.921–925, 2010. (in Japanese)
5.2.23 Jeng, D.-S., Ye, J.-H., Zhang, J.-S., and Liu, P.L.-F.: An integrated model for the wave-induced seabed response around marine structures: Model verifications and applications, Coastal Engineering, Vol.72, pp.1–19, 2013.
5.2.24 Ye, J., Jeng, D., Wand, R., and Zhu, C.: Validation of a 2-D semi-coupled numerical model for fluid–structure–seabed interaction, Journal of Fluids and Structures, Vol.42, pp.333–357, 2013.
5.2.25 Hsu, T.-J., Sakakiyama, T., and Liu, P.L.-F.: A numerical model for wave motions and turbulence flows in front of a composite breakwater, Coastal Engineering, Vol.46, pp.25–50, 2002.
5.2.26 Ye, J., Jeng, D., Liu, P.L.-F., Chan, A.H.C., Wang, R., and Zhu, C.: Breaking wave-induced response of composite breakwater and liquefaction in seabed foundation, Coastal Engineering, Vol.85, pp.72–86, 2014.
5.2.27 Ye, J., Jeng, D., Wang, R., and Zhu, C.: A 3-D semi-coupled numerical model for fluid–structures–seabed interaction (FSSI-CAS3D): Model and verification, Journal of Fluids and Structures, Vol.40, pp.148–162, 2013.
5.2.28 Ye, J., Jeng, D., Wang, R., and Zhu, C.: Numerical simulation of the wave-induced dynamic response of poro-elastoplastic seabed foundations and a composite breakwater, Applied Mathematical Modelling, Vol.39, pp.322–347, 2015.
5.2.29 Zhang, C., Zhang, Q., Wu, Z., Zhang, J., Sui, T., and Wen, Y.: Numerical study on effects of the embedded monopile foundation on local wave-induced porous seabed response, Mathematical Problems in Engineering, Vol.2015, 13 pp., 2015.

5.2.30 Zhang, C., Sui, T., Zheng, J., Xie, M., Nguyen, V.T.: Modelling wave-induced 3D non-homogeneous seabed response, Applied Ocean Research, Vol.61, pp.101–114, 2016.
5.2.31 Zhang, J.-S., Zhang, Y., Zhang, C., and Jeng, D.-S.: Numerical modeling of seabed response to combined wave-current loading, Journal of Offshore Mechanics and Arctic Engineering, ASME, Vol.135, 031102, 7 pp., 2020.
5.2.32 Lin, J., Zhang, J., Sun, K. Wei, X, and Guo, Y.: Numerical analysis of seabed dynamic response in vicinity of mono-pile under wave-current loading, Water Science and Engineering, Vol.13, No.1, pp.74–82, 2020.
5.2.33 Chang, K.-T. and Jeng, D.-S.: Numerical study for wave-induced seabed response around offshore wind turbine foundation in Donghai offshore wind farm, Shanghai, China, Ocean Engineering, Vol.85, pp.32–43, 2014.
5.2.34 Li, Y., Ong, M.C., and Tang, T.: A numerical toolbox for wave-induced seabed response analysis around marine structures in the OpenFOAM® framework, Ocean Engineering, Vol.195, 106678, 18 pp., 2020.
5.2.35 Fujisawa, K. and Murakami, A.: Numerical analysis of coupled flows in porous and fluid domains by the Darcy–Brinkman equations, Soils and Foundations, Vol.58, pp.1240–1259, 2018.
5.2.36 Gotoh, H.: Ryushiho [Particle Method], Morikita Publishing Co., Ltd., 304 pp., 2018. (in Japanese)
5.2.37 Gotoh, H., Sakai, T., Tominaga, K., and Toyoda, Y.: Hendou haatsu wo ukeru kaitei jiban no kyodou no suchi mogi heno kobetsu yousohou [Application of discrete element method to numerical simulation of seabed response under fluctuating wave pressure], Proceedings of Coastal Engineering, Vol.41, pp.596–600, 1989. (in Japanese)
5.2.38 Gotoh, H. and Sakai T.: Numerical simulation of surface-sheared-sand layer, Doboku Gakkai Ronbunshu, No.521, II-32, pp.101–112, 1995. (in Japanese)
5.2.39 Sakai, T., Gotoh, H., Harada, E., Hama, Y., and Imoto, Y.: Effect of wave-induced liquefaction of seabed on sediment transport rate, Proceedings of Coastal Engineering, Vol.48, pp.981–985, 2001. (in Japanese)
5.2.40 Chen, G., Zen, K., Kasama, K., Takamatsu, K.: Numerical simulation of the movements of sand particles under wave-induced seabed liquefaction, Proceedings of Coastal Engineering, Vol.48, pp.476–480, 2001. (in Japanese)
5.2.41 Gotoh, H., Harada, E., and Sakai, T.: Generalized movable bed simulator by using 3D distinct element method, Proceedings of Hydraulic Engineering, Vol.46, pp.613–618, 2002. (in Japanese)
5.2.42 Harada, E., Gotoh, H., Sakai, T., and Tei, T.: 3D granular material model of caisson-side cavern formation, Proceedings of Coastal Engineering, Vol.50, pp.891–895, 2003. (in Japanese)
5.2.43 Honda, A., Nagao, T., Yoshioka, K., Kyono, T., Yasuda, K., and Nakase, H.: Analysis of destructive mode of bearing capacity by DEM, Proceedings of Civil Engineering in the Ocean, Vol.21, pp.981–986, 2005. (in Japanese)
5.2.44 Takayama, T., Higashira, K., and Kim, T.M.: Computed behaviors of a composite breakwater by discrete element method, Annual Journal of Coastal Engineering, Vol.51, pp.756–760, 2004. (in Japanese)
5.2.45 Takayama, T. and Takahashi, M.: Computed behaviors of a composite breakwater by DEM representing a caisson as a single element, Annual Journal of Coastal Engineering, Vol.53, pp.841–845, 2006. (in Japanese)
5.2.46 Tsujio, D., Takayama, T., Oosato, M., Yamaguchi, Y., Suzuki, N., and Sera, K.: Numerical evaluation on breakwater ductility by discrete element method, Journal of Japan Society of Civil Engineers, Series B2 (Coastal Engineering), Vol.68, No.2, pp.I_841–I_845, 2012. (in Japanese)
5.2.47 Gotoh, H., Sakai, T., Hayashi, M., Oda, K., and Ikari, H.: Gridless analysis of toe scouring of seawall due to tsunami return flow, Proceedings of Coastal Engineering, Vol.49, pp.46–50, 2002. (in Japanese)

5.2.48 Ikari, H., Gotoh, H., and Yodoshi, H.: Development of fluid-elastoplastic hybrid particle method for tsunami simulation due to slope failure, Journal of Japan Society of Civil Engineers, Series B2 (Coastal Engineering), Vol. B2-65, No.1, pp.46–50, 2009. (in Japanese)
5.2.49 Gotoh, H., Ikari, H., Komaguchi, T., Mishima, T., and Yodoshi, H.: Numerical analysis on cave-formation process behind seawall by particle method, Journal of Japan Society of Civil Engineers, Series B2 (Coastal Engineering), Vol.66, No.1, pp.821–825, 2010. (in Japanese)
5.2.50 Ikari, H., Gotoh, H., and Yodoshi, H.: Fundamental study on large deformation analysis of composite breakwater by improved elastoplastic MPS method, Journal of Japan Society of Civil Engineers, Series B2 (Coastal Engineering), Vol.67, No.2, pp.I_731–735, 2011. (in Japanese)
5.2.51 Gotoh, H., Tsuruta, N., Harada, E., Ikari, H., and Kubota, H.: High-resolution DEM-MPS methods for solid–liquid multi-phase flow, Journal of Japan Society of Civil Engineers, Series B2 (Coastal Engineering), Vol.68, No.2, pp.I_21–I_25, 2012. (in Japanese)
5.2.52 Gotoh, H., Ikari, H., Haraguchi, K., Nakajima, H., Tonomo, K., and Ishii, T.: Development of numerical wave flume by particle method for simulating tsunami-overtopping induced failure of composite breakwater and its prevention work, Journal of Japan Society of Civil Engineers, Series B2 (Coastal Engineering), Vol.69, No.2, pp.I_881–I_885, 2013. (in Japanese)
5.2.53 Ikari, H., Gotoh, H., Yoshinaga, K., and Tanbo, T.: Development of improved higher order Laplacian model and numerical simulation on scouring due to submerged jet, Journal of Japan Society of Civil Engineers, Series B2 (Coastal Engineering), Vol.70, No.2, pp.I_36–I_40, 2014. (in Japanese)
5.2.54 Ikari, H., Gotoh, H., Tanbo, T., and Ejiri, T.: MPS-based simulation of scouring due to submerged vertical jet with sub-particle-scale suspended load model, Journal of Japan Society of Civil Engineers, Series B2 (Coastal Engineering), Vol.71, No.2, pp.I_19–I_24, 2015. (in Japanese)
5.2.55 Ikari, H., Gotoh, H., Ejiri, T., and Konishi, K.: Numerical modeling on caisson breakwater failure induced by scouring due to tsunami overtopping using particle-based fluid-elastoplastic interaction model, Journal of Japan Society of Civil Engineers, Series B2 (Coastal Engineering), Vol.73, No.2, pp.I_1033–I_1038, 2017. (in Japanese)
5.2.56 Ikari, H., Gotoh, H., Matsushima, R., and Niwa, G.: Numerical simulation on caisson breakwater failure due to tsunami overtopping using particle-based elastoplastic model with modification of stress by MLS, Journal of Japan Society of Civil Engineers, Series B2 (Coastal Engineering), Vol.74, No.2, pp.I_157–I_162, 2018. (in Japanese)
5.2.57 Shimizu, Y., Harada, E., Ikari, H., Gotoh, H., and Iga, S.: Hydraulic experiment and numerical simulation of mound scouring behind breakwater induced by tsunami overtopping, Journal of Japan Society of Civil Engineers, Series B2 (Coastal Engineering), Vol.74, No.2, pp.I_349–I_354, 2018. (in Japanese)
5.2.58 Imase, T., Maeda, K., Miyake, M., Tsurugasaki, K., Sawada, Y., and Sumida, H.: Instability of caisson-type breakwater due to seepage of tsunami into rubble mound and seabed, Journal of Japan Society of Civil Engineers, Series B2 (Coastal Engineering), Vol.67, No.2, pp.I_551–I_555, 2011. (in Japanese)
5.2.59 Imase, T., Maeda, K., Miyake, M., Sawada, Y., Tsurugasaki, K., Sumida, H., and Zhang F.: Destruction of caisson-type breakwater due to bearing capacity by tsunami with seabed soil damage by earthquake and overflow, Journal of Japan Society of Civil Engineers, Series B2 (Coastal Engineering), Vol.68, No.2, pp.I_866–I_870, 2012. (in Japanese)
5.2.60 Oie, T. Wang, D., Takatani, T., Araki, K., Li, S., Goto, H., Arikawa, T.: Numerical simulation of scouring behind the seawall caused by tsunami overflow with accurate ISPH method, Journal of Japan Society of Civil Engineers, Series B2 (Coastal Engineering), Vol.71, No.2, pp.I_253–I_258, 2015. (in Japanese)
5.2.61 Gotoh, H., Khayyer, A., Ikari, H., Arikawa, T., and Shimosako, K.: On enhancement of incompressible SPH method for simulation of violent sloshing flows, Applied Ocean Research, Vol.46, pp.104–115, 2014.

5.2.62 Mitourida, N. and Arikawa, T.: Incompressible SPH method for consideration of phenomenon of the toe scour of a seawall, Journal of Japan Society of Civil Engineers, Series B2 (Coastal Engineering), Vol.72, No.2, pp.I_595–I_600, 2016. (in Japanese)

5.2.63 Uejima, H. and Sato, S.: A study on countermeasure against scour behind coastal dike due to tsunami overflow, Journal of Japan Society of Civil Engineers, Series B2 (Coastal Engineering), Vol.74, No.2, pp.I_355–I_360, 2018. (in Japanese)

5.2.64 Gotoh, H., Suzuki, K., Ikari, H., Arikawa, T., Khayyer, A., and Tsuruta, N.: Development of particle-based numerical wave flume for multiphase flow simulation, Journal of Japan Society of Civil Engineers, Series B2 (Coastal Engineering), Vol.73, No.2, pp.I_25–I_30, 2017. (in Japanese)

5.2.65 Tsuruta, N., Khayyer, A., Gotoh, H., and Suzuki, K.: Simple and effective sedimentation technique for scour simulation by the projection-based particle method, Journal of Japan Society of Civil Engineers, Series B2 (Coastal Engineering), Vol.74, No.2, pp.I_151–I_156, 2018. (in Japanese)

5.2.66 Maeno, S., Ogawa, M., and Bierawski, L.G.: Modeling submerged breakwater using VOF-DEM-FEM, Annual Journal of Coastal Engineering, Vol.53, pp.886–890, 2006. (in Japanese)

5.2.67 Yamaguchi, A., Maeda, K., Matsuda, T., and Takagi, K.: Fluidity and pore water pressure variation in the sandy ground induced by surface layer flow using DEM-CFD, Journal of Japan Society of Civil Engineers, Series B2 (Coastal Engineering), Vol.73, No.2, pp.I_517–I_522, 2017. (in Japanese)

5.3 Treatment of interface between fluid and porous domains

5.3.1 Çeşmelioğlu, A. and Rivière, B.: Existence of a weak solution for the fully coupled Navier–Stokes/Darcy transport problem, Journal of Differential Equations, Vol.252, pp.4138–4175, 2012.

5.3.2 Chidyagwai, P. and Rivière, B.: A two-grid method for coupled free flow with porous media flow, Advances in Water Resources, 34,1113–1123, 2011.

5.3.3 Badea, L., Discacciati, M., and Quarteroni, A.: Numerical analysis of the Navier–Stokes/Darcy coupling, Numerische Mathematik, Vol.115, pp.195–227, 2010.

5.3.4 Girault, V. and Rivière, B.:DG approximation of coupled Navier–Stokes and Darcy equations by Beaver–Joseph–Saffman interface condition, SIAM Journal on Numerical Analysis, 47(3), 2052–2089, 2009.

5.3.5 Chidyagwai, P. and Rivière, B.: On the solution of the coupled Navier–Stokes and Darcy equations, Computer Methods in Applied Mechanics and Engineering, Vol.198, pp.3806–3820, 2009.

5.3.6 Cai, M., Mu, M., and Xu, J.: Numerical solution to a mixed Navier–Stokes/Darcy model by the two-grid approach, SIAM Journal on Numerical Analysis, 47(5), 3325–3338,2009.

5.3.7 Urquiza, J.M., N'Dri, D., Garon, A., and Delfour, M.C.: Coupling Stokes and Darcy equations, Applied Numerical Mathematics, Vol.58, pp.525–538, 2008.

5.3.8 Mu, M. and Xu, J.: A two-grid method of a mixed Stokes–Darcy model for coupling fluid flow with porous media flow, SIAM Journal on Numerical Analysis, 45(5), 1801–1813, 2007.

5.3.9 Beavers, G.S. and Joseph, D.D.: Boundary conditions at a naturally permeable wall, Journal of Fluid Mechanics, 30(1), 197–207, 1967.

5.3.10 Saffman, P.G.: On the boundary condition at the surface of a porous medium, Studies in Applied Mathematics, 50, 93–101, 1971.

5.3.11 Neale, G. and Nader, W.: Practical significance of Brinkman's extension of Darcy's law: Coupled parallel flows within a channel and a bounding porous medium, The Canadian Journal of Chemical Engineering, 52, 475–478, 1974.

5.3.12 Bars, M.L. and Worster, M.G.: Interfacial conditions between a pure fluid and a porous medium: Implications for binary alloy solidification, Journal of Fluid Mechanics, 550, 149–173, 2006.

5.3.13 Ochoa-Tapia, J. and Whitaker, S.: Momentum transfer at the boundary between a porous medium and a homogeneous fluid: I. Theoretical development, International Journal of Heat and Mass Transfer, 38, 2635–2646, 1995.
5.3.14 Ochoa-Tapia, J. and Whitaker, S.: Momentum transfer at the boundary between a porous medium and a homogeneous fluid: II. Comparison with experiment, International Journal of Heat and Mass Transfer, 38, 2647–2655, 1995.
5.3.15 Yu, P., Lee, T.S., Zeng, Y., Low, H.T.: A numerical method for flows in porous and homogenous fluid domains coupled at the interface by stress jump. International Journal for Numerical Methods in Fluids, 53, 1755–1775, 2007.
5.3.16 Fujisawa, K. and Murakami, A.: Numerical analysis of coupled flows in porous and fluid domains by the Darcy–Brinkman equations, Soils and Foundations, 58(5), 1240–1259, 2018.
5.3.17 Kobayashi, H.: The subgrid-scale models based on coherent structures for rotating homogeneous turbulence and turbulent channel flow, Physics of Fluids, 17, 045104, 2005.

5.4 Bottom sediment transport by particle method

5.4.1 Cundall, P.A. and Strack, O.D.L.: A discrete numerical model for granular assemblies, Géotechnique, Vol.29, No.1, pp.47–65, 1979.
5.4.2 Kobayashi, T.: Suchi ryutai hando bukku [Numerical Fluids Handbook], Maruzen Publishing, 723pp., 2003. (in Japanese)
5.4.3 Gotoh, H.: Suchi ryusa suirigaku [Computational Mechanics of Sediment Transport], Morikita Publishing, 223pp., 2004. (in Japanese)
5.4.4 Yeganeh, A., Gotoh, H., and Sakai, T.: Applicability of Euler–Lagrange coupling multiphase-flow model to bed-load transport under high bottom shear, Journal of Hydraulic Research, Vol.38, No.5, pp.389–398, 2000.
5.4.5 Drake, T.G. and Calantoni, J.: Discrete particle model for sheet flow sediment transport in the nearshore, Journal of Geophysical Research, Vol.106, No.C9, pp.19859–19868, 2001.
5.4.6 Calantoni, J. and Thaxton, C.S.: Simple power law for transport ration with bimodal distributions of coarse sediments under waves, Journal of Geophysical Research, Vol.113, C03003, 2008.
5.4.7 El Shamy, U. and Zeghal, M.: Coupled continuum-discrete model for saturated granular soils, Journal of Engineering Mechanics, 131(4):413–426, 2005.
5.4.8 Suzuki, K., Bardet, J.P., Oda, M, Iwashita, K., Tsuji, Y., Tanaka, T., and Kawaguchi, T.: Simulation of upward seepage flow in a single column of spheres using discrete-element method with fluid–particle interaction, Journal of Geotechnical and Geoenvironmental Engineering, Vol.133, No.1, pp.104–109, 2007.
5.4.9 Peskin, C.S.: Numerical analysis of blood flow in the heart, Journal of Computational Physics, Vol.25, pp.220–252, 1977.
5.4.10 Kajishima, T. and Takiguchi, S.: Interaction between particle clusters and particle-induced turbulence, International Journal of Heat and Fluid Flow, Vol.23, pp.639–646, 2002.
5.4.11 Tsuji, T., Narutomi, R., Yokomine, T., Ebara, S., and Shimizu, A.: Unsteady three-dimensional simulation of interactions between flow and two particles, International Journal of Multiphase Flow, Vol.29, No.9, pp.1431–1450, 2003.
5.4.12 Akiyama, M. and Aritomi, M.: Atarashi kiekinisoryu suchi kaiseki – tajigen ryudo kaiseki [Advanced Numerical Analysis of Two-phase Flow Dynamics – Multi-dimensional Flow Analysis], Corona Publishing, 261pp., 2002. (in Japanese)
5.4.13 Ushijima, S., Takemura, M., Yamada, S., and Nezu, I.: Computational method for multiphase incompressible flows (MICS) and its applicability to particle-laden liquid flows, Doboku Gakkai Ronbunshu, Vol.2003, Issue 74, pp.121–130, 2003. (in Japanese)
5.4.14 Yamada, S., Ushijim, S., and Nezu, I.: Numerical prediction for incipient motion of bodies under shielding effects with MICS, *Proceedings of Hydraulic Engineering*, Vol.49, pp.757–762, 2005. (in Japanese)

5.4.15 Harada, E., Tsuruta, N., and Gotoh, H.: Large eddy simulation for vertical sorting process in sheet-flow regime, *Journal of Japan Society of Civil Engineers, Series B2 (Coastal Engineering)*, Vol.67, Issue2, pp.I_471–I_475, 2011. (in Japanese)

5.4.16 Fukuoka, S.: Prediction of three-dimensional movement of gravel particles in a movable-bed numerical channel, Thesis 2013, France, 2013.

5.4.17 Ginglod, R.A. and Monaghan, J.J.: Smoothed particle hydrodynamics: Theory and application to non-spherical stars, Monthly Notices of the Royal Astronomical Society, Vol.181, pp.375–389, 1977.

5.4.18 Koshizuka, S. and Oka, Y.: Moving-particle semi-implicit method for fragmentation of incompressible fluid, Nuclear Science and Engineering, Vol.123, pp.421–434, 1996.

5.4.19 Ohta, M., Sakai, M., Shimada, N., Honma, S., and Matsukuma, Y.: Konsoryu no suchi simyureshon [Numerical Simulation of Multi-phase Flows], Maruzen Publishing, 149pp., 2015. (in Japanese)

5.4.20 Gotoh, H.: Ryushiho [Particle Method], Morikita Publishing, 289pp., 2018. (in Japanese)

5.4.21 Inman, D.L.: Wave-generated ripples in nearshore sands, U.S. Army Corps of Engineers, Beach Erosion Board, Tech. Memo 100, 1957.

5.4.22 Bagnold, R.A.: Motion of waves in shallow water, interaction between waves and sand bottoms, Proceedings of the Royal Society of London Series A, Vol.187, pp.1–15, 1946.

5.4.23 Mogridge, G.R. and Kamphuis, J.W.: Experiments on bedform generation by wave action, Coastal Engineering Proceedings, ASCE, Vancouver, Canada, pp.1123–1142, 1972.

5.4.24 Sleath, J.F.A.: On rolling grain ripples. Journal of Hydraulic Research, Vol.14, pp.69–81, 1976.

5.4.25 Dingler, J.R. and Inman, D.L.: Wave-formed ripples in nearshore sands, Coastal Engineering Proceedings, ASCE, Honolulu, Hawaii, pp.2109–2126, 1976.

5.4.26 Harada, E., Tazaki, T., and Gotoh, H.: Numerical investigation of ripple in oscillating water tank by DEM–MPS coupled solid–liquid two-phase flow model, Journal of Hydro-environment Research, Vol.32, pp.26–47, 2020.

5.4.27 Gotoh, H., Tsuruta, N., Harada, E., Ikari, H., and Kubota, H.: High-resolution DEM–MPS methods for solid–liquid multi-phase flow, Journal of Japan Society of Civil Engineers, Series B2 (Coastal Engineering), Vol.68, Issue 2, pp.I_21–I_25, 2012. (in Japanese)

5.4.28 Khayyer, A. and Gotoh, H.: Modified moving particle semi-implicit methods for the prediction of 2D wave impact pressure, Coastal Engineering, Vol.56, No.4, pp.419–440, 2009.

5.4.29 Khayyer, A. and Gotoh, H.: A higher order Laplacian model for enhancement and stabilization of pressure calculation by the MPS method, Applied Ocean Research, Vol.32, No.1, pp.124–131, 2010.

5.4.30 Khayyer, A. and Gotoh, H.: Enhancement of stability and accuracy of the moving particle semi-implicit method, Journal of Computational Physics, Vol.230, No.8, pp.3093–3118, 2011.

5.4.31 Tsuruta, N., Khayyer, A., and Gotoh, H.: A short note on dynamic stabilization of moving particle semi-implicit method, Computers & Fluids, Vol.82, No.15, pp.158–164, 2013.

5.4.32 Harada, E. and Gotoh, H.: Computational mechanics of vertical sorting of sediment in sheetflow regime by 3D granular material model, Coastal Engineering Journal, Vol.50, No.1, pp.19–45, 2008.

5.4.33 Ergun, S.: Fluid flow through packed columns, Chemical Engineering Progress, Vol.48, pp.89–94, 1952.

5.4.34 Wen, C.Y. and Yu, Y.H.: Mechanics of fluidization, Chemical Engineering Progress Symposium Series, Vol.62, pp.100–111, 1966.

5.4.35 Ayrton, H. and Ayrton, W.E.: The origin and growth of ripple-mark, Proceedings of the Royal Society of London. Series A, Vol.84, No.571, pp.285–310, 1910.

5.4.36 Kawano, M., Tazaki, T., Harada, E., and Gotoh, H.: Koseido ryushiho ni yoru sindoryuka rippulu no syoki keisei katei no kento [Investigation of the Initial Formation Process of Ripple Under Oscillatory Flow by Enhanced Particle Method], 2020 JSCE Kansai Branch Annual Academic Lecture, II-49, 2020. (in Japanese)

5.5 Wave-induced seabed response by grid method

5.5.1 Jiang, Q., Takahashi, S., Muranishi, Y., and Isobe, M.: A VOF-FEM model for the interaction among waves, soils and structures, Proceedings of Coastal Engineering, Vol.47, pp.51–55, 2000. (in Japanese)
5.5.2 Arikawa, T., Yamada, F., and Akiyama, M.: Study of the applicability of tsunami wave force in a three-dimensional numerical wave flume, Annual Journal of Coastal Engineering, Vol.52, pp.46–50, 2005. (in Japanese)
5.5.3 Nakamura, T. and Mizutani, N.: Development of coupled fluid–sediment–seabed interaction model and its application, Journal of Japan Society of Civil Engineers, Series B2 (Coastal Engineering), Vol.69, No.2, pp.I_1026–I_1030, 2013. (in Japanese)
5.5.4 Bear, J.: Dynamics of Fluids in Porous Media, American Elsevier Pub. Co., New York, p.166, 1972.
5.5.5 van Gent, M.R.A.: Wave Interaction with Permeable Coastal Structures, Ph.D. thesis, Delft University, 177 pp., 1995.
5.5.6 Mizutani, N., Maeda, K., Mostafa, A.M., and McDougal, W.G.: Tousuisei kouzoubutsu no teikou keisuu no hyouka to nami-sensui tousuisei kouzoubutu no hisenkei sougosayou no suchi kaiseki [Numerical simulation of evaluation of drag coefficients of permeable structure and nonlinear interaction between waves and submerged permeable structure], Proceedings of Coastal Engineering, Vol.43, pp.131–135, 1996. (in Japanese)
5.5.7 Sarpkaya, T.: Vortex shedding and resistance in harmonic flow about smooth and rough circular cylinders at high Reynolds numbers, Naval Postgraduate School, Monterey, CA, 1976.
5.5.8 Kajima, T., Sakakiyama, T., Matsuyama, M., Sekimoto, T., and Kyoya, O.: Antei genkai wo koeru harou ni taisuru syouhakou no henkei to bouha kinou no henka ni tsuite [Deformation of wave dissipating blocks and change in their performance against waves over stability limit], Proceedings of Coastal Engineering, Vol.39, pp.671–675, 1992. (in Japanese)
5.5.9 Sakakiyama, T. and Ogasawara, M.: Sentei ni yoru syougeki saiharyoku no teigen to jikkenn suke-ru kouka [Reduction of impact breaking wave force due to submerged breakwater and scale effects], Proceedings of Coastal Engineering, Vol.40, pp.746–750, 1993. (in Japanese)
5.5.10 Gu, Z. and Wang, H.: Gravity waves over porous bottoms, Coastal Engineering, Vol.15, pp.497–524, 1991.
5.5.11 Takahashi, S., Suzuki, K., Muranishi, Y., and Isobe, M.: U-π form VOF-FEM program simulating wave–soil interaction: CADMAS-GEO-SURF, Proceedings of Coastal Engineering, Vol.49, pp.881–885, 2002. (in Japanese)
5.5.12 Nakamura, T., Cho, Y., and Mizutani, N.: Development of VOF-based numerical model applicable to unsaturated soil and application to seepage and erosion phenomena, Journal of Japan Society of Civil Engineers, Series B3 (Ocean Engineering), Vol.75, No.2, pp.I_229–I_234, 2019. (in Japanese)
5.5.13 Yoshikawa, T., Noda, T., Kodaka, T., and Cui, Y.: Kuuki-mizu-tuti kokkaku rensei yuugen henkei kaiseki wo motiita fuhouwa shintou mokei jikken no suchi simyure-syon [Numerical simulation of a unsaturated seepage model tests using soil–water–air coupled finite deformation analysis], Proceedings of the 26th Symposium on Geotechnical Engineering, pp.13–18, 2014. (in Japanese)
5.5.14 Sugii, T., Yamada, K., and Okumura, Y.: Kouhouwa ji ni okeru suna no fuhouwa tousui keisuu ni kansuru kousatsu [Discussion on hydraulic conductivity of unsaturated sand at high saturation], Proceedings of the JSCE Chubu Branch Annual Meeting 2001, III-6, pp.267–268, 2002. (in Japanese)
5.5.15 Yoden, T.: Kasen teibou no etsuryuu shinsyoku no mekanizumu ni kansuru kenkyuu [Study on mechanism of overflow-induced erosion of river banks], Doctoral dissertation, Kyoto University, 148pp., 2014. (in Japanese)

6

Future Challenges

Taro Arikawa

Presently, we are in a period where the safety of the living environment in the future is uncertain because of global warming, climate change, etc. Thus, long-term predictions by science and technology and examination of the adaptation measures based on predictions are beginning to play a more important role than ever. To predict the behavior of the living environment, it is essential to improve the prediction technology of complicated structures that comprise various materials, including soil, water, concrete, and steel. At present, examinations are being performed on the basis of various ideas because a system that considers the prediction of such complicated structures in a unified manner has not yet been established. Typical examples include model experiments and numerical calculations.

However, when handling large structures, particularly civil engineering structures, it is difficult to conduct experiments on an actual scale. Thus, experiments are usually conducted using models smaller than the actual size. Moreover, methods of coupling multiple basic equations to perform calculations are mainly used in numerical calculations at present. Thus, in either case, the results acquired from some sort of assumption or constraint conditions are used to estimate what will occur in the real world.

Accordingly, in the case of a model experiment, problems with the law of similarity occur, such as how to return the scaled-down experiment to the actual scale. In the case of a numerical calculation, besides the validity of a basic equation, handling of the boundary conditions and handling of the coefficients in equations become problems. The problems handled are extremely complicated in the first place. Even with the same equation, if the condition differs, the law of similarity also differs. On top of that, the constitutive basic equations may also be different.

Based on such viewpoints, this book contains a summary of the model experiments, the law of similarity used in the experiments, and the numerical calculation methods for the handling method of the interactions in complex fields of waves and ground of coastal areas. The scouring phenomenon under breakwaters and beach nourishment were covered as detailed case studies. The concept of the model scale for each phenomenon was examined. The challenges are listed below. Please refer to each chapter for details.

Chapter 2 on model experiments summarizes the hydraulic model experiments on the basis of the viewpoint of both the coast and ground. It shows that the way the experimental methods develop differs according to the difference in the phenomenon focused on. This is because the constraint conditions change when the size of the model is reduced. Conversely, since standard experimental methods have not yet been established, the characteristics of the ground materials have not been sufficiently considered.

DOI: 10.1201/9781003424758-6

Thus, a future task is to make further improvements using ingenuity. When conducting a model experiment, it is necessary to recognize that the similarity of the progression rate and time required for geographic features to reach the equilibrium state accompanying the interaction of scouring, erosion, etc. has not yet been clarified.

Chapter 3, which covers the examination of the law of similarity, summarizes the law of similarity appropriate for different experimental methods, the law of similarity based on the purpose of the experiment, and the law of similarity suitable for the quality of the material of the experiments. The difficulty of standardization can also be understood in this summary. The law of similarity when the final shape of the geographic features becomes settled also differs depending on the case study. Nevertheless, in the case where the deformation process and hysteretic characteristics become essential, there are not enough case studies for the similitude ratio at such time. Thus, it is necessary to handle it carefully. Regarding this point, conducting an experiment in a centrifugal force field may be one of the promising key solutions on which expectations are pinned.

Chapter 4 elucidates how to effectively apply the law of similarity through a detailed case study. It became clear that the effect of the model scale on the behavior of the sand in the surface layer is that the scouring shape can be expressed accurately by applying the Dean number to the particle size of the sand, like the scouring phenomenon. Nevertheless, in the case of a beach nourishment filling, it is necessary to determine the similitude ratio of the time in the deformation process according to the mechanism of each phenomenon that occurs during the deformation process of a beach nourishment filling. Additionally, application of the law of similarity in consideration of the characteristics of unsaturated ground is required for unsaturated soil. Since this is a complex phenomenon that includes a gas phase, it is necessary to perform an examination in the future, including the experimental method.

The problem of movable bed experiments such as the deformation of beaches can be mentioned as a subject that could not be published in this book because the discussions did not result in a conclusion. The committee discussed the matter of which the scour depth and scouring time sometimes did not meet Froude's law, although Froude's law was applied to the water level and flow velocity, which are external forces when reproducing a scouring phenomenon of a site by a movable bed model experiment. Therefore, it is necessary to continue examining how to convert the deformation process acquired in a model experiment into the scale of the site. By clarifying these matters, it is expected that the solutions are applied to the sedimentation and erosion process of sand on beaches, resulting in the clarification of mechanisms and improvement of sand drift prediction.

Finally, Chapter 5 summarizes the numerical calculation method. Various types of verifications are now available accompanying improvements in the handling of boundary conditions and computers themselves. Although issues in the unified handling method remain, it seems that there are still great limitations on the computational resolution, including grid size and the performance of computers. If the use of a grid size of 1 mm or less becomes available, the accuracy can be improved, including that of the behavior of sand. However, in the present stage, high calculation costs are required to calculate the reproduction on a laboratory level, even for the calculation of a grid size of approximately 10 mm, i.e., 1 cm. If 1,000 times or more is desired, the calculation of such a level is attainable after 15–20 years according to Moore's law, where the understanding of the phenomenon is expected to advance further.

Although research on the complex phenomenon of waves and ground has just started, future challenges seem to have come into view by organizing the laws of similarity, etc. to present the current condition of experimental methods and computation methods. In short, we believe that the challenges for the next ten years are to elucidate the mechanisms while skillfully utilizing experiments, etc. that use a centrifuge modeling and to promote the sophistication of numerical calculations to establish a general-purpose law of similarity, which can consider the deformation process, hysteretic characteristics, etc. The sophistication of experimental techniques can be expected by examining what kind of experiment data shall be acquired through considerations via backcasting based on the future aspects of a computer. We truly hope that the number of engineers who grapple with such research will increase further by reading this book.

Index

Page numbers in **bold** refer to tables and those in *italic* refer to figures.

A

Accelerometer, 46, *46*
Amplitude domain reflectometry (ADR) method, 44
AV equipment, 46

B

Beach nourishment filling
  bore-type run-up waves, 142
  collapsed sand, 153
  deformation of, *150*
  erosion process, 141–142
  erosion situation, 156, **157**
  grain size distribution of, *148*
  hydraulic model experiments, 142
  initial shoreline, 152
  material composition of, **149**
  perpendicular slope surface, 155
  sand dune and beach scarp, 142–146, *143–146*
  slope gradient, 150–151
  static and dynamic, 141
  structure of, *149*
  topographical change, 151–152
  washed-away pebbles, 150
  wave run-up point, 153–155
Beam-type failure, 143, *145*
Beam-type geotechnical centrifuge, *36*
Beavers–Joseph condition, 178, *178*
Bottom sediment transport
  active momentum exchange, 187–188
  description, 187
  flow velocity, porous medium, *188*
  multiphase flow analysis, 188–195
  numerical analysis, 188–195
Boundary element method (BEM), 164
Breaking waves, 79–80
Buried low stiffness pipe, *33*

C

Caisson breakwater model, *16*
Calibration
  acquisition and organization of, 52–54
  defined, 20–21
  example of, *21*
  measuring instruments, 52
  test waves, 21–23, *22*
Capacitance wave gauges setup, *18*
Centrifugal force field
  advantages and disadvantages, 37–42
  drum-type devices, 36
  geotechnical centrifuges, 36
  slope stability and bearing capacity, 35
  Technical Committee 104 (TC104), 35–36
Centrifugal gravity field, 81, **81**, 87–88
Centrifuge model tests, 56–60
  acceleration, 132
  drum-type geotechnical centrifuge, *132*
  fall speed, 133
  grain size distribution curves, *135*
  Ng centrifugal field, 133
  test cases and conditions, **134**
  wave–soil–structure, 131
Chinese Academy of Sciences (CAS), 169
Coastal Development Institute of Technology, 165
Cohesive soil, 49
Complex wave motion and ground problems, 60–62
Complex wave–structure–soil interactions problem, 63
Composite-type breakwater, wave breaking, *170*
Computational conditions, 197–198
Computed pressure distribution, 185
Constant acceleration curve, *39*
Coriolis effects, 135–136, *136*
Coriolis force, *40*, 40–41
COSMOL Multiphysics, 171
Coupled level set and volume of fluid (CLSVOF) method, 205
Cross-hole seismic survey method, 111
Culmann-type failure, beach scarp, *146*
Current meter, 12–13, *13*, 17–18, *18*

D

Darcy–Brinkman equations, 182, 183
Darcy flow velocity, 182
Darcy/Navier–Stokes coupled flow, 177, 183
Data logger, 47
Dean number, 3, 110, 229
Dean's Law (Dean Number), 137–140, *139*
Deformation process, 230
DEM-type spring–dashpot model, 173
Detached breakwaters, 111
Dimensionless numbers, **2**, **3**
Direct numerical simulation (DNS), 162
Discontinuous deformation analysis (DDA), 172
Displacement meters, 45, *45*
Distinct element method (DEM), 172
Drag-type interaction force, 175
Drum-type geotechnical centrifuge, *132*
Dummy wooden caissons, 25
Dupuit–Forchheimer equation, 87–88
Dynamic beach nourishment, 141

E

Earth pressure gauge, 43, *43*
Earth's gravitational field, 85–87, **86**
Earth's gravity field, 80, *81*
Elastoplastic MPS method, 173
Electromagnetic current meters setup, *19*
Euler–Lagrange coupling, 189–191
Excess pore water pressure, *218*
Experimental wave conditions, **121**
Experimental wave test, 21

F

Fabrication method, 124–125
Finite difference method (FDM), 164
Finite element method (FEM), 164
Finite volume method (FVM), 164
Fixed bed experiment, 11–12
Flow velocity, 12–13, *13*
  normal direction, 179
  tangential direction, 178–179
Fluid (wave) and ground interaction, 1, *2*
Fluid and porous domains, 177–178
Fluid flow field, 180
Fluids
  breaking waves, 79–80
  centrifugal gravity field, 81, **81**
  earth's gravity field, 80, *81*
  fluid–structure interaction, 81–82
  kinematic viscosity coefficient, 77
  Navier–Stokes equation, 77–79
  Weber number W_e (surface tension), 78
Fluid–structure–seabed interaction (FSSI), 169
Fluid–structure–sediment–seabed interaction model (FS3M), 169
Fluid *vs.* ground model tests
  sand drift, 93–98
  scouring phenomenon, permeable structure, 100–102, **101**
  seabed topography changes, 98–100, **100**
Froude law of similarity, 3

G

Generation and dissipation processes, *58*
Geosynthetic clay liners (GCLs), 34, *35*
Geotechnical engineering approach, 1
Geotechnical model test
  boundary setting, 47
  categories of, 29
  centrifugal force field, 35–42
  gravitational force field, 32–35, *33*, *35*
  gravity and centrifuge, 56–60
  implementation order, 54
  model test method and design, 30–32, *31*
  purpose of experiment, 30
  sandy and cohesive soil ground, 48–49
  scaling ratio, 30
  static model test, 54–56, *55*
  structure of model test, *29*
Gotoh model, 172–173
Gravitational force field, 7, 32–35, *33*, *35*
Gravity model test, 56–60
Grid method
  CADMAS-SURF, 164–165, 168
  pressure around caisson-type seawall, *168*
  two-dimensional VOF-FEM model, 169
  type 1 and 2, 164, **165–167**
Grid *vs.* particle method, 176–177
Ground
  B matrix, 83
  centrifugal gravity field, 87–88
  dimensionless quantities, 83
  earth's gravitational field, 85–87, **86**
  prototype and model, 84
  similitude ratio, geotechnical model test, **83**, 85
  soil–structure interaction, 88–89, **89**
Gyro sensor, *46*

H

Hagen–Poiseuille flow, 177, 181
Horizontal wave force, *26*, *27*
Hudson formula, 111
Hydraulic model experiments, 196–197, 228–229
Hydraulic model test
  experimental data, 23–24
  fixed bed experiment, 11–12
  flow velocity, 12–13, *13*
  installation and calibration, instruments, 16–21, *18–21*
  model-making, 15–16, *15–17*
  movable bed experiment, 11–12
  plan
    experimental facility, 7, *8*, 9, *9*
    model scale setting, 6–7
  suspended sediment concentration meter, 13, *14*
  wave basin experiments, 11
  wave-dissipating concrete blocks, *27*, 27–28, *28*
  wave flume experiments, 9–11
  wave force acting on structure (caisson), 24–27, *25–27*
  wave force and pressure, 13, *14*
  wave height, 12, *12*

I

Incompressible SPH (ISPH), 176
Inertia force, 7
Interfacial conditions, 181–183

K

Karman-type slope collapse analysis, 144
Keulegan–Carpenter (KC) number, 209, 210
Kinematic viscosity coefficient, 77
Kozeny–Carman equation, 208

L

Lagrange–Lagrange coupling, 194–195
Large Eddy simulation (LES), 169, 204
Large-scale experiment (1/16 scale of the site)
  characteristics of flow within mound, 117, *118*
  cross section of, 115, *115*
  generation status of scouring, 116–117, *117*
  mound with scouring, 119
  sand suspension situation, 115, *116*
  scouring of each cross section, 120, *120*
  wave conditions and scouring, 117, 119
Law of similarity, 229; *see also* Fluids; Fluid *vs.* ground model tests; Ground; Unsaturated ground; Wave *vs.* ground model test
  dimensionless quantities, 73
  Earth's gravity field, 72
  Froude and Reynolds numbers, 74
  large-scale civil engineering and building structures, 72
  π numbers, 73
Laws of physics, 42
Level set (LS) method, 205
Liquefaction, initial ground, 127–128
Load cell, 43–44, *44*
L-type scouring, 112

M

Main laws of similarity, **2**
Manning roughness coefficient, 96
Maximum shear stress around composite-type breakwater, *169*
Measurement time, 23
Microelectro-mechanical system (MEMS) technology, 46
Model bed making, 15–16, *15–16*
Model experiments, 228
Modeling of models, 31
Modeling of prototypes, 31
Model tests; *see also* Geotechnical model test; Hydraulic model test
  actual object/large-scale, 5
  geomaterial and earth's, 6
  waves, currents and beach deformation, 5
Modified incomplete Cholesky conjugate gradient (MICCG) method, 207
Monotone upstream-centered schemes for conservation laws, total variation diminishing (MUSCL TVD), 207
Moore's law, 229
Movable bed development process, 199–201
Movable bed experiment, 11–12
Moving particle semi-implicit (MPS) method, 172
Multicomponent load cell, wave pressure gauge, 18–20, *19*, *20*
Multi-interface advection and reconstruction solver (MARS), 169, 207

Multiphase flow analysis
Euler–Lagrange coupling, 189–191
Lagrange–Lagrange coupling, 194–195
Lagrangian tracking, sand particle motion, 189
particle–fluid DNS and LES, 191–194
single-phase/solid–liquid phase turbulence, 188–189

N

Navier–Stokes equation, 79, 171
Numerical calculation methodology, 228
economical efficiency, 162
Froude's law, 163
Navier–Stokes equation, 162
Rubey's equation, 163
Numerical simulation method, 197–198

O

Offshore wind power generation, *172*

P

Particle image velocimetry (PIV), 52–54, 53
Particle implemented simulator for physical and engineering research (PARISPHERE), 176
Particle method; *see also* Bottom sediment transport
advancement of particle methods, 172, *173*
elastoplastic MPS method, 173
Gotoh model, 172–173
Passing wave test, 21
Permeability coefficient, 87–88
Piecewise linear interface calculation (PLIC), 205
Pilot experiments, 30–31
Pore pressure gauge, 43, *43*
PORO-WSSI II, 169
Pressure gauges setup, *20*, 43, *43*
Production experiments, 30–31

R

Reflectivity, 82
Relative density adjustment, 49–50, *50*
Resistance force
linear component, **208**
nonlinear component, **209**
Reynolds-averaged numerical simulation (RANS), 170, 204
Reynolds law of similarity, 3
Reynolds number, 210
Reynolds stress, 80

S

Sampling interval, 23
Sand drift
dimensionless quantity, 93–95
suspension, 97–98
traction, 95–97, **97**
Sand dune and beach scarp
beach scarp formation, *147*
beach scarp migration and eventual destruction, *147*
beam-type failure, 143, *145*
damage patterns, road embankment, 142, *143*
ground conditions, 142
maximum beach scarp height, *146*
relative density, collapsed sediment volume and shear strength, 142, *143*
sand dune failure mechanisms, *144*
shear-type failure, 143, *145*
test model and gauses, *148*
Sand hopper, *48*
Sand level gauge, 46, *47*
Saturation control, 50–52, *51*
Scale effect, 41–42
Scaling ratio, 30
Scouring phenomenon, *see* Subsidence phenomena
Seabed deformation model, 175
Seabed model, 211–212
Seabed topography changes, 98–100, **100**
Sediment movement phenomena, 63
Seepage flow, crushed stone ground, 82
Seepage flow velocity, 182, 184
Sensor body and lead wire, 42
Shear stress, *175*
Shear-type failure, 143, *145*
Shields number, 3
Similitude ratio, **56**, **59**
Simple linear interface calculation (SLIC), 204
Simplified marker-and-cell (SMAC) method, 206
Single-domain method, 179, *180*
Slope instability problem due to waves, *103*
Small-to-medium-scale model experiments, 113–115, **114**

Smoothed particle hydrodynamics (SPH) method, 172
Soil container, 42
Soil moisture meter, 44, *44*
Static beach nourishment, 141
Strain gauge, 44, *44*
Structural model making, 16, *16*
Submerged breakwaters, 111
Subsidence of breakwaters, 111
Subsidence phenomena, 111–113, **112–113**
Super-large-scale experiment (1/4 scale of the site)
  experimental flume and condition, 120, *121*
  reproductive experiment, movable bed, **121**, 121–123, *123*, *124*
Super Roller Flume for Computer-aided Design of Maritime Structure (CADMAS-SURF), 164–165, 167–169
Suspended sediment concentration meter, 13, *14*, 20

T

Tangential flow velocity, **179**
Test cases and condition, **57**
Three-component load cell setup, *19*
Time domain reflectometry (TDR) method, 44
Transmitted wave height, 82
Turbulent diffusion coefficient, 79
Two-dimensional cavity flow, 183
Two-domain method, 179, *180*

U

Unsaturated ground, 89–93
  description, 89
  effective stress equation, 90
  gravitational and centrifugal force fields, **92**
  permeability coefficient, 90–91
  unsaturated infiltration characteristic model, 91
Unsaturated seepage experiments
  dike, *215*
  overflow-induced erosion, *216*, *217*
  sand column, *214*

V

van Genuchten–Mualem model, 91, 207
Vertical stress distribution, *38*
Volume-averaged Reynolds-averaged Navier–Stokes (VARANS) equation, 169
Volume-of-fluid (VOF) method, 165, 204

W

Water pluviation method, 127
Water–soil coupled analysis, 171
Wave basins, *8*, *9*, 11
Wave-dissipating blocks, 81–82
  experimental wave flume, 126, *127*
  Hudson formula, 111
  large-scale experiment (1/16 scale of the site), 115–120
  liquefaction, initial ground, 127–128
  small-to-medium-scale model experiment, 120–123
  small-to-medium-scale model experiments, 113–115, **114**
  subsidence phenomena, 111–113, **112–113**
  super-large-scale experiment (1/4 scale of the site), 120–123
Wave-dissipating concrete block model, *17*, *27*, 27–28, *28*
Wavefield analysis, 171
Wave flumes, *8*, 9–11
Wave force, 13, *14*
Wave gauge, 16–17, *18*
Wave generator, 7
Wave height, 12, *12*
Wave-induced local scouring, *219*
Wave-induced sediment movement, **61**
Wave-induced stress changes, seabed soil, **62**
Wave model
  computational grid, 206
  continuity equation, 205
  Dupuit–Forchheimer resistance law, 206
  FS3M, 206–207
  GS component, 207
  Navier–Stokes equation, 205
  wavefield model, 204
Wave overtopping
  description, 129
  experimental wave flume, *130*
  progress of scouring, *129*
  rate, 82
Wave pressure, 13, *14*
Wave–seabed interaction
  grid method, *see* Grid method

numerical calculations, classification of, 163, **164**
particle method, *see* Particle method
Wave–structure–soil interaction in Japan, **64**
Wave *vs.* ground model test, 102–105, *103*, **104**
Wave *vs.* seabed model, 212–214
Weber number We (surface tension), 78
Wire-type displacement meters, 45

Y

Young researchers, 1

Z

Zero down-crossing method, 24
Zero up-crossing method, 24